TO DIE FOR

THE PHYSICAL REALITY OF CONSCIOUS SURVIVAL

James E. Beichler, Ph.D.

Trafford
PUBLISHING

Order this book online at www.trafford.com/07-2808
or email orders@trafford.com

Most Trafford titles are also available at major online book retailers.

© Copyright 2008 James Edward Beichler.
All rights reserved. No part of this publication may be reproduced, stored in a retrieval system, or transmitted, in any form or by any means, electronic, mechanical, photocopying, recording, or otherwise, without the written prior permission of the author.

Note for Librarians: A cataloguing record for this book is available from Library and Archives Canada at www.collectionscanada.ca/amicus/index-e.html

ISBN: 978-1-4251-6167-5

We at Trafford believe that it is the responsibility of us all, as both individuals and corporations, to make choices that are environmentally and socially sound. You, in turn, are supporting this responsible conduct each time you purchase a Trafford book, or make use of our publishing services. To find out how you are helping, please visit www.trafford.com/responsiblepublishing.html

Our mission is to efficiently provide the world's finest, most comprehensive book publishing service, enabling every author to experience success. To find out how to publish your book, your way, and have it available worldwide, visit us online at www.trafford.com/10510

Trafford PUBLISHING™ www.trafford.com

North America & international
toll-free: 1 888 232 4444 (USA & Canada)
phone: 250 383 6864 ♦ fax: 250 383 6804 ♦ email: info@trafford.com

The United Kingdom & Europe
phone: +44 (0)1865 487 395 ♦ local rate: 0845 230 9601
facsimile: +44 (0)1865 481 507 ♦ email: info.uk@trafford.com

10 9 8 7 6 5 4 3 2 1

CONTENTS

Introduction		9
Chapter 1 Some things never change(Some things do)		11
	The boy cried	11
	Science, Religion and Nature	17
	Science does mind	29
Chapter 2 Supernatural Perceptions of Death		42
	The religious pot of stew	42
	Eastern Religious Perspectives	48
	Western Religious Perspectives	57
	The religious stew boiled down	74
Chapter 3 Natural Perspectives of Death		84
	The road not so well traveled	84
	From Thanatology to NDEs	92
	The nature of consciousness	106
Chapter 4 Paranormal Perceptions of Death		120
	Science beyond the normal	120
	The spirit of science	127
	The loss of soul and spirit	143
	Apparitions and Ghosts R Us	164
	Small, Mediums and Large	189
	There and back again	198

Chapter 5	The SOFT Life	**205**
	That's Life	205
	Physical Reality	209
	Imagining the un-imaginable	215
	Our five-dimensional Life	224
	LIFE: The body inside out	240
	Memories are made of this	246
	The Neuron	249
	The Microtubule	251
	The Neuron as an LRC Circuit	257
	Microtubules in an Axon	260
Chapter 6	**The Nature of Death**	**264**
	IRIMI – Entering Death	264
	We enter death SOFTly	270
	SOFT NDEs	280
	Post NDE SOFT landings	292
	NDLEs	306
	'Where' is death?	319
Chapter 7	**A Universe of Purpose**	**328**
	Everything that has a beginning has an end	328
	Natural Purpose	334
	Some things never change and some things do	347
	To live and let die	357
Chapter 8	**Epilogue: (Some things never change, but …)**	**365**
	Some things do	365
	Play it again Sam	366
	What goes around, comes around	377
	And goes around again	385
	Solving the Universe	395
	Buddha, Jesus, Human Enlightenment and the dawn of the Mysphyts	409
	The Kaballah—a Warning	424
	The Boy Cried	426
Bibliography		**428**
General References		**430**

*There is a dimension, beyond that which is known to man.
It is a dimension as vast as space
and as timeless as infinity.
It is the middle ground between light and shadow,
between science and superstition,
and it lies between the pit of man's fears
and summit of his knowledge.
This is the dimension of imagination.*

—Rod Serling – *The Twilight Zone*

*For the mysphyt
in all of us*

INTRODUCTION

> *If we do discover a complete [unified] theory of the universe, it should in time be understandable in broad principle by everyone, not just a few scientists. Then we shall all, philosophers, scientists, and just ordinary people, be able to take part in the discussion of the question of why it is that we and the universe exist. If we find the answer to that, it would be the ultimate triumph of human reason – for then we should know the mind of God.*
>
> Stephen Hawking – 1988 – *A brief history of time*

To Die For is a science book. It not only offers new science, but it chronicles how that science has developed. It is about the science of death and afterlife. *To Die For* is a book about physics. In the broadest sense of the word, physics is nature so this book is about the very nature of death and how death fits within the context of nature and our natural world. Only one other book that I know of, *The Unseen Universe* by Peter G. Tait and Guthrie Stewart, has ever attempted to develop a theory that explains death and what happens to us when we die. It would seem that *The Unseen Universe* should have sent lasting ripples through the scientific and scholarly communities and challenged science to answer one of the most important questions ever asked; "What happens to us when we die?" But *The Unseen Universe* was largely forgotten as the second scientific revolution changed the basic assumptions and context for all of science and culture in the twentieth century.

The Unseen Universe was a product of its time and the questions that were important within the context of its time. Two other books on death that explain what happens when we die were also products of their times and helped to shape their own cultures: The Egyptian and Tibetan books of the dead. When these books were written, the afterlife was placed within the cultural and religious context of the times and reflected the cultural norms and perspectives of the societies from which they emerged. So, in a sense, they were forerunners to *The Unseen Universe* since the nineteenth century culture from which *The Unseen Universe* emerged was a science-dominated culture just as the earlier cultures were religion-dominated cultures. All three books are forerunners for the book *To Die For*.

It is hoped that *To Die For* will teach people not to fear death, but to embrace it when it comes. Nor should people ever force death to come when it is not due. Death is natural and should be viewed as a celebration of life, neither welcomed nor forced before its time, but accepted for what it is when it comes calling. So what is death? Death is a passing onward to another state of being or even a passing into an unseen part of our universe if you would prefer that explanation. The truth of the matter is simple and will be verified by science in the not too-distant future – each individual's mind and consciousness survives when the material body dies. The greater an individual's knowledge of the true nature of the universe when they die, the greater that individual's awareness of the surviving consciousness as an individual being within the context of the universe will be after death. This realization will enlighten and free each and every one of us from the shackles that bind us to our material life and that are the fundamental goal of our very existence.

CHAPTER 1
SOME THINGS NEVER CHANGE
(SOME THINGS DO)

[My advice] will one day be found
With other relics of 'a former world'
When this world shall be former, underground,
Thrown topsy-turvy, twisted, crisped, and curled,
Baked, fried or burnt, turned inside-out, or drowned,
Like all the worlds before, which have been hurled
First out of, and then back again into Chaos,
The Superstratum which will overlay us.

—Lord Byron–1821

The boy cried

The scene is all too familiar and the event that is playing out is as common as birth. A young boy, several days past his tenth birthday, wakes up on a chilly February morning and rushes downstairs for breakfast at the start of a new day. However, his mother meets him at the bottom of the stairs. He doesn't really notice the tired sad look in her eyes or the fact that her face is already puffy and red from crying. He doesn't even suspect what has happened or what he is about to be told. He'd slept soundly, without disturbance, throughout the night and his mother had let him sleep the night away peacefully, thankful that he had been spared the ordeal.

His mother takes him by the hand and leads him into the living room by the couch. She sits down on the couch and beckons him to sit on her lap. Once there, she hugs him and begins crying again. She then tells the boy that his father died suddenly and unexpectedly the night before. The boy had slept through all of the noise and commotion of ambulances and police as well as the frantic effort to transport his father to the hospital. But his father had died of a probable heart attack long before the ambulance had even arrived at the house. His father passed on without any warning or preparation, having had no prior heart problems, at the age of fifty-five.

The boy cried as much because his mother was crying as he cried because of his father's death. The boy felt grief, but wasn't completely sure of just what death meant. His step-grandmother had died a few years before, but he doesn't even remember meeting her and he felt no real grief at her passing. His uncle, his father's brother, had died from cancer the year before, and that was the only real case of death that the boy had ever encountered. The boy certainly missed his uncle, but his uncle's death had not directly affected the boy's life. So his father's death was essentially a new experience for him. The boy would cry more, and learn more of what his father's death meant over the next few years, as he slowly came to deal with his grief, the anguish, the deep loss and loneliness that he would come to feel.

In later years, the boy could not remember crying or feeling any special grief at his father's funeral. Nor does he remember viewing his father. Those memories seem to have been blocked out of his mind. He does, however, remember the other people who came to his father's funeral and how they thought he was so brave for not crying. He also remembered the people that looked at him then turned away and whispered to each other when he returned their gaze. But the boy just did not fully realize what death was, and by not crying then his sadness and grief was actually extended over the following years, to crop up later in his life in the most unlikely of places and times.

A few years later, in Junior High School, the boy took French classes. During one lesson on the family, the teacher asked him what his father did, of course in French, expecting a simple answer to her inquiry. Instead, the boy burst into tears, uncontrollably and spontaneously. He stayed after class and the concerned teacher asked why he was so sad and starting crying. She had not expected to hear that the boy's father had died two years earlier. The teacher was visibly shaken by the incident and quite apologetic. For many years thereafter, the boy had dreams that his father was still alive and would wake up in the mornings expecting to see his father, only to learn that his father's only presence was in his dreams. The father was still, and would always be, absent from the boy's waking life.

The boy has a sister, several days short of three years older. Their father died on the eve of his sister's thirteenth birthday. She had expected to become a full-grown teenager on that fateful morning, an accomplishment of some distinction and responsibility. It was supposed to be a day of happiness and great expectations for her. But instead she took her first steps on the road to adult life having to cope with the tragic loss of her father. She never fully recovered from that tragic event and blamed their father for deserting her just as she expected to embark upon the most important years of her life. Many years later, after two unhappy marriages for the sister ended in divorces, she talked with the boy, then grown into a young man, all night long during a rare visit home. She admitted to her brother that she had always blamed their father for dying when he did, as if it was a personal insult to her, and that his death was an important factor in her poor choices to love and marry the wrong men.

This story has probably been scripted and played out millions of times over the millions of years since humans first walked this earth. Only the details are different for different people. But the general characteristics are the same in all the cases. Under similar circumstances, everyone feels grief and the painful sense of loss when a loved one or friend dies, because we do not understand death. Each death in each family has long and profound effects on each and every member of the family. Death is among the most profound and important personality shaping events, if not the single most profound event that can occur during a person's lifetime. It is also the most memorable event due to the agonizing grief and anguish that is felt by those who are forced to carry on with their own lives after the death of a loved one.

Such profound and overwhelming grief is not just a human characteristic. Elephants are known to grieve the death of their kin as well as other members of their herd. Dogs have been known to mourn the loss of their masters, sometimes for years after the death occurs, just as we sometimes mourn the loss of our pets. Given the circumstances, it is not beyond possibility that many other animals and species also feel grief and loss upon the death of one of their own. So, some knowledge or recognition of death as an end and what death means for the living would seem to be universal.

Death is so much a part of our lives that whole mythologies and religions have sprouted up around the concept. Yet religion and mythology only represent the hopes, desires, wishes and expectations of what might happen to us after death, while we don't yet have any real verifiable scientific evidence that some part of us survives death. Nor is there any real evidence that we don't survive death in some manner since science has no real theoretical explanation of life, and not knowing what life is science cannot say with complete confidence what death is or isn't. Under these circumstances, the question of survival and what survives is completely open. We presently accept the survival of death on faith and faith alone, depending on what we have been raised to believe. These realities beg questions which demand immediate answers, if only provisional answers, but the questions either go completely unanswered by science or we only get excuses in place of real answers because of our lack of knowledge about what happens when and after we die. In a world such as ours, where death is so universally felt, even among different species, why is so little known about death?

As if the personal tragedies associated with death were not enough to occupy our thoughts and emotions, we also have to deal with other factors including the cultural responses to death. Religions have been known to use the promise of life after death to control those who believe in the tenets of that religion. Death, in one form or another, is often a volatile political issue. Some governments use death to move the people governed toward some end that the government thinks desirable or necessary. Terrorists and criminals use death and threats of death to manipulate their victims toward their points of view, to adopt their policies and belief systems or just to gain riches. Our very laws and senses of morality speak mountains on the concept of death, from dictums such as "thou shalt not kill" to medical and ethical clashes over the question of when a person should be considered dead and should life support be continued past the time when the person is incapable of continuing his or her own life without it. Some people just cannot accept the idea of death so they try to force others to live past the time when nature would have them die.

In still other cases people see a death as a sign of the will of a supreme all-knowing being. Perhaps they are just trying to find some meaning for what could have been a senseless death, but they might also be making excuses because they believe that death must have some greater reason for occurring than just the plain and simple fact that the death happened, however inconvenient and unexpected the death may have been. Some people need to place blame somewhere, so they blame whatever god they worship and state that the death was "God's will". Some might even believe that a particular death has some higher purpose and fulfills some unspecified 'destiny'. And others abhor death so much that they are willing to commit murder in their attempt to prevent deaths that they claim will occur. For a person to commit murder because that person does not believe in the general principle of 'killing others' is illogical. All told, a great amount of needless waste results from our basic lack of understanding of the nature of death and dying. There is no facet of our lives that is not touched or tainted by our deaths; that is the curse of our existence as conscious and sentient beings. We experience death, but we also seem doomed to know death only incompletely, at least until we can understand the nature of life itself or until we die, and that may be too late.

In medical and legal terms, we cannot even answer important questions concerning when the moment of death occurs. It was recently thought that death occurred when all brain activity ceased, but people who have died in extremely cold environments have been revived after brain activity had stopped and yet suffered no brain damage, which might indicate that 'something' more than the material brain is associated with life and dying and that 'something', whatever it is, could continue past the end of life. So now a person is not considered dead until the person is both warm and dead.

On the other hand, our most advanced sciences seem poised to conquer the universe even though science really knows little or nothing about the concept of death, beyond the fact that it happens. So we are forced to invent science fiction and fairy tales to explain what happens after death, if only to satisfy our immediate needs and curiosities about death. This situation reinforces and raises still other questions about death, in spite of the advances of science: What is death? Why do we grieve and mourn? Do we, or does some part of us, survive death? Are grieving and mourning necessary? Is there a better, less traumatic way to deal with death? Is there a scientific basis for survival? Members of the scientific, religious and scholarly communities may not like these questions, but the questions cannot be ignored forever without some acknowledgment of their existence and importance.

Quite frankly, science deals with our physical/material universe, the natural world in which we live. Religion deals with the supernatural, that which goes beyond or transcends nature, and that is considered to be the realm of spirit and soul. During life, religion deals with the spiritual aspects of a being, but after death religion prescribes the habitat and existence of spirit, of the continuance of a soul that had inhabited the living body before death and abandons the body when life ends. But there is an 'in-between place' in our thoughts and thought processes that may eventually help to explain the nature of death. Between normal science and religion, the natural and supernatural, there is a region of knowledge that can be called metaphysics within a philosophical context and the paranormal within a more experiential context. The paranormal exists between the supernatural and the normal as prescribed in our present scientific paradigms, but the paranormal is also a branch of science, albeit a somewhat neglected branch of science by the larger scientific community.

When push comes to shove, there are only two institutions that legitimately deal with the concept of death: science and religion. And in many cases, science capitulates or surrenders its right to investigate and understand the concept of death in favor of religion, while retaining only a practical hold on death and the biological act of dying for medical and psychological reasons. It is only at the very periphery of science that questions on the concept of death can be asked and answered without prior prejudice and that is in the scientific study of the paranormal. This dichotomy of approach toward such a profound conceptual subject can be traced back to the very foundations of science as laid down during the first scientific revolution of the seventeenth century.

Science, Religion and Nature

From the time of Aristotle in the fourth century BCE to the work of Kepler, Galileo and Newton in the seventeenth century, a philosophical understanding of nature slowly grew into what we now call science. Against this newly developed understanding stood various superstitions, mythologies, prejudices and beliefs in the supernatural, all of which were associated with the religious aspects of our society in some manner. But slowly, reason and logic, under the name of Natural Philosophy overtook superstitions and made successful incursions into the other aspects of religious belief, even though religion claimed to be philosophical in nature. Although a logical construct, religious philosophy was based on beliefs as well as misbeliefs, not on the rules that nature establishes, but the rules that the sometimes-infallible mind of man establishes, for better or worse. Over the centuries, humankind began to make logical sense of the natural world, against some of the teachings of religion, and science evolved.

But science and the first simple understanding of nature did not emerge out of ignorance without serious misgivings, pain and limits. The French philosopher and mathematician, Rene Descartes, set the basic limits for science when he separated our universe into two major categories, MIND and MATTER. This separation still exists under the guise of the observer versus the observed in modern physics or in less specific terms, the subjective part of nature versus the objective part of nature.

The overall study and investigation of matter became the realm of science, while the process by which matter was studied was institutionalized as logical reduction. In order to explain our world, it was reduced to its most common and fundamental elements, matter and motion and science thereby developed upon the basis of reduction. In order to understand nature and the logic of nature, various portions of nature were defined and measured, emphasizing only the quantitative aspects of nature. The mind could not be quantified, defined and measured so it was cast out of science. Quite simply mind was not the business of science.

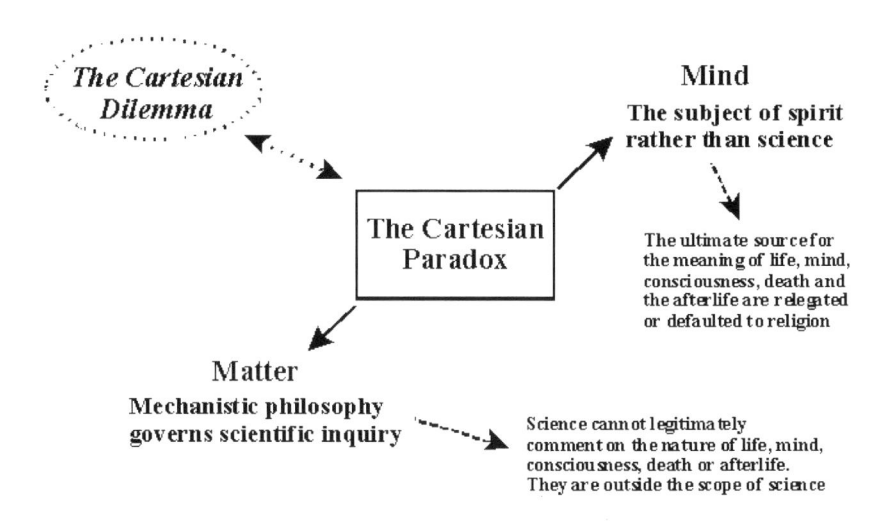

```
         ........
      .·          ·.
     : The Cartesian :              Mind
     ·.  Dilemma   .·       The subject of spirit
       ·.        .·         rather than science
          ······
              ◤    ┌─────────────┐    ▲
                ╲  │ The Cartesian│  ╱
                   │   Paradox    │            The ultimate source for
                   └─────────────┘             the meaning of life, mind,
              ╱                   ╲            consciousness, death and
            ▼                       ◥          the afterlife are relegated
                                               or defaulted to religion
       Matter
   Mechanistic philosophy        ╲     Science cannot legitimately
   governs scientific inquiry     ──▶  comment on the nature of life, mind,
                                       consciousness, death or afterlife.
                                       They are outside the scope of science
```

The Cartesian Dilemma is to define the boundary between MIND and MATTER

Life also fell into this category, but not completely for the same reason. There was no standard or method by which life could be reduced. In the intervening centuries, the objective method of reduction has proven to be extremely successful, but it has said little or nothing about the subjective qualities of nature as well as neglecting mind and life.

Even today, one of the major criticisms of science is its obsession with objective reduction and the subsequent neglect of the subjective and qualitative aspects of our world. But objective reduction has been absolutely necessary for the growth of science and our understanding of the nature of reality, a fact that the critics of science seem unable to understand. On the other hand, science has yet to accept the simple fact that any understanding of nature based solely on objective reduction will always be incomplete and thereby lead back to the subjective aspects of nature. So science must eventually transcend the split between the subjective and objective in nature to realize its own ambitions of true and total knowledge. Science is only now beginning to accept this new attitude toward discovery, but then this dual aspect of nature has always haunted science even though scientists would never admit to that fact.

In spite of its reliance on reduction, or perhaps because of it, the span and scope of science has greatly expanded over the centuries, partly through its constant incursion into new realms of phenomena as perceived by humans, including phenomena sometimes considered occult or supernatural and occasionally religious. Science is always expanding the domain of its influence by redefining the range of phenomena that it considers 'scientific'. This progression raises questions concerning the reality and scientific legitimacy of certain humanly perceived phenomena and the extent to which they reflect real and verifiable phenomena in our commonly shared physical environment. Because there is no objective measure of when a human mind is or is not misinterpreting sensory input from the environment, the misinterpretation of phenomena due to bias, culturization or any other reason, whether real or imagined, poses a real problem for science. So, science may neglect some phenomena that may be valid, simply to err on the side of caution, and science might ignore certain questionable phenomena if there is no pressing and timely reason for science to recognize and study them.

It is sometimes difficult to distinguish which phenomena are merely the results of our faulty and perhaps biased perceptions of the world and which phenomena actually represent occurrences and events in our world. For this reason, the difference between subjective and objective is sometimes overly emphasized in science, as the distinction blurs with regard to some phenomena. What is real and what is imaginary in many cases, science must shed light into the darkness and shadows of our misperceptions and subsequent misconceptions of the world, before it can progress. So science must precisely identify and define the quantities with which it must deal in our physical environment, both beyond and independent of our perceptions of the world, before it can proceed with its quest of explaining the world in which we live. So questions like mind, life and death pose special problems for science. They can be inferred from physical evidence, implied by physical concepts, but they are extremely difficult to isolate on their own terms. In spite of this difficulty, science still has a history of trying to explain them.

During the first scientific revolution, science successfully challenged religion as the main expression of the human perception of our world. Many of the earliest successes of science were as much or even more about overcoming religious interpretations of nature, superstition and religious scholasticism, than just carefully observing the world alone. Carefully and accurately observing the world is the scientific ideal, not taking the word of earlier scholars, including religious scholars, as absolute truth as was the religious tradition. The successes of science in this regard have promoted a bias in science, not always wrongly applied, against many types of phenomena dealing with the occult and psychical aspects of nature. This bias, which is in many cases a protective mechanism, fairly well precludes the possibility of any part of a person surviving death, in any form, although the full reasons why science seems to ignore the concepts of death and survival actually run much deeper.

While the scientific revolution institutionalized the separation between mind and matter, that separation implied a further separation between life and mechanism as well as spirituality and natural philosophy. This separation was part and parcel to Newtonianism, since the science that it was based upon reached its fullest development in Isaac Newton's *Philosophiae Naturalis Principia Mathematica* of 1687. And the *Principia*, as it is commonly called, marked the culmination of the revolutionary period. After the *Principia*, science was all about applying the principles and ideas set forth in the Newton's *Principia*.

By carefully defining mass in the *Principia*, thereby differentiating between mass, weight and density, Newton was able to derive the correct laws of motion and gravitation. This success where others had failed allowed Newton to delegate the whole study of our physical world to science and reason. But the Newtonianism that followed did not offer the final explanation for all of the phenomena in nature as perceived by humans. Physical quantities such as heat, energy, electricity, and magnetism fell outside of the original Newtonian worldview at first, as did life and mind, leaving them open to both scientific and non-scientific speculations. Yet in spite of his initial successes, Newton's work was openly and severely criticized for leaving God out of science. So Newton was forced to find a place for God in his overall scheme of the world in rebuttal to his critics. In so doing, Newton set a new standard for the interaction of science and religion by establishing boundaries for the evaluation and investigation of natural phenomena that had not yet been incorporated into the new scientific worldview.

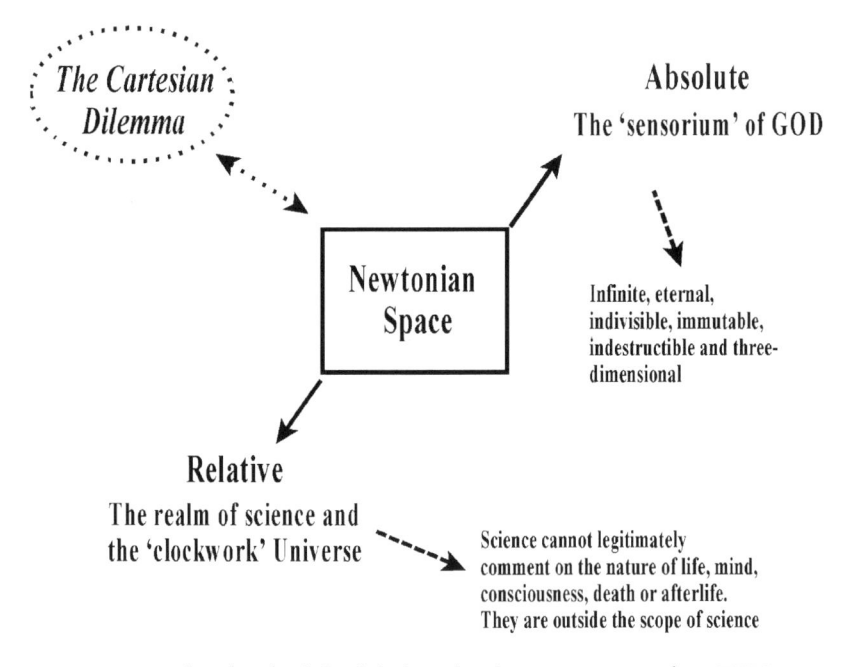

The Cartesian Dilemma

Absolute
The 'sensorium' of GOD

Infinite, eternal, indivisible, immutable, indestructible and three-dimensional

Newtonian Space

Relative
The realm of science and the 'clockwork' Universe

Science cannot legitimately comment on the nature of life, mind, consciousness, death or afterlife. They are outside the scope of science

Newton altered and redefined the boundary between MIND and MATTER

By defining what was scientific, all else was open for religion to claim within its sphere of influence, albeit a shrinking sphere as science continued to progress.

Newton separated the world as investigated by science into absolute and relative portions. He developed concepts of absolute space and time as opposed to relative space and time, but then acknowledged that relative space and time alone constituted the realm of scientific investigation. Although he did not explicitly compare mind to the absolute, Newton did associate God with the absolute in his famous "General Scholium" in a later edition of the *Principia*.

> He is not eternity and infinity, but eternal and infinite; he is not duration and space, but endures and is present. He endures forever, and is everywhere present; and, by existing always and everywhere, he constitutes duration and space. Since every particle of space is *always*, and every indivisible moment of duration is *everywhere*, certainly the Maker and Lord of all things cannot be *never* and *nowhere*. ... In him are all things contained and moved; yet neither affects the other; God suffers nothing from the motion of bodies; bodies find no resistance from the omnipresence of god. It is allowed by all that the Supreme God exists necessarily; and by the same necessity he exists *always* and *everywhere*. (Newton, 545)

For Newton, only the world of relative space and time was subject to scientific scrutiny, analysis and investigation since only relative space and time were necessary for a complete description of matter in motion. So life and mind, which were not explicitly placed within the realm of science, were incorporated, by default, into the realm of religion. God gave life to humans and created mind for them to think and worship God. Since death meant an end of life, and religion already claimed any state of being after death as a religious priority, the meaning of life as interpreted by religion was not so different from the nature of life, even though the search for and discovery of the nature of life should have been considered part of science.

In Newtonian science, the absolute could only be inferred from certain phenomena such as centripetal acceleration during circular motion, which could not be explained relative to any point in Euclidean space. The absolute only existed in science to give such accelerations a philosophical validity and could not be reduced and measured in any manner. The same was true for God. Material motion, as explained within a relative framework, did not affect God or God's existence in any way. Nor did God's existence offer any physical resistance to material motion. The same characteristics were used to describe absolute space, but God was not equivalent to absolute space and time, they formed his 'Sensorium'. Newton did not, and would not, define God, but he could still talk about that realm of being within nature through which God acted. Neither God nor absolute space and time were prone to any scientific measurement, but were abstractions from our common senses. According to Newton,

> And so, instead of absolute places and motions, we use relative ones; and that without any inconvenience in common affairs; but in philosophical disquisitions, we ought to abstract from our sense, and consider things themselves, distinct from what are only sensible measures of them. For it may be that there is no body really at rest, to which the places and motions of others may be referred. (Newton, 8)

It is also evident from these statements that Newton was seeking both a philosophical validation of both God and an unseen scientifically motivated yet non-demonstrable (absolute) world. So, while Newton established science as the only way to directly know and understand our material world, he left room for the existence of God and a justification for religion and religion's interpretation of God. Others would conclude that he also left a place for life, mind, death and the occult in between the two extremes of God and science.

Obviously, life could not be denied as a scientific subject, but it could not be reduced either, at least in the same manner as normal material reality was reduced. In the late 1600s, the German philosopher Gottfried Leibniz, one of those who criticized Newton for leaving God out of science, tried to explain the essence of life in his system of 'monads'. He used a 'philosophical reduction' as opposed to a 'physical reduction'. Monads were like small particles of life, sharing some of the characteristics of material atoms. But Leibniz' ideas were not accepted by other scholars or scientists, who were otherwise familiar with the concept of purely material atoms. The atomic theory of ancient Greece was well known and many early scientists speculated that atoms constituted the smallest pieces of matter. However, no modern atomic theory was announced until Dalton's theory in the early 1800s, and even then the atomic theory became part of the new chemistry, rather than a subject in physics.

In spite of such attempts, there still remained unsuspected gaps and holes in the Newtonian worldview, due in large part to the same separation of mind and matter that had been necessary at the earliest stage of the evolution of science. The separation of mind and matter had been crucial to guarantee the identification and definition of quantities that would eventually fall under the eye of science and this left room for occult concepts and ideas to exist at the far edge of the scientific world. Even Newton was something of a mystic through his alchemical researches, while society and culture in general could not abandon their occult trappings as easily as science could. Occult disciplines such as practical witchcraft remained, and perhaps even thrived, in those corners of society where science was slow to make inroads. The occult continued a tenuous coexistence with science during this period, even as science groped to bring the forces of electricity and magnetism into its fold.

Mind and life, and thus death, still remained outside of the Newtonian structure of the physical universe in spite of concurrent developments in science that would eventually lead to a science of biology. For all intents and purposes, life seemed to be 'something extra', more than a living body that appeared to be just another Newtonian mechanism. Leibniz' failure to develop his system of monads as a basis of life only exacerbated the problem. Since the true nature of electricity and magnetism were not discovered until the 1800s, these physical forces were equated to life and mind in the speculations of scientists, non-scientists and occultists. On the other hand, scholars who wished to make life sound more Newtonian, and thus more scientific, began to call that 'something extra' a 'life force'. But before electricity and magnetism could be fully understood within the context of the new Newtonian worldview, and explained as 'fields', scientists and non-scientists alike mistakenly used them to explain life. The early concept of a 'life force' evolved into such exotic concepts as bioelectricity and biomagnetism in the late eighteenth century.

Mesmerism was the best known of the occult practices that utilized the force of magnetism to explain life as 'something extra' beyond the concept of a Newtonian force. Mesmerism first took root in Paris before the French Revolution. While magnetism certainly has a place in the science of life, even today, it cannot completely account for life as in Mesmer's concept of 'animal magnetism'. On the other hand, electricity also plays an important role in biology and today's science of life and that role was also misinterpreted and misrepresented by early scientists. During the eighteenth century, equating life directly to electricity only emphasized what little Newtonian science knew of both life and electricity. This association came under the guise of such concepts as 'animal electricity,' 'neuro-electric fluid' and 'bioelectrogenesis.' However wrong the notion was, it did lead to advances in science. In the 1780s, Luigi Galvani conducted a series of experiments on 'bioelectrogenesis', but his discoveries led instead to Volta's discovery of the battery and current electricity, rather than a true basis or explanation of life.

Another perspective on the relationship between electricity and life came in a rather non-scientific package, although it accurately reflected the current scientific concepts. Mary Shelley's Romantic novel *Frankenstein,* published in 1818, romanticized the scientific speculations of a relationship between electricity and life. The Frankenstein monster was a mishmash of inanimate body parts gathered from different cadavers, made whole by Dr. Frankenstein, but still without life until the monster was brought to life by electricity. The complete but lifeless body of Frankenstein's monster represented a Newtonian view of a mechanism that is the sum of its parts, while the introduction of electricity provided the 'something extra' that was life. However, as Newtonian physics progressed and incorporated electricity and magnetism into the overall structure of science, leading to Maxwell's electromagnetic theory in the 1860s, electricity and magnetism lost their 'occult' status, thereby rendering these physical fields unavailable for a complete explanation of life.

Newton's original work dealt with the physical quantities of matter, motion, forces and momentum, yet these quantities were eventually found insufficient for a complete description of physical reality. By the late 1700's, new concepts of power and energy were developed and added to the Newtonian worldview. These new advances were made in conjunction with the industrial revolution and the development of the steam engine to run early industries. The physical concept of energy was eventually applied to describe the random motions of the hypothetical atoms and molecules that were thought to constitute gross material objects. This application allowed the use of Newtonian mechanics to explain heat, temperature and the states of matter themselves. In fact, by equating heat energy to mechanical energy during the 1840s, physics evolved out of, or perhaps separated from natural philosophy with the birth of thermodynamics. This separation allowed the other sciences of biology, chemistry and geology to evolve within their own domains of knowledge. Science was already becoming too complex to be dominated by one discipline under one heading.

At nearly the same time as the other sciences were beginning to evolve, both electricity and magnetism became viable subjects for scientific study. The first natural laws governing these forces were modeled on the Newtonian world structure and Newton's explanation of the only other natural force known, gravity. Electricity and magnetism were first explored and explained as separate forces, but the more that science discovered about their characteristics and interactions with matter, the more science came to realize they were actually related, like two sides of the same coin. The development of the battery in 1800 as a simple source of electrical current allowed science to discover and understand the true nature of electricity and magnetism as different forms of the single electromagnetic field by the 1840s, just as thermodynamics and physics left their original home in Natural Philosophy.

When all of these changes and advances in Newtonian science are taken into account, Newton's accomplishment begins to look as much like a program to determine and define what quantities and concepts in nature were scientifically valid as it does a single scientific theory. Any phenomenon that could be explained within the Newtonian worldview was deemed scientific. Even as other scientific disciplines began to develop their own methods and principles they still remained Newtonian in their overall worldview. With new advances in mathematics and methods of mathematical analysis, the Newtonian laws of nature became ever more accurate in describing all forms of mechanical motion, including what seemed to be the inner workings of the world. Very early on, Newtonian science had been proven so accurate that God was not needed to intervene in the daily running of the universe. It was as if the universe was a gigantic Newtonian mechanical device, a clock, and God had only been necessary to design and initially wind the clock, and perhaps occasionally intervene to keep the mechanism running correctly. Upon this basis a new form of religion, called 'theism,' evolved under the Newtonian banner during the 1700s. So Newtonianism and science, in spite of their many successes, did not supplant religion even though the 'clockwork' universe became the basis for the 'theism' religious movement.

Many scholars and scientists continued to believe in private that life was something more than a complex Newtonian mechanism, but such notions did not become a normal part of Newtonian science. If life were truly 'something extra' then it would be reasonable to assume that that 'something' could survive death, but scientists could not state this because such an admission was outside the scope of science as determined by Newtonianism. Besides, such a statement would trespass on religious beliefs. Biology slowly began to emerge as the science of life in the early 1800s against this background of belief, when reduction was successfully applied to living organisms. The reduction of life was finally accomplished through the study of chemical processes rather than the physics of machines. Early biology dealt with anatomy, physiology and medical knowledge of the human body and other living organisms, but the early development of chemistry allowed a new method for thinking about life itself and changed biological practices. Life was reduced to chemistry and was found to be a process rather than a 'thing' or a 'quantity' that could be defined, categorized and measured, or rather than 'something extra' beyond the Newtonian mechanism. But in spite of the new chemical perspective on life, the main influence on the concept of life after the 1840s came from a completely different and unexpected quarter, the newly forming science of geology, and this new perspective on life directly challenged religious beliefs.

Science does mind

By the 1840s, when physics finally emerged from natural philosophy, the physical sciences of geology and chemistry had also developed to the point where they were ready to emerge as separate fields within science. So they also broke away as new scientific disciplines with their own methods, practices, set of questions and standards as well as relevance under the same Newtonian umbrella. Geology, in particular, had been influenced by Newtonianism and strictly adhered to Newtonian principles. Geologists determined that our earth had evolved over millions of years through a physical mechanism called 'uniformitarianism' as opposed to the Christian 'theory' that God created our earth several thousand years before in the exact form and shape that we now find it. According to 'uniformitarianism,' the geological features of the earth had slowly changed over thousands of millennia by slow mechanical processes such as erosion and uplift, yielding the present day geological structure and form of the earth.

The new assault on the concept of life did not, however, come from biology, but instead from this new view of geology. In his 1859 book *The Origin of Species*, Charles Darwin, a geologist and adherent of uniformitarianism, applied the principles of the theory to living organisms including human beings, thereby developing the theory of evolution. Evolution progresses through a process called 'natural selection.' Darwin's theory was directly opposed to the notion of 'teleological design' as implied in religion and predicted that the earth was at least a few hundred million years old. But Darwin's form of human evolution could not account for the evolution of the human mind, and, by extrapolation, it could not account for the mind (or spirit) after the death of the human body. To many, especially those in the religious community, mind was so specialized that evolution cold not possibly account for its 'evolution' and development, and thus the hand of God was made evident by the very existence of the human mind.

If the human body was a Newtonian mechanism that had evolved through 'natural selection,' and life and mind were something extra for which Newtonianism and Darwinism could not account, then there was no reason why the human spirit could not survive death of the host body, which now seemed no more than the cessation of mechanical and chemical activity. Neither Newtonian theory nor Darwinian evolution, nor any other form of accepted science, addressed this issue, so the question of survival of the mind and/or spirit remained open for the spiritualists to fill in the blanks as they saw fit. Newtonianism had found a place for God and religion by associating God with the absolute, but Darwinism challenged the last and only place that Newton had found for religion to coexist with science, one of religion's last philosophical strongholds, by demonstrating that humans had evolved, not been created.

Coincident with this new scientific challenge to religion, another assault on the concept of life emerged at the end of the 1840s. However, this other challenge was more specifically concerned with the concept of death and the afterlife. Modern spiritualism, as it was called, was not a scientific movement per se, but a popular movement that was directly related to science and scientific attitudes as they were interpreted, and sometimes misinterpreted, by society as a whole. It has been commonly thought that the modern spiritualism movement began in a New York cabin in 1848 when the Fox sisters heard a rapping sound that they could not immediately explain. The sisters began asking the 'source' of the rapping sounds questions and eventually discovered it was the discarnate spirit of a traveling salesman who had been murdered and buried in the cellar of the house before the Fox family moved in. The news of this phenomenon spread quickly across the United States and Europe, quickly evolving into the international social movement of modern spiritualism.

The evolution of modern spiritualism, especially in its scientific aspects, was certainly not due to any failure of science to cope with the occult and religion, but instead it was due to the success of science in its expansion into previously questionable territories of human perception and thought. The theory of evolution was just the instrument of this expansion. The historical fact that science first attacked the problem of mind indirectly, through spirit, rather than directly through a science of mind, is also easy to explain: Science was being driven to some extent and motivated to action by popular beliefs within the general population. The modern spiritualism movement coincided with the first attempts of science to define itself relative to the human mind, which was forced upon it by the theory of evolution as well as the recent successes of Newtonianism, and science was forced to reevaluate the separation of mind and matter.

Spiritualism in itself was not new since there were many older forms of the practice. But this new strand of spiritualism was quite different from previous forms. Modern spiritualism differed from other forms in its vast scope, its rapid and overwhelming rise as a vast cultural movement, in the fact that it made broad and sweeping statements about matters such as reality and death, it also reflected some scientific ideals at least as they were commonly perceived, it was a popular semi-religious movement that was not affiliated with any specific religious faction or established religion, it dealt specifically with the spirits of the dead as opposed to older forms of spiritualism whose spirits were religious in nature, and it depended heavily on the practice of mediumship and various methods of communicating with the dead. In this respect, modern spiritualism directly challenged the primacy of the church and established religions as the sole authorities on an afterlife.

It is certainly no random coincidence that the spread of modern spiritualism throughout civilized society occurred at the same time as fundamental changes that resulted in the reordering of science. The movement spread too rapidly and too far afield to have 'originated' in just a single incident or event, such as the Fox family legend contends. More realistically, the basic tenets and intellectual ideals upon which the modern spiritualism movement was built already existed in society before the Fox incident. The rise and success of modern spiritualism actually depended upon a confluence of factors, many of which predated the Fox incident. Mesmerism had undergone radical changes in the early nineteenth century and withdrew fundamental dependence on the concept of 'animal magnetism' just as science began to make real progress in understanding electricity and magnetism. Mesmerism then concentrated more on psychic and other mental phenomena such as telepathy. Other forms of spiritualism, such as that practiced by the Swedenborgians also helped form a basis for the development of modern spiritualism, while superstitions and popular witchcraft still existed at the very edges of a rapidly growing culture of science. A revival of romantic attitudes in culture that focused on the non-mechanistic aspects of nature and the macabre was also developing. The study of phrenology emerged and became quite popular between the 1820s to the 1840s. Phrenology was as an early precursor to more scientific studies of the human mind. Each of these elements was an important factor and played a role in the development of modern spiritualism.

However, the important role of science can be neither overlooked nor underestimated in this respect. The successes of Newtonian science fostered a growing skepticism toward established religions and religious beliefs. So the religious view of an afterlife in heaven or hell, depending on an individual's actions and choices during his or her life, was now being questioned. On top of this, society as a whole was becoming more educated and worldly as the industrial revolution changed the social mores, habits and attitudes of large portions of human civilization. Between the industrial revolution and Newtonian science, the old ideas associated with established religion were dying and individual people were seeking new ideas regarding their role in the universe with respect to human existence in general. The Darwinian concept of evolution was as much a product of some of these changes in attitude as it was an intellectual stimulant for further changes. Upon its development, the concept of evolution became pert of this complex milieu of new ideas and helped shape new questions about the role of humankind, life itself, the development of mind, and by implication the possibility that some form of our living existence might survive death of the body. More precisely, the modern spiritualism movement was at least an indirect result of the successes of Newtonian science, as adjusted by social and cultural factors, if not a direct effect of Newtonianism.

In turn, modern spiritualism did not just affect the common masses of people in society. It could count among its members many of society's best-educated professionals, precisely those people who could not be so easily duped by common superstitions. Society was searching for meanings to its new style and conditions of life and modern spiritualism seemed too many to be the product of that search for meaning. Even some scientists joined the movement and those scientists who did not become spiritualists certainly had to pay heed the message presented by the movement. The very fact that noted scientists developed an interest in modern spiritualism is a matter of historical importance and may seem to pose a dilemma for science, but the answer to this dilemma is not hard to find. Newtonian science had pushed its boundaries so deeply into the realms of physical reality formerly claimed by religion and believers in the occult, that science was forced to face the challenge posed by the existence of mind as well as the role of the human mind in our common physical reality. Quite simply, science had matured enough to reexamine the fundamental role that had been given to the separation of mind and matter and question that unnatural limitation to its view of nature. Through evolution, science came to realize that mind was as much a part of nature as matter. However, the first attempts of Newtonian science to deal with mind and its relationship to physical reality brought science into direct contact with the occult and common psychical experiences. So modern spiritualism and science were both headed in the same direction at the same time in human history, toward the end of the eighteenth century, at least until science could recast mind and psychical experiences in its own terms.

Within this historical context, a whole new variety of psychical experiences and phenomena, based on the possibility of survival of the spirit, came to characterize the modern spiritualism movement. Mediumship became a growth industry in America and Europe. What had begun as the rappings of a dead salesman in a small house in New York in 1848 evolved into full-blown séances, characterized by such paradoxical events as apportments of flowers and falling rain inside the Victorian living rooms and studies of well known and respected scientists. Spirits made visible during séances oozed ectoplasm and other otherworldly forms of matter. In one famous incident, a well-known and well-studied medium by the name of D.D. Home floated out of the window in the home of an investigator and into the window of another room under the watchful eyes of a group of distinguished scientists. Acts of levitation were also reported during other séances. All of these phenomena were decidedly physical in character and quite flagrant and spectacular in practice, flying in the face of Newtonian science, or so it seemed.

The scientific establishment at large would not give the scientific investigation of the spiritualistic phenomena a fair hearing, in spite of the scientific stature and reputations of the scientists who became interested in the phenomena. Scholarly research and proper scientific papers on psychical matters were shunned by the larger part of the scientific and academic communities. The SPR was formed in part so that scientists would have a forum to raise questions about the newly reported phenomena. According to Alfred Wallace,

So strong was the feeling against the paper ("On some Phenomena associated with Abnormal Conditions of the Mind" by W.F. Barrett) in official scientific circles at the time that even an abstract was refused publication in the *Report of the British Association*, and it was not until the Society for Psychical Research was founded that the paper was published, in the first volume of its *Proceedings*. It was the need of a scientific society to collect, sift and discuss and publish the evidence on behalf of such supernormal phenomena as Prof. Barrett described at the British Association that induced him to call a conference in London at the close of 1881, which led to the foundation of the Society for Psychical Research early in 1882. (Quoted in Marchand, 425)

It is quite evident that those scientists who appeared to control segments of the scientific establishment would 'a priori' discount spiritualistic phenomena, which would seem an unscientific position to take, but such a position regarding parapsychology is still taken by many scientists today. This does not imply that those scientists did not believe in the spiritual world. They could deny that the reported psychical phenomena had any scientific merit, while still keeping intact their personal beliefs of a separate spiritual (in a religious sense) world.

Still, there was enough support in the scientific community to warrant the development of professional organizations for the 'systematic investigation of this subject by science.' Scientific societies were formed to study these new phenomena while individual scientists weighed in with their opinions, pro and con, on the validity of the new varieties of psychical phenomena. The Society for Psychical Research (SPR) was the foremost of these organizations and the first to be formed. This organization was formed in 1882 by a group of British scientists including William F. Barrett, F.W.H. Myers, Henry Sidgwick, Lord Rayleigh, William Crookes, Sir Oliver Lodge and Alfred R. Wallace. The inclusion of Wallace in this group is especially significant because Wallace was a co-founder of the theory of evolution, although his role in that respect is often overshadowed by Darwin's work.

Other scientists were influenced by the same questions and cultural factors that gave rise to modern spiritualism and the SPR, but responded in a completely different manner. In one such case, they approached the question of survival from a purely scientific and decidedly non-spiritualistic direction. In 1875, P.G. Tait and Balfour Stewart published *The Unseen Universe: or Physical Speculations on a Future State*. This popular book held a peculiar position with respect to the spiritualist movement. It was not a work on spiritualism, and had nothing to do with spiritualism, yet it so closely paralleled the attitudes of spiritualists that they could not ignore it. Yet the book was popular among many scientists who were openly antispiritualistic. It must be remembered that the movement of modern spiritualism dealt to some degree with attempts to put older forms of spiritualism on a scientific basis. Any work which was spiritualistic or shared common fundamental characteristics with spiritualism, while being authored by so widely known a scientist as Tait, deserved special attention. However, Tait and Stewart left it to no one's imagination that they were not spiritualists and strictly qualified their opposition to spiritualism. In this book, they argued for the survival of the human spirit after the death of the human body upon strictly thermodynamical principles. They were clearly influenced by the same desire to find a place for mind in science as others. Mind could not survive death of the mechanical body as spirit unless it was itself a real physical 'thing' before death.

William James and other American scientists founded an American Society for Psychical Research (ASPR) in 1884 by on the other side of the Atlantic Ocean. James was interested in the human mind and consciousness as well as telepathy and other psychic phenomena. His wide range of interests dealing with the human mind exemplified the suspected connections between these aspects of the human mind. Ironically, James is considered the a father of the science of psychology, but modern psychology has very little or nothing to do with modern parapsychology in spite of its psychical roots. This may be due in large part to the fact that soon after psychology split from philosophy, as an independent science, it was given over to a different theoretical basis independent of any general concepts of consciousness. On the other hand, James thought psychology should be based upon a study of consciousness. Instead, the concept of consciousness was relegated to an intellectual position behind another important aspect of the human mind, behavior.

The development of psychology as an independent science during this same era is not a coincidence, but further emphasizes the fact that science in general, and Newtonian science in particular, had evolved to the point where they could finally address the concept and role of mind. But Newtonian science had also evolved to the point where it could begin to question science itself, a subject, which paradoxically, is related to the development of the science of psychology. In science, and especially in physics, the human perception of nature and our physical environment had always been taken for granted as representing true physical reality, but that assumption also became suspect in the last decades of the nineteenth-century. In his book the *Science of Mechanics*, Ernst Mach made the first successful philosophical argument refuting the Newtonian concept of absolute space. He also criticized, quite severely, the use of non-Euclidean geometries and extra-dimensional spaces as a haven and home for spirits.

Few concrete scientific theories emerged to explain these new varieties of psychical phenomena, which seemed fairly impervious to explanation within the Newtonian paradigm. So explanations outside of the Newtonian paradigm were sought. J.K.F. Zöllner, a German astrophysicist, thought that spirits existed in a fourth dimension of space, beyond our normal three-dimensional space of experience. He hoped to demonstrate both the existence of spirits and the fourth dimension in a long series of experiments with the American medium Henry Slade. Slade performed tricks that defied normal logic and three-dimensional Newtonian physics. However, he was caught cheating on one occasion by investigating scientists and exposed as a charlatan, very nearly ruining Zöllner's scientific reputation and career.

So, without naming Zöllner, Mach was certainly attacking Zöllner's work with mediums. The concept of non-Euclidean geometries was admittedly valid in mathematical speculations, but that did not mean that our common space was anything other than three-dimensional Euclidean, as commonly perceived. Yet some scientists were conducting astronomical observations with the hope of determining if space was non-Euclidean or Euclidean, so not all scientists agreed with Mach on the strictly *a priori* three-dimensional Euclidean nature of space. The mere existence of mathematically consistent geometries other than Euclid's geometry immediately raised questions about the fundamental structure of space as studied by physics.

On the other hand, some scientists conducted theoretical research into the existence of a luminiferous aether that was thought to convey electromagnetic waves through an otherwise empty space and other scientists postulated that vortices in this aether were, in fact, the very atoms that make up matter. In many respects, the luminiferous aether came to be the embodiment of Newton's absolute space. Mach argued that it was ridiculous to adopt such hypothetical concepts as physical realities since we could not really know physical reality, we could only know our sensations and thus our perceptions of reality rather than reality itself. Mach attempted to redefine the physical world we experience in terms of the sensations that our minds receive from the physical reality outside of our minds.

Reality beyond our sensations was not legitimately questionable by science according to Mach; therefore speculations on concepts such as an 'Unseen Universe' and spiritualism were superfluous to science. Mach's philosophy offered an alternative solution to the dichotomy of mind and matter by circumventing the reality of the physical world. Although Mach did not believe in the possibility of an 'Unseen Universe,' his philosophy was a product of the same influences that were driving the spiritualists in their search for the 'Unseen Universe.' Mach was reacting to the temper of the time no less than were the spiritualists, Tait and Stewart, Zöllner, James and the members of the SPR, but he chose to interpret the synthesis of mind and matter in a different manner than the spiritualists and many other scientists.

Two of Mach's contemporaries, J.B. Stallo and Karl Pearson, shared strikingly similar views of science with Mach. If we were to ask Mach, Stallo and Pearson what constituted science, they would all agree that science deals with the sensations or sense impressions that humanity as a whole derives from the physical world in which it exists. The sensations are grouped, categorized, abstracted and conceptualized in the simplest or most economical terms to give us Natural Laws. But these Natural Laws are products of our minds and cannot be imposed on the physical world since our sensing faculties limit us in our knowledge of the physical world. In this way, these three men, and especially Mach, differed from Tait and Stewart who sought to set up Natural Laws as independent of man's mind in a universe that had both sensible (material and physical) and non-sensible (spiritual) components.

Obviously, not everyone thought of mind as something separate from physical reality and thus fulfilling a role as an interpreter of physical reality, as did Mach. But all scientists were beginning to recognize importance and necessity or understanding the interactions of mind with the world. Mind was now recognized as a player in the game of reality that was described by science, rather than a separate observer or spectator of the game. Science needed to find the place of mind within its ranks instead of ignoring mind as prescribed by the old Cartesian dictum separating mind and matter. These various opinions offer further evidence that Newtonian science had at last progressed far enough, been successful enough, to tackle the subject of the human mind, and out of this philosophical introspection, psychology evolved. But the direction that psychology took as the new science of the mind was different from what would have been expected. Another scientific revolution was under way and the varieties of psychical research were about to change once again.

In far more concrete rather than philosophical ways, the successes of the Newtonian worldview opened the floodgates for the downfall of the Newtonian worldview. The Newtonian worldview was notoriously deterministic in its mechanisms. Given the position and velocity of all material particles in a system, or even the universe as a whole, the complete future of that system, or the universe, could be predicted or determined. But the new electromagnetic theory, originally based upon Newtonian principles, gave different predictions than Newtonian mechanics for phenomena involving the interaction of light waves and material bodies. This discontinuity in the theoretical basis of physical reality posed a serious problem for science and the solution to this looming crisis came in the form of a Second Scientific Revolution, which changed the primary worldview from determinism to indeterminism.

The second scientific revolution began in 1900 and is generally thought to have concluded in 1927 when the philosophical and scientific foundations of quantum theory were finalized. Otherwise, relativity theory, the other leg upon which the revolution stood, was developed between 1905 (special relativity) and 1916 (general relativity). However, this revolution was about more than just physics. Other sciences as well as society as a whole were changing. The science of psychology finally emerged during the first decades of the twentieth century, but the psychology that developed was different from that envisioned by James and others due to the influence of the new physics. Quantum physics was based upon the probabilistic nature of physical reality, which was wholly different from the Newtonian conceptions of physical reality as fixed. The scientific study of the human mind was reduced to statistical analyses and theories of human behavior while the broader questions of mind; consciousness and their relation to physical and material realities were again pushed to the bottom of the basket of scientific priorities.

The new science of the mind would occupy those scientists concerned with such matters for nearly seven decades before the role of mind and consciousness would again surface. In a similar fashion, and following a similar timetable, the quantum ideal of a discrete reality, as opposed to a continuous basis for physical reality, would dominate physical thought for the first several decades of the twentieth century. And, while modern spiritualism did not die out, the scientists concerned with it were unable to verify either the existence of spirits or communication with spirits, so the movements most of the small amount of scientific support that it gained.

Most of the practicing mediums were eventually found to be cheating at one time or another by faking their communications with the dead. Once a medium was caught, that medium was tainted and ignored by science. So those few scientists who were willing to work with the mediums slowly came to the realization that any possibility of scientific verification of their work was hopeless. However, there was circumstantial evidence for other forms of mental communication such as telepathy, so scientific interest in the psychical aspects of the mind did not completely die away. As a social movement, modern spiritualism faded but did not die out as political realities and world wars filled the thoughts of the civilized world, only to be slowly replaced by parapsychology after the 1930s.

Both relativity theory and quantum theory were originally developed to account for discrepancies between electromagnetism and Newtonian mechanics. Yet each took a new life and evolved in unexpected directions that put them at odds with the Newtonian worldview. Relativity theory rendered absolute space and time irrelevant in science. Relativity changed the basic framework of physical reality from a three-dimensional space with a separate time to a four-dimensional space-time continuum. On the other hand, quantum theory introduced the concept of indeterminism in science, based on the discrete probabilistic interpretation of physical reality at the most fundamental level, as opposed to the deterministic views of the relativists. The Newtonian concepts of absolute space and time just disappeared from science. The loss of the absolute in the face of the relative caused almost no stir in other quarters. Any association that the absolute might have had with religious thought, the concepts of life, mind or death, had been forgotten over the previous half century or more, as science and philosophy moved on to greener postures and more important and timely questions.

CHAPTER 2
SUPERNATURAL PERCEPTIONS OF DEATH

Beneath multiple and specific and individual distinctions, beneath innumerable and incessant transformations, at the bottom of the circular evolution without beginning or end, there hides a law, a unique nature participated in by all beings, in which this common participation produces a ground of common harmony.

—Chuang Tzu – c. 250BCE

Religion [is] the crystallization, brought about be a scientific process of cooling, of what mysticism had poured, while hot, into the soul of man. Through religion all men get a little of what a few privileged souls possessed in full.

—Henri Bergson – 1932

The religious pot of stew

We generally accept the fact that religion is based upon our belief in the existence of a god, gods or a supreme being, but religion is just as closely associated with birth, death and the afterlife. In religion, the afterlife is as real as life itself and entering the afterlife is an important event in the progression of the soul or spirit. Although these three 'events' mark specific milestones in the existence of individual human beings, religions apply the concepts themselves to the whole of the universe resulting in various stories of god's creation of the universe (birth), the expected destruction of the universe (death) and in many cases the rebirth of the universe (afterlife) in a new form. Many religions carry the notion of god as creator still further and view our material universe as a battleground in an eternal struggle between good and evil. Just as mortals can choose between good and evil deeds in their daily lives, and thereby face judgment by god upon their death, god is also the final judge of the fate of the universe and will eventually destroy or allow the universe to be destroyed, so that the universe can be reborn or renewed.

Gods do not have to follow or adhere to normal physical laws. In other words, gods transcend natural reality and are thus supernatural beings. In fact, as a creator of the universe god also creates the physical laws by which the universe operates as well as the laws and rules by which human beings and other species must live. So god is the law-giver. The basic concepts of morality, the very fact that there are natural or god-given laws that regulate human life by which humans must live and by which they will be judged upon death to determine their destiny in the 'afterlife', are also a religious affair because that morality derives from god the law-giver. Religions therefore postulate a supernatural origin for human morals and thereby assume that it is their responsibility to determine right from wrong in human affairs and thus what is good or evil. Accordingly, all religions make it their business to estimate and judge human worth and destiny in the name of their god or gods, in so far as their religion is based upon a concept of god. However, a few religions are not based upon the concept of god or creator, but instead on human worth and destiny without the intervention of a god. These religions are not necessarily godless, but rather they presume that human destiny and the afterlife are based upon human actions rather than any judgment by god or in the name of god. In either case, the afterlife would be considered supernatural, beyond physical and material reality, as opposed to an afterlife that is subject to the physical or material explanations of science.

Within this context there are two major approaches to religious thought. These approaches cover the majority of people in our world and thereby affect the whole world both culturally and materially. The separation between these two approaches is not just philosophical, but was originally geographical in nature. The two different philosophical approaches developed to a large degree independent of one another due to their geographical isolation from each other's influence. One approach was developed in the Far East (East and Southeast Asia) and the other in the Middle East (Southwest Asia and Northeast Africa), relative to the traditional European perspective of the world. The eastern oriental religions and philosophies developed in China and India. From their earliest inception, Chinese religions have been both animistic (a belief in spirits) and superstitious in orientation. There is no central god, although there are various legends of gods, demons, immortals and great heroes. Chinese religious beliefs are also full of ghosts. People literally 'pay' homage and tribute to the dead while some ancient rulers of Chinese kingdoms built vast 'terra cotta' armies to protect them in death. The Chinese people respect their dead and ancestors and perform specific rituals ('li') for the dead because they believe that the dead still exist and can directly affect our material world. The Chinese cultural attitudes toward the existence of spirits and souls comes at the expense of a centralized belief in a god that has allowed a more philosophical, pragmatic and less dogmatic evolution of religion in China. Religion in China was more personal than elsewhere without an over bearing religious system to control individuals.

Through its culturally formative years, China was geographically isolated from the rest of the world by seas and oceans, deserts (the Gobi), Mountains (the Himalayas) and vast regions of Arctic tundra. However, there has been a long tradition of trade with China for specialized goods such as spice and silk, so China has not been completely isolated from outside influences however much those influences have been minimized by the isolation. Commercialized trade occurred only after China had already reached a moderate and stable level of civilization and the basic religious principles unique to Chinese culture had already been established. Trade moved to the West along the silk trade routes to the Middle East and over passes through the Himalayan Mountains to India. The trade with India formed a small cultural bridge whereby the other great esoteric and philosophical religious traditions of the East could influence the further development of Chinese religions.

The Hindu religion of India is a polyglot of different ideas. In fact, any religion practiced in India which does not go by its own name is considered part of the Hindu religion, so there are many different forms of worship under the Hindu umbrella. Again, this situation arose from the isolation of the early Indian civilization from outside influences and provided a cauldron of sorts for different local varieties of worship to thrive and develop as part of a greater Hinduism. The esoteric and philosophical nature of Hinduism is very probably a long held tradition carried over from the original inhabitants of the Indus River region in the Northwestern corner of the subcontinent. Little is known about the religion and religious practices of the Indus River people, but archeological artifacts depict these people as practicing ritual meditation in their religious observances while meditation and inward introspection have developed as an integral part of the Hindu religions that still survives today. In fact, similar meditative and ascetic practices in Chinese spiritual philosophies were very probably influenced by the Hindu example. However, very early in the history of human civilizations Aryan migrations brought new traditions to India as the Indus River civilization was overrun and conquered. The Aryan influences forced the development of a greater diversity within Hinduism, as the different religious ideas were adapted to local and regional variations resulting in the Hinduism that exists today.

On the other hand, the primary thrust of western religions has been the development of monotheism, the belief in a single all-powerful god. The western traditions began in ancient Sumeria (a pre-Babylonian civilization of the Tigris-Euphrates delta) and

Egypt. Not much is known about the Sumerian religions beyond some early legends, especially the epic of Gilgamesh. Portions of the epic of Gilgamesh seem to be the origin of some common stories found in the Judaic/Christian Bible. However, the first form of writing (cuneiform) and science (in the form of Astrology) seem to have been developed in the environs of Sumeria, while the Babylonian civilization that followed ancient Sumeria gave the West its first form of written laws in the Hammurabi Codes. So the overall influence of the Sumerian beliefs and their immediate religious descendants is assured.

By contrast, a great deal is known about Egyptians forms of worship and religion as well as their influence on the development of other religions throughout the Middle East. The first stirring of monotheism came from Egypt with the short-lived practices of the pharaoh Akhenaton, who worshiped the god 'Aton', while the Egyptians developed and documented very elaborate rituals and funerary practices for the dead. The Egyptians thereby set the tone for later spiritual developments in the Middle East through their strong political, scholarly and social presence in the region.

The intricate web of influences and interactions between all of these religions would be hard to reconstruct even if good sources of information were available, but information regarding the actual development of religions in the far distant past is not readily available. So a lot of the story of religion can only be pieced together by logical deduction based on archeological evidence while the rest is mere conjecture and speculation. The greatest amount of conjecture deals with the very origins of the concept of religion itself. The earliest origins of religion are lost to antiquity, but religions do seem to fulfill some type of psychological need in the minds and psyches of humans, so we can assume that religion in some form or other has existed far into the human past in some rudimentary fashion. Religion could then be interpreted as a natural development of the human mind to fulfill some basic psychological need.

The earliest development of religions and religious rituals may have resulted from no more than a simple human psychological need to belong to something greater than early societies. People surely would have questioned their role in the world when confronted with the vast expanses of the heavens on a clear dark night. Perhaps early societies invented gods to explain the relationship between the society and the perceived universe. Or possibly the development of early religions amounts to a desire to fill an emptiness generated by our intuitive notions of our role and participation in a greater physical reality that was anthropomorphized by some abstract notion of god. But then again, the origin of religions might not even depend on any concept of god or gods, which might have only come after the first religions were already formed around rudimentary beliefs in an afterlife. In this case, animism predates formal religions that postulate a god or supreme being to rule over the universe. At the very least, this explanation sounds more plausible then many others. There is certainly enough archeological evidence that indicates the human belief that something continues after death has existed for tens if not hundreds of thousands of years, long before any evidence that humans have believed in an abstract concept such as god. Whatever the case may be, our various religions and religious beliefs form an important portion of our cultural heritage and existence today.

Even though we know little about the actual origins of religion, the later development of religion does become much clearer as early written records first became available in the early centuries Before the Common Era (BCE), so our picture of religions in the early years of the Common Era (CE) is more accurate. By the beginning of the common era (originally the Christian era and denoted as BC until the western calendar became the international standard and BC became BCE and AD became CE) the basic religious principles behind our present religious systems solidified as we understand them today while the differences between Western and Eastern religions became set in stone. The eastern religions were basically Taoism, Buddhism and Hinduism with spatterings of other religions such as Shintoism in Japan, while the western standards were Judaism, Christianity and Islam. Each grouping is interrelated by its basic characteristics and the historical parameters of their development.

Eastern Religious Perspectives

The eastern religions are marked more by philosophical and mystical introspection than their western counterparts. In fact, Buddhism is not a true religion if we define religion in such a way that it must be based upon a belief in god, gods or a supreme being. Nor does Taoism have a real god in which to believe, and if it does then one is not even to think about it. Striving to realize the Tao (the way) will only render it harder to grasp. The Tao can only be realized intuitively and not be expressed in words or described by logic. Taoism teaches harmony with nature as a way of acting without any dissipation of our vital energies, or 'chi'. However, there is a striking difference between the philosophical tenets of Taoism and practical Taoism, which is full of gods, immortals, demons and other fantastic beings. Older folktales of such beings have been adapted to Taoism, but they do not form any special religious doctrine. On the other hand, Hinduism has a number of different Gods of various ranks, or just three gods, or in an esoteric sense only one Supreme Being. Hinduism even has an atheistic branch in which no gods are needed. There are no set rules or absolute standards by which each of these religions could be judged or completely defined, so all religions should be judged within their own historical and cultural context. Hinduism was developed long before the Chinese religions and the circumstances of that development give Hinduism it unique character.

In essence, Hinduism would be the oldest of the established eastern religions, beginning anywhere between 3500 and 1500 BCE with the Indus Valley culture. Little is known about the religious practices of this early society but physical artifacts of the civilization show costumed people practicing meditation. So the meditative aspects of Hinduism date back four thousand years and more in some form. Two of the oldest books in the Hindu religion are the *Mahabharata*, an epic poem that includes the *Bhagavad Gita*, and the *Ramayana*. Both are tales of war and conquest probably based on older Aryan traditions and legends of conquest. But there is another group of religious texts called the *Veda* dating from around 1500 BCE. The last books in the *Veda* are called the *Upanishads* and date from around 600 BCE. The *Upanishads* deal with the nature of self (atman) in relation to the reality that underlies the world (Brahman). The interplay of these two basic concepts forms the philosophical backbone of the Hindu religion, in spite of the many varieties of gods worshiped. Of the various schools that interpret the Veda, the Vedanta school is the largest. The *Veda* are the most sacred books in Hinduism, even though other books play a more direct and important role in Hindu worship. Technically any Indian religion that is not another religion in its own right is Hindu, so Hinduism is actually a classification of many different groups and beliefs, all mixed together.

The main gods worshiped within the Hindu faith are Brahma (the creator), Vishnu (the preserver) and Shiva (the destroyer), along with their consorts Saraswati, Lakshmi (or Sri) and Parvati. Vishnu often appears in the religious literature as one of his avatars such as Krishna or Rama or even as Buddha and Jesus in some minor Hindu sects. The majority of temples in modern India are dedicated to Shiva, quite of few less to Vishnu and/or his various avatars and other lesser gods within the pantheon, but only a very few to Brahma, the creator. If asked why, you could be told that Shiva represents destruction, war and death, so people fear and want to appease Shiva and pacify his anger through worship, while Brahma has already finished the creation so there is no need to build temples to worship him. Worship of the other gods falls in-between these two extremes with many lesser gods worshiped locally within India. This approach to worship may seem a bit awkward from the western point-of-view, but it is a more pragmatic if not realistic approach from the Hindu point-of-view. Yet it also emphasizes some philosophical discrepancies in the Hindu religion because Brahma is also associated with Brahman, the underlying reality of the universe. So Brahma is actually closer to a western concept of a single all-powerful god who both created and rules the universe. But Brahma is not directly worshiped to any extent in Hinduism in spite of his obvious philosophical importance as the underlying and unifying concept that binds all of Hinduism together as a single religion. The seeming significance of Brahma to all of Hinduism and the lack of respect paid to his significance through direct worship is paradoxical.

In spite of the wide-ranging differences between the many different sects of the Hindu religion, there is still a common basis to all of Hinduism. Hinduism is based upon a belief in a simple structure of life and the universe as well as rebirth after death and these concepts are wedded to the ideas of Atman and Brahman as described in the *Upanishads*. The conditions of rebirth are decided by the actions of people while alive (karma), so what ensues after death is neither predetermined by god nor random in nature. Even the universe is subject to the cyclical changes of birth, death and rebirth. Each cycle is called a kalpa and lasts about 16 billion years long. In their most philosophical and esoteric religious practices, the Hindus strive for release (moksha) from the cycle of birth and death (samsara). These last beliefs are incorporated in the asceticism that characterizes Hinduism. Throughout the history of India, many people have given up everything to practice various forms of asceticism, ranging from wandering pilgrims to those who would practice various forms of meditation in order to reach moksha, while they do little else with their lives. Society tolerates and supports these ascetics, demonstrating the importance of Hinduism within the culture.

It was within this ascetic tradition that early reformers tried to change Hinduism. Early on, the ascetic tradition was strong in the north of India where two such reformers emerged to establish new independent religions. Buddhism and Jainism both emerged in Northern India during the sixth century BCE. Siddhartha Gautama (born in the 560s BCE) was a prince of the Shakya clan of the Kapilavesta people. Siddhartha became an ascetic at the age of thirty-nine, giving up his royal titles, riches and family to take up the ascetic life of a searcher after truth. Some years later he sat down under a Bo tree for forty-nine days, vowing to overcome the cycle of birth, death and rebirth to finally attain enlightenment. The Buddha, as he is now called, then taught others how to reach enlightenment and break the cycle of birth, death and rebirth, thus founding the religion of Buddhism. Mahavira, another ascetic of the same period of time, discovered how to conquer the forces that keep people tied to rebirth through his own ascetic practices, thus founding the religion of Jainism. Both of these religions were originally reform movements within Hinduism, but both later evolved as separate religions. For example, Buddhism strictly rejects Hindu rituals so it could never be considered a separate sect within the Hindu fold although some believe that the Buddha was an avatar of Vishnu. Jainism never really spread beyond the Indian subcontinent, but Buddhism became a world religion as it spread in all directions from India. Ironically Buddhism was eventually wiped out within India, its country of origin.

The first surge of Buddhism came in the third century BCE when the emperor Ashoka conquered a large part of the Indian subcontinent. Ashoka is credited with spreading Buddhism throughout southern India and Sri Lanka. In the following centuries, Buddhism found its way west to Afghanistan, north to Nepal, Tibet and into China as well as southeast into the lower portions of Asia (Burma, Thailand, Cambodia, Vietnam, and Malaysia) and into Indonesia (Bali). At one time Buddhism cut a large swath throughout eastern and southeastern Asia. As Buddhism fanned out across the world, two different versions of the religion developed. The Mahayana seems to have located more in the East Asian regions while the Theravada Buddhists are more centralized in Southeast Asia. Although the differences between these two approaches to Buddhist doctrine are quite complex, Theravadan Buddhism does seem to be more devotional while the Mahayana brand of Buddhism seems to be directed more toward enlightenment through individual efforts and practicing meditation.

Also, Mahayana Buddhists believe in various heavens where Buddhas and Bodhisattvas (people on the way to becoming a Buddha) reside, while the Theravadan Buddhists feel that there is no permanent self, so a person may well just cease to exist after reaching the ultimate nirvana or enlightened state. But once again, there are intimate subtleties to the concepts that are not satisfied by describing the principles in such simple terms. Yet there does remain a single underlying principle in all of Buddhism that unites everyone and that is the singular fact that in each and every person there is a Buddha-nature and each and every person can break the cycle of birth and rebirth, attain nirvana or enlightenment, by realizing their own inherent Buddha-nature.

During the sixth century of the Common Era, Bodhidharma (called Daruma by the Japanese) officially introduced Buddhism into China. It is possible that the Chinese already knew a great deal of Buddhism by this time because Bodhidharma was brought to China specifically to teach Buddhism. It is also at this time that the native Taoist beliefs and Buddhism successfully merged to form Ch'an Buddhism. Ch'an is interpreted as recovering the 'original mind' through meditation. The word itself is a Chinese form of the Indian Sanskrit word 'dhyanna'. Ch'an Buddhism thrived over the next century and was studied in many monasteries, including the Shao Lin monasteries famous for their Kung Fu fighting techniques. However, the Ch'an tradition split in the seventh century when Hui Neng (638-713), the sixth patriarch, was forced to flee from the Shao Lin temple with his followers.

Hui Neng was a poor illiterate beggar boy who was naturally and spontaneously enlightened before he came to the temple to study Buddhism. Only the patriarch of the temple understood and knew of Hui Neng's enlightenment so Hui Neng held no special position within the Temple political structure. When it came time for the fifth patriarch to pick his successor, he knew that Hui Neng was the only person who could truly follow him, but the politics of the temple would not allow such a thing to happen. So, the patriarch initiated a poetry contest to choose his successor. A poem written by the chief monk won the contest because Hui Neng could neither read nor write.

> The body is the Bodhi-tree,
> The heart is like a bright mirror stand,
> Strive to clear it at all times,
> And allow no dust to cling.

The poem was philosophically correct, but obviously did not reflect the intuitive understanding of enlightenment that comes from first hand experience. When the winning poem was read to Hui Neng, he disagreed with the sentiments of the poem, and had a friend secretly write a rebuttal in the form of another poem on the wall of the temple.

> Fundamentally no Bodhi-tree exists,
> Nor the stand of a mirror bright.
> Since all is empty from the beginning,
> Where can the dust light?

Hui Neng's poem reflected his first hand experience of enlightenment. The patriarch knew immediately who the author of the new poem was because only Hui Neng had reached the level of enlightenment that was expressed in the new poem. So the patriarch gave Hui Neng the official insignia of the office and the Patriarch's robe and bowl and told him to run for his life. Thus, the Ch'an Buddhist movement split into different feuding factions.

Ch'an Buddhism was introduced into Japan in the thirteenth century where it came to be called Zen Buddhism, so named because the practitioners practice zazen or sitting in meditation without purpose or goal. In Japan, Buddhism took a strange new twist when it became entangled with the fighting philosophies of the warrior caste. The samurai warriors of Japan so respected the discipline of mind exhibited by the Zen Buddhists that they based their Bushido warrior's code on Buddhist thought. In this manner, some of the basic Buddhist principles were incorporated into various forms of Japanese martial arts. In a sense, this completed the adaptation of Taoism and Buddhism that began in China. The syllable 'do' in the word Bushido is just Japanese for the word 'tao', which can be translated as the way or the path.

By the time that Taoism and Buddhism first began to merge together in China, Taoism was already more than a thousand years old. Taoism had first emerged during the days of the one thousand philosophies in the sixth century BCE. Different philosophies and schools of thought were spread throughout China, with no single clear direction on how religion would further develop. Out of this entanglement of ideas, two different philosophies (religions) emerged to dominate the other philosophies: Taoism and Confucianism. Taoism is the sole product of the mind of Lao Tzu. Although he was very probably a real person, the name Lao Tzu actually means 'old master' so that was probably not his real name. Confucius or K'ung-fu-tzu, however, was a real person and a great deal is known about him. Confucius lived from 551 to 479 BCE. Lao Tzu's Taoism played to the esoteric nature of the Chinese scholars while Confucianism offered a far more practical panacea for the philosophical confusion of the times.

Confucianism became the religion of the Chinese state for a millennium because of its practical values. Confucius taught that every person should cultivate his or her inner humanity. This inner humanity was not just a characteristic of the rich and noble, but it was a characteristic that was shared by all people. Perhaps this human trait is why the Chinese liked Buddhism. There is a striking similarity between this inner humanity ('jen' or genuine humanity) and the Buddhist concept that every person has a Buddha-nature. But Confucius took his ideas in another direction and applied the concept to education rather than individual enlightenment. He taught that all qualified people deserved an education, not just the rich and noble. It is this aspect that made Confucianism so practical since a large number of educated people were needed to run the vast bureaucracies of the Chinese imperial government.

Confucius also stressed the importance of family values and incorporated them into his basic principles of how to live. This factor also fit the notion of an imperial government quite well. He taught respect for all people and redirected the traditional respect the Chinese had for the dead toward the living, including the rituals ('li') performed for the dead. All people were to observe rules of propriety ('li') toward one another. This respect was the basis of Confucius' golden rule of conduct, "Do not impose on others what you yourself do not desire", which he published in his *Analects*

(15.24). Although not really a religion in the normal sense of the word, Confucianism did evolve as a religion as older religious rituals, beliefs and myths were adapted to the new philosophy.

On the other hand, Taoism was an esoteric philosophy that was readily adapted to religious beliefs from the very beginning. Taoism began with the book *Tao Te Ching* as dictated by Lao-Tzu. The Tao is the way of nature, but it is also a vague intuitive thing that cannot be described by words. According to the *Tao Te Ching*, the Tao cannot be given a name it is indescribable. Yet the Tao is the mother of all things, including humans. The Tao takes up no space or volume and it is timeless, yet the Tao exists and is real. The Tao exists nowhere; at least it does not occupy a volume of space, yet it is in all things. And people should strive to be like the Tao, yielding to the forces of nature by not acting with deliberate intention. Acting with deliberate intention against nature and the natural forces wastes the vital life force that is in all humans. There are three vital forces, the most important of which is 'chi' ('ki' in Japanese), the original breath. A person's 'chi' must work in harmony with nature. It is the principle of 'chi' that brings Taoist philosophy into the realm of medicine and the martial arts. Good health depends on a proper flow and balance of 'chi' and the other vital life forces throughout the body. So controlling 'chi' became the basis of Chinese medicine, as in the practice of acupuncture. But the flow and control of 'chi', the inner force, is also the goal of the martial artist. According to legend, Bodhidharma not only brought Buddhism to China, but he also developed the first form of kung fu at the Shao Lin temple of Ch'an Buddhism. Through practice and meditation, martial artists learn to live and act in harmony with nature and control their 'chi', in the Taoist manner, and thus to control the concentration and activation of 'chi' for power when needed during fighting.

From this short look at Eastern religions, which is far from thorough or comprehensive, a developmental pattern is already beginning to emerge. Strangely enough, during a period of about a century coinciding roughly with the sixth century BCE, a number of significant changes were made in the eastern religions. Not only did Confucius and Lao Tzu develop their philosophical and religious principles during this time in China, but Siddhartha Gautama (the Buddha) and Mahavira were also active in northern India at about the same time. The *Upanishads* also date from this same period, so it would seem that the Eastern religions all took a sharp turn toward a more philosophical and abstract interpretation during this period. Even more astounding is the fact that this same period of time covers the Jewish captivity in Babylon. Daniel was reputed to have lived during the Babylonian captivity and supposedly authored the book in the Bible by that same name. The book of Daniel is unique and significant because of its apocalyptic nature and predictions. The Jews were subsequently set free when Cyrus the Great of Persia conquered Babylonia, allowing them to return to Judea although many Jews instead migrated to Persia where they came into contact with Zoroastrianism, a religion that had only been founded by Zarathustra a short time before. So these were also formative years and a period of change for the development of western religions.

This same period of time also saw the philosophical work of Thales on the island of Miletus and the work of Pythagoras of Samos that initiated the development of Greek philosophy as well as the long march toward science. The Greek philosophies would later influence the development of Christianity. So it would seem that the sixth century BCE was the most productive century for the concept of religion that has ever been because religion as a whole turned toward philosophy and philosophical introspection during this short period of time. The significance of so many philosophical changes over so large a geographic area almost simultaneously is unimaginable, but perhaps it marks a specific period of development in human consciousness as a whole. However, any allusion to this notion would be speculative at best even though the probability of so many changes happening within such a short span of time is extremely small.

Western Religious Perspectives

The major western religions of Judaism, Christianity and Islam developed along different lines than their eastern counterparts although they do share some characteristics with their eastern cousins. The emphasis in the west is more on the individual and group worship of god than in the east where the development of the individual mind and consciousness is emphasized. And, as in the Far East, the characteristics of the western religions are due to their historical development, only in this case beginning with the first religious concepts in Sumer. The Sumerian civilization developed in Mesopotamia, the area between the Tigris and Euphrates Rivers, in about 3500 BCE. The area included many different city-states, a new political and social construct. Sumer was the first important city-state in the area. The city-states originally led independent existences, but after about 2300 BCE different rulers of the city-states began to seek dominance over the other city-states. The Assyrian and Babylonian empires eventually came to dominate the area until 539 BCE, when Cyrus led the Persian conquest Babylonia. Each city-state originally worshiped its own god, but each city-state respected the gods of the other city-states. These gods formed a hierarchy of their own. The gods dwelled in heaven, but also took form on the earth. The details of this religion are somewhat sketchy, but what is known comes from clay tablets with cuneiform, the first writing system of the world. It is from such clay tablets that modern scholars have learned of the epic of Gilgamesh.

Gilgamesh was the king of Uruk in about 2700 BCE. He was reportedly the fifth king after the flood. In the epic tales, Gilgamesh was extremely strong and quite handsome and mingled with the gods. His brother Enkidu, with whom he shared his adventures, lost his own strength and died forcing Gilgamesh to realize his own mortality. Gilgamesh then sought to find the secret of life and the answer to immortality, but let it slip through his hands. So Gilgamesh finally came to accept his own mortality. During his quest to find immortality, Gilgamesh met Utnapishtim, whose city was destroyed by a great flood that was sent by Enlil, the king of the gods, because humans were too noisy. However Ea, the god of water, took pity on Utnapishtim, warned him of the impending flood and instructed him how to build an ark on which he could save two of every animal that lived. The gods rewarded Utnapishtim with everlasting life after the flood. Obviously, this story is the origin for the biblical story of Noah's ark. Other early stories were also adapted to the needs of the newly forming Jewish religion that emerged from Mesopotamia several centuries after Gilgamesh. Gilgamesh's city of Uruk has even been identified with the city of Erech in the Judeo-Christian Bible.

Abraham, the patriarch of both the Israeli and Arabic cultures, emigrated from the city of Ur in Mesopotamia somewhere between 2000 and 1600 BCE. He fathered Jacob by his wife Sarah and Ishmael by his concubine Hagar. Jacob's offspring then became the Israeli nation and Ishmael fathered the Arabic race. Abraham worshiped a single all-powerful god, which became Jehovah to the Jews and Allah to the Arabs over the next two millennia. Abraham's god promised both Jacob and Ishmael that they would found their own nations and indeed they did. Not only did they father nations, but each also inspired their own religions several centuries later, Judaism and Islam. Both religions claim Abraham as their forefather.

It is at this point that the Egyptian religion enters the picture. By whatever means, the Jewish people came to be slaves or workers within Egypt a few hundred years after Abraham, even though God had promised them the land of Canaan as their own. Moses, Jewish by birth but raised in the Egyptian royal family, killed an Egyptian to protect a Jewish slave when he was a young man. Fearing retaliation and his own life, Moses fled Egypt in about 1500 BCE. He spent the next forty years of his life in exile in Midian. During his exile, he found and communicated with the God of Abraham on mount Horeb in the Sinai desert. God appeared as a burning bush that was not consumed by the flames and told Moses to return to Egypt and free the Jews from captivity. After many trials and tribulations, Moses led the Jews from Egypt and into a forty year exile in the Sinai Desert. Moses again met God (JHVH or Jehovah) on the mountain where he was given the Ten Commandments as well as the Torah, the first five books of the Judeo-Christian Bible. The Torah tells the early history of the world relative to the Jewish experience and also establishes a complicated set of laws, rituals and rules by which the Jews should live. Moses eventually took his followers to Judea where they settled and became the Jewish nation with a land and a religion all their own. The Jews now had their own history, a present life and a future as well as a destiny to fulfill. The Jews were God's chosen people. Ironically, Moses was not allowed to enter the 'promised land'. God forbade it. So the manner and circumstances of Moses death are unclear and based on legends within legends, not really a fitting end for so important a Jewish icon, but perhaps a fitting end for a person who was trained and educated as a member of the pharaoh's family.

No verifiable evidence of Moses' life has ever been discovered other than the biblical stories, so a great deal of scholarly interpretation and speculation has been made about Moses. Some scholars believe that the 'Moses' story is an adaptation from earlier legends and myths, like the flood story, while some have proposed that Moses was an Egyptian monotheist who still worshiped 'Aton' as Akhenaton had done. But whatever the reality turns out to be, Moses is an important pivotal figure in western religion, not just in Judaism. While Abraham brought the religion of a single God out of Mesopotamia, Moses gave the Jewish people a religious doctrine to follow in the Ten Commandments and the Torah. Yet Moses was raised as an Egyptian and the Jews in captivity would have been subject to Egyptian religious practices. While the Jews may have wandered around the Sinai for forty years in an attempt to cleanse themselves of what could be considered heretical Egyptian religious practices, such as idol worship, there would still remain some deep-rooted philosophical parallels between Judaism and the Egyptian belief system. For example, the Egyptians practiced mummification of the dead. The bodies of the dead were sanguinized, and then washed in honey and oil. The brain was removed as well as the internal organs, which were sealed in Canopic jars to be placed in the tombs with the other remains. Then the bodies were chemically treated, wrapped and mummified. The whole body was saved because it was needed for the journey into the afterlife. When the wrapped bodies were placed in the sarcophagus, a 'book of the dead' or 'tomb book' was also placed nearby to help the dead person on his or her journey to the afterworld.

By the same token, it was Jewish custom to carefully cleanse the body of a dead person with oils and wrap the body in clean linen before placing it in the tomb. Even in today's world, after terrorists murder Jews on the streets of Israel, orthodox Jews will come and clean up every drop of the victim's blood to be buried with the victim, according to Jewish tradition and law. This practice is reminiscent of placing all of the remains of an Egyptian to be mummified in separate jars that were then placed with the mummified remains. In the Egyptian religion, the whole body needed to be present in the tomb for the soul to make the journey to the afterworld. It seems that each and every material drop of a human's life is sacred, which is not so outrageous or strange an idea given the modern fact that each and every cell of a person contains that person's genetic code.

Large parts of the Jewish Torah are complex rituals, laws and rules that the Jews must follow during their daily lives. Whether these were God-given dictums or not, they are to be followed to prepare Jews to enter their death when it comes. After a manner, these rules are similar to the rituals and rules in the Egyptian tomb books that were meant to guide the dead to the afterlife. The individual rules themselves are not similar; the similarity lies in the purpose behind the rules. Rituals for the living have replaced the rituals for the dead to prepare for the journey after the living die. Whereas the Egyptian priests prepared the dead for the journey, Jews prepare themselves for their concept of the journey during life. In his own time and place, Confucius altered the perspective on life and death of his followers by taking the respect and rituals ('li') that were reserved for the dead and showing that the same respect and proprieties should be given the living. Likewise, Moses turned the journey and preparation for the afterlife that Egyptians reserved for the dead and made it mandatory through the laws and rules in the Torah for the living. Moses did not necessarily copy the Egyptian tomb books for use in the Torah, but he did change the ideas that they represented for the dead and applied them to the living. Moses rendered life itself and the manner in which the Jews lived their lives as the journey toward an afterlife when they died. This practice of living a 'good' life in order to enter 'heaven' on death is still followed today in religions throughout the modern world. This change in perspective marks Moses' real accomplishment as a founder of the Jewish religion.

The Jewish religion continued to grow and prosper over the ensuing centuries until Judah was conquered by the Babylonians and its people carted back to Babylonia in 586 BCE, an event called the Diaspora. The Jews remained in Babylonia until 539 BCE when the Persians conquered Babylonia and Cyrus allowed the Jews to return to their homeland in Judea. During the Babylonian captivity, the Book of Daniel was written and added to the Bible. The book of Daniel is different from other books in the Bible in that it is not a history. The book of Daniel is an apocalypse. The book is meant to inform the people that they must remain faithful and true to their Jewish heritage and religion in spite of their captivity, oppression and persecution. But Daniel also addressed the issue of resurrection from the dead, an event that the Jews can look forward to if they remain true to their Jewish heritage and laws. Daniel's visions of a spiritual realm reinforced the hidden meaning of the Jewish faith and laws, so they offered hope to the Jews during hard times and gave meaning to their lives such as they were during captivity. Daniel's visions renewed the meaning of Judaism as a journey toward death followed by a resurrection in an afterlife in a spiritual realm. It has been argued by scholars that Daniel was not a real person but a pseudonym, and the book of Daniel was written at a much later date than the Babylonian captivity, but that fact has never been established. Even if the book was actually written at a later date by someone else, there is still no reason to believe that the intellectual and philosophical ideas presented in the book are any less valid. Scholars also speculate that the very idea of an apocalypse originated in the Zoroastrian religion of Persia, so Daniel could have been written as a result of the Zoroastrian influence.

Cyrus the Great of Persia was considered a Messiah by some of the Jews that he freed and they followed him back to Persia instead of returning to Judea. In Persia, the Jews learned of another monotheistic religion like their own, but different, Zoroastrianism. Zarathustra (Zoroaster in Latin) founded Zoroastrianism sometime during the first half of the first millennium BCE. Modern scholars tend to push the date of Zarathustra's accomplishment back into earlier times, about 1000 BCE, but Plato placed Zarathustra's lifetime about 600 BCE. There is also some evidence that Zarathustra lived during the *Rig Veda* period of the Hindu religion and was thus influenced by the Aryan before they invaded northern India. Characters in the *Rig Veda*, like the god Mithra, are archangels in the Zoroastrian religion. Given the fact that Zarathustra's first converts to his new religion lived in the province of Bactria (northern Afghanistan) and that the Aryans came out of the Steppes north of Bactria and down into India, there is little doubt of a very early pre-Hindu connection between Zarathustra and the Aryans. The Aryans had earlier swept south into Persia and become the Persians. Rather ironically, the modern name if Iran is Persian (Farsi) for Aryan. Whichever the case may be, Zoroastrianism was already established in Persia (Iran) when the Jews migrated to that country and parts of the Zoroastrian religion influenced the development of Judaism. Darius, Cyrus' successor, proclaimed Zoroastrianism the official religion of the Persians.

The religion that Zarathustra taught was strictly monotheistic. A great deal of information about the details of Zoroastrianism has been lost through the persecution of Zoroastrians during later invasions of Persia by Alexander the Great of Macedonia and the Islamic Arabs, but the general tenets of the religion are still preserved. Precise ideas are vague except for story of the worlds' creation by Ahura Mazda the 'wise lord'. Everything emanates from Ahura Mazda and in the end returns to Ahura Mazda. Ahura Mazda has no form, but does have six aspects, three male and three female. The first aspect is divine hope, which is beyond human comprehension although its purity is represented by fire. The second aspect is wisdom, the third creative power, the fourth is devotion and faith, and the fifth is harmony and beauty, while the sixth and last aspect is called Ameretal. The final aspect, one of a female, represents immortality and knowledge of this aspect will dispel the fear of death. This aspect is associated with the tree of life and immortality, which can only be found in heaven.

The physical world is a combination of both good and evil, a product of two opposite spirits. The good spirit is in all people while the evil spirit is an illusion by which we realize what is good in life. Good will eventually triumph over evil, as Ahura Mazda destroys and renews the universe in a final conflict. In this final conflict, there would be a final judgment where friends of 'the Lie' would be condemned to Hell and the pious were allowed to enter Heaven. These elements of Zoroastrianism still survive in Judaism, Christianity and Islam. In addition to the concepts of a final judgment, resurrection, the supernatural struggle between good and evil and the participation of humans in that struggle, Zoroastrianism also introduced the concept of angels as well as the devil into the mix of western religions.

Coincident with the Diaspora and other political events in the Middle East, a completely new perspective of the world began to emerge farther to the west in Greece. No one has determined what is so special about the Greek culture and religious concepts for the new philosophies to emerge in Greece rather than elsewhere, but a new understanding of the world and a new worldview through which the world could be described and explained by physical principles rather than gods was just beginning to evolve in the Greek culture in the sixth century BCE. Thales (624-547 BCE) of Miletus, a small island near the coast of Asia Minor, initiated this new worldview. Greek popular religion was based upon a polytheistic structure. The Greeks worshiped a number of different gods and goddesses who lived on Mount Olympus and regularly interacted and interceded in human affairs. But why the new philosophies evolved out of these religious views is unknown. Thales certainly traveled to Egypt where he learned geometry, so the Egyptian influence cannot be discounted, but Miletus was also a crossroad for trade and other philosophical influences cannot be discounted. Whatever the unique influences on his philosophies, there is no doubt that Thales ushered in a whole new era of thought.

Thales explained the world by reason and logic, not as the product of the will of a god or gods. Thales is also thought to have placed geometry on a logical basis for the very first time by proving several basic theorems of geometry. He also predicted a solar eclipse, but more importantly he tried to explain earthly natural events logically and rationally rather than supernaturally. Thales believed that water was the fundamental element out of which the whole universe was created, thereby initiating a long search for the most fundamental and common concept to explain physical and material reality. That search continues to this day in modern theoretical physics.

Other Greek philosophers followed Thales example, although complete credit cannot be given to Thales. The fact that others so readily listened to Thales and accepted his ideas indicates that Greek thought was ready and waiting for some kind of a change. Thales only primed the pump and then lit the fuse. Pythagoras of Samos (569 to about 475 BCE) was also a central figure in this development. Pythagoras is considered the first pure mathematician, although much of his original work in mathematics is unknown because he started a secret semi-religious cult and his discoveries were kept secret within the group. It is known however that he traveled extensively and probably studied in other countries including Egypt. He also studied under Thales and Thales' student Anaximander on Miletus. Thales thought that the basic element of the Universe was water, while the material world was both a mixture of sand and water or water also gave rise to sand and thus water was the basis of all material existence. On the other hand, Anaximander thought it was the 'indefinite' (apeiron) and the later philosophers Anaximenes and Heraclitus thought that the basic element of reality was air and fire, respectively. In each of these cases, the single element represented the fundamental all-pervading principle that contained everything else in the universe. But Pythagoras thought the basic element was pure numbers.

Pythagoras was a musician and played the lyre and noted that the pitch or tone of a plucked string changes in the same ratio as the length of the string. If the middle point in the string were held down, then the pitch would double. If the string was held down at the one-third point, then the pitch of the tone would be tripled, and so on. From this simple principle, Pythagoras deduced that the whole world could be explained by the whole numbers and ratios of the whole numbers. The universe had harmony. This abstraction of physical phenomena to mathematical terms became the fundamental conceptual basis of the Pythagorean cult as well as the conceptual basis of all science over the next three millennia. Eventually the Pythagoreans learned that the length of a diagonal of a square was a transcendental or irrational number, which could not be expressed as the ratio of two whole numbers and the cult disappeared. Such a simple notion as the square root of two destroyed the philosophical basis upon which the Pythagorean cult was structured. Even so, the diagonal of a square could still be measured geometrically even though it could not be expressed rationally, such as by the ratio of two whole numbers, so the evolving science was based instead on measurement and geometry rather than counting and arithmetic. However, the fact that nature could be rationally and logically explained alone, without reference to a god or gods, would forever change the religious concepts that were emerging. In particular, the Greek philosophical concepts had a profound effect on Christianity. The Greek concept of an eternal soul was especially influential.

Empedocles of Acagras (492 to 432 BCE), a Greek city in Sicily, was the first to use the system of four elements; earth, air, fire and water. He saw our world of physical phenomena as a development within a cosmic cycle of eternal change, growth and decay ruled by the dual natures of Love and Strife, as opposed to the religious view of good and evil. Love (Philia) was a force of attraction and combination while Strife (Neikos) was a force of repulsion and separation. Empedocles also developed, it seems, the first mechanistic description of the origin of species. Living organisms were created by Love and Strife. Empedocles argued that the Olympian gods of the Greeks were a result of misinterpretations of nature and that the real 'gods' were the basic elements of nature and the cosmic forces that shaped and directed reality through the endless evolution of cycles. Within this context, the human soul could undergo purification through proper religious and ethical teachings to achieve a unity with perfect Love. The cycles through which the soul goes through its various incarnations parallel those of the cosmos itself, seeking the harmony of a past time when a golden age of universal harmony existed. The soul was thus eternal and intimately connected to the whole cosmos and the harmony of that cosmos.

The philosophies of Plato and Aristotle were especially influential in the later developments of western religion. Plato lived from 427 to 347 BCE, well after Thales, Pythagoras and the others. In his day, the Greek philosophical concepts had become more abstract in nature and Plato dealt first with issues of Being (the changeless ideals of the world) and Becoming (that which changes in the world) as well as what constituted our sensed reality. Plato argued that the real world of the changeless had a strong Being while the changing world of appearances displayed a form of weak Being. So for Plato, Being and Becoming formed the total reality of our world together. Upon this basis, Plato instituted a Doctrine of Forms. The unchanging real world was made of perfect Forms, ideal shapes and concepts against we judged our sensed world of change. For example, there might be an ideal Form of a tree against which we could determine which of all the plants that we sensed in our common world were trees and which were not trees. The concept of ideal Forms was especially helpful in geometry, since we all have an idea of what a perfect triangle or circle is, but the triangles we draw or see in our changing world are never that perfect even though we recognize them as triangles and circles. The world of Forms was the 'real' changeless world, while the changing world of our senses was a shadow of those Forms. The shadow analogy was made elsewhere in Plato's writings but we can assume that he was still referring to the world of Forms. Plato pictured our sensed world as shadows on the wall of a cave, whereas the ultimate reality was the Forms that produced those shadows. Virtues such as love and truth were also Forms and what we sense as love and truth in the common world are just shadows of those pure Forms. The ability to recognize the shadows in our common sensed world was an innate ability that humans acquired before the soul was placed in the imperfect material body.

Plato also spoke of a single ideal god, called the Demiurge (Demiourgos), in his explanation of the creation of the universe as portrayed in the *Timeaus*. The Demiurge was both the Father and begetter of our world, but the Demiurge was also a symbol or Form of the concept of a soul. The Demiurge did not create the universe from a void or from 'no-thing', as had the gods in common religions, but instead brought order out of elementary matter by imitating pre-existing eternal Forms. The universe itself was more than just material it was a living thing. The planets and heavenly bodies circled around a spherical Earth along perfect circular paths. This particular structure of the universe, with modifications, became the basis of both religious and scientific cosmology for the next two millennia. Plato's Demiurge was not like the Jewish God of Abraham, but rather the creator of good, an ideal Form, because good could only be created by good.

The human soul was an invisible and rational life principle that occupied the material body. Like all Forms, the soul was eternal, but it had three parts: the intellect, the spirit and the appetite, which dealt with the basic impulses that drive living beings. The influence of Plato's ideas on the development of early Christianity is hard to pinpoint, but Jewish and Christian philosophers in the first century CE associated Plato's concept of Forms as thoughts within the divine, perfect and ideal mind of God. In particular, the Christians developed a concept of an immutable God that was not originally part of the Jewish doctrine, but instead based on the Platonic concept of the Demiurge. The Christian concept of dualism in the form of mind versus matter or nature versus spirit was also influenced by Plato's concept of an ideal reality based on Forms as opposed to the sensed world of matter.

Aristotle's worldview was, however, quite different from that of Plato even though Aristotle was a pupil of Plato. Aristotle was primarily interested in the real world of senses and based his 'science' on direct observation of the world. Aristotle catalogued flora and fauna from all over the world as they were sent back to him from Alexander the Great, his pupil, while conquering the known world. So Aristotle developed concepts of species and an hierarchy of living things. But Aristotle also wrote the first book of *Physics* (Greek for 'nature'), defining the basic principles upon which science would be conducted for the next two millennia. In particular, Aristotle differentiated between cause and effect in the physical world and developed a notion of the 'first cause'. Aristotle was a religious man and spoke directly about God. God was not a creator because the universe was eternal, but Aristotle did offer a philosophical proof of God by extrapolating to the 'first cause' of motion. Aristotle elaborated on Plato's concept of the moon and planets circling the earth and developed a cosmology based on that fact. Everything on earth was made of the four basic elements; earth, air, fire and water. The natural positions of earth, water and air were at the center of the universe, which was the center of the earth, so the elements fell downward by gravity. But the moon and everything in the astronomical heavens was made of aether, the perfect element, whose motion was perfect circular. This cosmology of the universe was eventually adopted by the Roman Catholic Church as supporting their view of a Hell in the center of the Earth and the spiritual Heaven.

Christianity came more than three centuries after Plato and

Aristotle. It was built around the life and teachings of a Jewish carpenter named Jesus (Yeshua), the son of Joseph and Mary. Many people believe that Jesus was the Jewish Messiah; a figure that Jewish prophets had predicted would come and save the Jewish people. A great deal of legend has grown around Jesus' birth and life, in part because the written versions of his life, as appear in the *New Testament* of the *Bible*, give differing accounts of his birth, life and death. In fact, almost nothing is known about his life from shortly after his birth, to the time that he started his ministry at about thirty years of age, at least if we only read the books in the Christian *Bible*. But other books, called the *Apocrypha*, exist that tell of Jesus' earlier years. The *Apocrypha* were banned as religious texts and declared heretical by the Council of Nicea in 325 CE (or AD). The Council of Nicea also decided that reincarnation was impossible, setting the standard for a Christian afterlife for all future Christians. Very little is actually known of the real person of Jesus. Even his birth date is unknown in spite of the fact that we base our calendar on that historic date (zero BC or zero BCE), and his birthday is unknown even though we celebrate the birth of Jesus on the twenty-fifth of December. That date is actually the birthday of Mithra, a Zoroastrian deity or archangel of light who was born on the winter solstice. Such inaccuracies are the result of political decisions rather than true theology and history. Many of the legends of Jesus' birth and life may have been fabricated and/or adapted from other religions to give Jesus a more messiah-like appearance. However, no matter what the actual details are, Jesus was a true reformer of Judaism.

The teachings of Jesus are simple, but the tradition of what Jesus was (or is) and what he taught are quite complicated. Jesus tried to reform Judaism, which he thought had become corrupt through its own political dealings as well as through Roman influence. Jesus traveled around Galilee (northern Israel) where he lived, teaching and performing miracles. As his popularity and following grew, he moved on to Jerusalem to directly challenge the religious authorities in the capital, the Pharisees. The times were revolutionary as religious fanatics called zealots attacked the Roman occupiers and the puppet government of King Herod. In Jerusalem, Jesus continued to preach, but he also cast the moneylenders and changers out of the Temple of Solomon as an abomination to God and was questioned by the religious authorities for his heretical views. But they could find no evidence that he was anything other than a good Jew. Jesus was eventually arrested for sedition against the Roman occupiers, tried before Pontius Pilate the Roman governor and sentenced to die by crucifixion. Jesus was crucified, died and was placed in a tomb. But three days later, he reportedly rose from the dead and ascended to heaven.

This story of Jesus confirmed the concept of resurrection that had become part of the Jewish tradition, while Jesus was a Jewish reformer. However, political events and unfolding circumstances resulted in a cult of Jesus being elevated to a whole new religion, Christianity, rather than a reform movement within Judaism. The word 'christ', the honorific title given to Jesus, is derived from the Greek word 'christos', meaning savior, messiah or anointed one. While alive, Jesus had twelve confidants or disciples who traveled with him, learning at his feet. One of the twelve betrayed Jesus to the Romans, but another replaced that follower. So after the Resurrection, the disciples became the teachers and spread Jesus' message throughout the world. The problem is that Jesus' message has been garbled through so many translations in both time and location. All Christians claim that they know exactly what that Jesus' message is, or was, and that other Christians with other opinions are either misguided or outright wrong. So the exact details of Jesus' teachings and life as well as Christian doctrine are still debated.

There have been so many changes and interpretations of Jesus' teachings and his life story as well as the manner of his death that hardly anything can be said in general about Jesus and Christianity without raising criticism, and sometimes-volatile debate, from some corner. The minutest details of Christian beliefs are matters of intense emotion for many people, but some basic facts of

Christian belief are still discernable and held by the vast majority of Christians. According to the Christian belief system, Jesus is the son of the God of Abraham (Jehovah). There is a great deal of controversy over whether Jesus actually taught or believed this himself. Jesus testified before the Pharisees that he as the son of man, but elsewhere he very nearly claimed that he was the Son of God. Within this context, many Christians believe that Jesus was the Messiah foretold by Jewish prophets. Obviously, Jews do not believe that Jesus was the Messiah or they would be Christians instead of Jews.

Jesus taught love. Jesus taught that the classical sins are bad and committing them will get you committed to Hell when you die. Being good, such as not committing any sins, will get you into heaven. However, Jesus died on the cross to provide sinners with forgiveness that allows anyone who believes in Jesus, and accepts him as their savior or Messiah, to enter heaven when they die. Jesus died on the cross to free humankind from the stain of original sin. Jesus died for the sins of humankind, so they could be saved and spend an eternal life in heaven. Jesus' doctrine of love is stated in the Golden Rule: 'Love thy neighbor as you would have your neighbor love you.' This rule is very nearly the same as the Golden Rule taught by the Buddhists and Confucius, which leads some to claim that there was a Buddhist influence in the life of Jesus, but that is only a speculative idea. Jesus committed the ultimate act of love as he died on the cross to save the world from sin and was resurrected three days later and ascended to heaven.

Beyond these basic tenets of Christianity there are so many differences of opinion and interpretation as well as approaches to worship, that the religion almost seems a quagmire. At this juncture the Greek philosophical heritage enters Christianity leading to many more philosophical interpretations that can be associated with Jesus and the different approaches to worshiping Jesus. For example, was Jesus the son of man, the Son of God, or both? Was Jesus God incarnate? Is God a trinity of father (creator), son and the Holy Ghost? Will those who are sinless but do not believe in Jesus go to heaven or Hell? When you take the Eucharist of wine and bread, are you actually consuming Jesus' body and blood (which could be construed as cannibalism) or is the Eucharist only a symbol for Jesus' body and blood? And the list goes on.

Questions such as these have led to both major schisms and minor splits in Christianity over the centuries. The first split came in 1054 CE when Christianity divided into the Roman Catholic Church and the Eastern Orthodox churches. Christianity split again in sixteenth century Europe, in an age called the Reformation. The Lutherans, Calvinists, Anglicans and other groups broke away from the dominance of the Roman Catholic Church. Today there are so many different Christian sects, each with its own interpretation of what Jesus said and did, that they would be difficult to precisely define and count. However, Christianity as a whole is the largest single religion in the world and it has contributed significantly to the development of the European and other cultures.

The last and most recent addition to the world's religions is Islam. Islam, an Arabic word for 'submission', is based upon the total submission to the will of God or Allah. Islam developed upon the teachings of Muhammad who died in 632 CE. Muslims, as followers of Islam are called, believe that 'There is no god but Allah and Muhammad is his prophet (or messenger)'. While in the desert (or a cave) meditating, the angel Gabriel appeared to Muhammad and dictated the *Qur'an*, the Muslim holy book to him. This book offered a radical reorganization of the Arabic society of that day, so Islam is deeply imbedded in the culture of its followers. Islam does not separate religion from other areas of daily life, so people must live Islam every moment of every day of their lives to be Muslims. This sentiment strengthens the ties between Islam and culture and strengthens the hold of Islam over its adherents and followers. The holy cities are Medina and Mecca in Arabia. The Kaaba in Mecca is the central object of worship to Muslims and it is strictly off limits to non-Muslims. Every Muslim is expected to make a pilgrimage to Mecca and visit the Kaaba at least once in their lifetime. Muslims everywhere are required to bow toward Mecca and offer prayers to

Allah several times a day when called to prayer. Jerusalem is also a holy city because Muhammad ascended to heaven from the Temple of the Rock in Jerusalem, while other cities are also holy to Islam for other reasons.

Allah is the sole God of Islam; there can be no others. So Jesus is just another prophet, the last before Muhammad, as are all the prophets in the Jewish religion going back to Abraham. Islam is built upon Judaism and Christianity, but is uniquely different from either one. Islam was originally a reform movement, but not to Judaism or Christianity. It was a reform movement to the polytheistic and animistic practices on the Arabian Peninsula at the time of Muhammad. Since God is Allah alone, Muslims reject the Christian Trinity and many of the other trappings of Christianity. Allah is the creator, not what Allah created, so the worship of any material object is strictly forbidden. Worshiping or appearing to worship anything other than Allah is a serious sin. Indeed, paintings, photographs and any images of a person are not allowed in the strictest of Muslim traditions because they amount to idolatry. Mistaking the creator for the created is a sin (shirk) while denying Allah and defying Allah's will is another sin (kufr). The *Qur'an*, the holy book of Islam, is a guide to every facet of Islamic life and thus all of society. All living beings must serve Allah and follow Allah's commandments. Those who follow the commandments go to a heavenly paradise or a garden of paradises, and those who do not will be judged by Allah and sent to Hell. This judgment will come at the end of time when Allah raises everyone from the dead and passes judgment.

Like Christianity and other religions, major splits occurred in Islam as time went on, but the greatest split came shortly after Muhammad's death. Since Islam is so closely associated with society and culture, the political leader of a society is also the religious leader and vice versa. When Muhammad died, he was a political as well as a spiritual leader, so a schism occurred because Muhammad appointed no successor and left no instructions on how he would be succeeded. Those who chose to follow one of Muhammad's generals or another community leader have become the Sunni sect of Islam. In Sunni Islam, the religious matters are left to religious scholars. However, another group claimed that Ali, Muhammad's cousin and son-in-law, the husband of Muhammad's daughter Fatima, had been appointed as Muhammad's successor. The direct descendants of Ali and Fatima are Imams in the Shiite sect of Islam, and they are revered by about 15% of the Islamic world. There are still other significant differences between the Sunnis and the Shiites. Those differences are mainly philosophical, however, the country of Iran (modern Persia) is the center of the Shiite sect and some ancient Persian traditions have colored the Shiite perspective of Islam, rendering this branch of Islam even more different from the Sunnis than normal political and philosophical differences would dictate.

Even though these three major western religions, Judaism, Christianity and Islam, are based upon the same fundamental principles and share the same ancestry, they have contended with each other throughout history. And sometimes that contention has been quite severe and deadly, as it still is. Even with each religion, the various factions and sects battle each other, sometimes for control and sometimes, it seems, for no other reason than just pure hatred. This paradox in this contention is that all of these religions are basically the same underneath the trappings of their rituals, down at the fundamental levels where things really count. Many practitioners are willing to compromise their religion's principles to save those principles from imagined and non-existent attacks. Too many people are just too willing to quibble over the minute details of differences and unwilling to consider the vast panorama of similarities where the real meaning and significance of the religions can be found, the ultimate truths are the same in all cases.

The religious stew boiled down

The study of how religions formed and how they are related to one another is called 'comparative religions', but the study of comparative religions also seeks to identify the underlying principles of all religions. While comparative religion is a logical study of religions, it does not represent a philosophy of religion in itself. Philosophies of religion are associated with the individual religions that are compared whereas a comparison of those philosophies comes closer to developing a 'science' of religions. In the broadest sense, science seeks to define and derive principles of comparison between physical quantities, while the study of comparative religions tries to define and derive the principles of comparison between the philosophical and qualitative aspects of religions. The different areas of comparison can be easily identified in religions. Most religions are based on the belief in a god, gods, or a supreme being, as was stated earlier. However, a more common feature of religions is the concept that some portion of each and every human survives death and that part of each of us that survives death is commonly called a soul or spirit. Some such concept of an afterlife, in all of its various manifestations, is part and parcel to all religions.

There is even a name for the study of the afterlife. This study is called eschatology from the Greek word 'eschatos', translated as the 'last' or 'furthest'. So eschatology is actually the study of 'last things', which could be interpreted as the end of an individual life or the end of the universe as a whole. On an individual basis, the study of eschatology includes stories of death and the afterlife, which could include, but would not be limited to, judgment, heaven, hell and reincarnation. Since the universe as a whole can go through the same cycles as individuals, eschatology also includes, but again is not limited to, the destiny of the universe, the final judgment, destruction of the universe and the possible rebirth of the universe in a renewed form. Whichever the case may be, local or global, specific or general, individual or universal, the key to understanding religions is to understand the concept of an afterlife and how a religion approaches the subject of death.

While it is commonly believed that the afterlife cannot be verified because it is beyond the experiences of most people, enough circumstantial and anecdotal evidence exists to give some credence to the existence of an afterlife. And that evidence has been persistent enough throughout history that the question of an afterlife cannot be simply ignored. In fact, some form of a belief in an afterlife has existed for countless millennia, so the afterlife concept seems fairly basic to the human condition, and therefore to all religions in one respect or another. Within religions, the afterlife can take on one of two different forms.

The first and simplest view of the afterlife postulates that a soul or spirit survives bodily death and is judged by the merits of the life that has just ended. The judgment is conducted by god or in the name of god by some other agency. Those people who led good lives are rewarded. Their souls will spend the rest of eternity, sometimes in the presence of god, in heaven, in a paradise or in some condition that is pleasurable. The reward of the righteous is happiness and love. On the other hand, if the person has been judged bad, immoral or corrupt, that person is punished or possibly even tormented. The punishment can last an eternity, with no possible redemption (some forms of Christian Hell) or just until the end of time when a final judgment for the whole universe is made. In Zoroastrianism, the soul walks across a narrow sword to heaven. The sword is turned broadside up for the virtuous and sinless and turned sharp blade up for sinners who must walk precariously across to heaven. The slightest imbalance in the favor of an overall sinful life and the soul plummets to Hell where it awaits the final judgment before it again has an opportunity to enter heaven. In the Egyptian tradition, the soul was weighed against a feather to see how virtuous the person had been during life. On the other hand, Jewish literature showed very little concern with an afterlife until after the Jewish sojourn in Iran where it came into contact with Zoroastrianism. So it is assumed that Jewish views of judgment and heaven were strongly influenced by the earlier Zoroastrian example. On the other hand, it is also possible that souls just wait for the final judgment, when they will be raised from the dead and judged, as in Islam. In either case, the corrupted soul or spirit is sent to a place that is decidedly unpleasant as its punishment for misdeeds in life.

According to some religious beliefs, entering heaven or hell, a good place or a bad place, does not completely depend on having lived a good or bad, moral or immoral life, but whether that person was completely faithful to god (Islam) and believed in or accepted god (or Jesus in Christianity). In one form of Buddhism, to have faith in the Amida Buddha will lead to a rebirth in a 'Pure Land', a paradise which is only a short distance from nirvana and ending the cycle of birth and rebirth, but that is another story. Although the issues are far more complex than indicated here, the concepts are correct in general. Even in Taoism, the dead can become immortals in a heavenly realm, while that immortality can also be gained through alchemy and virtuous living. Views of the afterlife based on a judgment of values are far more prevalent in the western religions of Judaism, Christianity and Islam, than in their eastern counterparts, but the concepts of heaven and judgment are not completely unknown in the eastern religions.

On the other hand, the afterlife could just be the next life down the line in an endless cycle of reincarnations. The soul or spirit can then be reborn into another living person after the death of an individual. It could even be reincarnated into another life form or species such as an animal or an insect. In Buddhism, reincarnation can occur in six different ways. A soul can be reborn into different hells, a realm of hungry ghosts, a realm of asuras and titans, and an animal, a human or even one of the heavens. The mechanics of reincarnation vary from religion to religion and sect to sect, but in general the eastern religions tend more toward reincarnation of some kind than an eternity in heaven (or hell) as an explanation of the afterlife. However, there is some small tradition of reincarnation in the west. Empedocles claimed to be a reincarnated 'daimôn', a person to whom a long life had been granted, but who committed the sin of eating flesh and whose punishment for that sin was to undergo various reincarnations for 'three myriads of years'. Plato also talked about reincarnation in different places in his writings and there were forms of reincarnation in some early Christian beliefs, but reincarnation was purged from Christian doctrine at the Council of Nicea.

While there is normally no element of active judgment according to a person's past deeds or misdeeds where reincarnation is concerned, there is a passive concept of 'karma' whereby good deeds during a lifetime merit a better reincarnated life the next time around and bad deeds detract from the next life in a form of cosmic balance. Although no specific judgments are made to guide or condemn a soul or spirit, the soul or spirit is still held responsible for having committed good or evil acts during a lifetime. Events in our current lives are therefore consequences of choices made and acts committed in our own past lives. One surprising consequence of this belief is that it offers a new form of evolution, an evolution of the soul or consciousness such that the soul or consciousness can progress and learn through successive lives. Under these circumstances, the soul would eventually evolve into a higher state of existence, breaking the cycle of reincarnation. The goal of all Buddhists is to reach nirvana, a form of unconditioned reality beyond either life or death. In those religious systems where rebirth and reincarnation are common principles (Buddhism and Hinduism), it is also possible to break the cycle and attain a higher state of being such as nirvana at any time during any lifetime, not necessarily and only after death. The Buddha taught that 'enlightenment' could break the cycle of birth and rebirth, but what happens after the cycle is broken then presents a new problem. If the cycle of birth and rebirth is broken, then the soul might just dissipate into nothingness, or go to some alternative type of heaven and afterlife, or even be absorbed into some unspecified 'something' that is indeed greater than physical reality itself such as a cosmic soul, a universal intelligence or a supreme (state of) being. In any case, and in the case of any of these various religions, we all continue on some larger stage of reality after our deaths.

Within the context of these two traditions, there are two well known works of religious literature that describe the processes of attaining the afterlife: The Egyptian and Tibetan books of the dead. In either case, the title of a 'book of the dead' is of recent popular vintage. The Tibetan book of the dead is actually titled the *Bardo*

Thodol, which translates as *The Great Liberation upon Hearing in the Intermediate State*. The Tibetan book is traditionally thought to be the work of Padma Sambhava in the eighth century CE. On the other hand, the Egyptian book is not really just one book. There are many different variations of the book, many containing the same passages with slight differences. The oldest passages in the book seem to date to a time before the first Egyptian dynasty, over 5000 years ago, and then those passages may have been old at that time, as they exist only as remaining fragments of what seems to be a still older and larger work. Since it was the Egyptian custom to place copies of the book in the tombs of the dead so that the dead could more easily find their way into the afterlife, these books are often called 'coffin texts' or 'tomb books' instead of the commonly used 'book of the dead'. In any case, the purpose of the Tibetan and Egyptian books was similar, to help the dead navigate their transition into the next step of the afterlife.

In the ancient Egyptian belief system there were six parts to the human being: The physical body, its shadow, the individual's name, the ka (spirit, guide or soul), the 'ba' (the individual's personality or id), and the 'akh' (the quality of immortality). The Egyptians followed very careful mummification procedures because each of the aspects of the person was necessary for the dead person to move on to the afterlife, so each had to be preserved through the funerary and mummification procedures of the Egyptians. The ka (soul or spirit) was a duplicate of the body, so it had the same needs in the tomb as the body had while living, so the Egyptians put food and other offerings in the tomb for the ka to use on its journey. On the other hand, the 'ba' or personality had existed since the first breath of life of the person and could freely travel between the underworld and the physical world. But the 'akh' was immortal and unchangeable.

If the funerary rights were performed correctly, they would bring forth (or liberate) the 'akh' so that the 'akh' could join the gods in the underworld. Since the 'akh' was immortal, the successful release of the 'akh' would guarantee that the person could not die a second time, which would cause the end of the person's existence.

The 'book of the dead' was placed in the tomb near the mummified body so that the soul could use the information contained in the book to overcome any obstacles along its path to the underworld and gain admittance to the underworld. The elaborate procedures carried out by the Egyptians were meant to guarantee the continued existence of the person after his or her death. The very fact that variations of this book were used for thousands of years is a testament to the seriousness of the Egyptian belief in the afterlife as well as the influence that their beliefs in an afterlife had over their normal lives. Over the years and centuries of use, the book changed and reflected new perspectives on death and the afterlife.

On the other hand, the Tibetan book is read aloud to the person after his or her death. The very fact that the book is read emphasizes the unspoken assumption that some part of the person that can still 'hear' and understand the reading has survived material death. The soul, or whatever name is given to that part of the person that has survived, can still interact with our physical and material world. The book is not a description of an elaborate ritual, not is it a compendium of magical incantations and spells to help the soul find its way into the afterworld and gain entrance. The *Bardo Thodol* is a guide for the soul to navigate through the reality that exists between death and reincarnation. If the reading is successful, it is believed that the soul will recognize its true nature and gain release from the cycle of rebirth, which is the goal of all Buddhist activities. However, should this not happen the soul can at least be directed into one of the higher realms rather than the lower realms and attain a better position after reincarnation.

When a person dies, the surviving soul (or consciousness) is freed from its natural and earthly restraints, but it is also disoriented and confused from the loss of its connection to the material world of the body. The soul (or consciousness) enters a self-styled dream-like state where it creates its own reality. That reality can include wonderful visions or terrible sights, which all occur in three specific phases or 'Bardos'. Reading of the *Bardo Thodol* instructs and helps the soul to recognize the heavenly realms that it encounters and guide the soul toward a higher state of being and reality through which it will be liberated.

Upon death, the soul first enters the 'Clear Light of Ultimate Reality'. It is within this reality that the soul should remain, but only if it can recognize this reality as the place to remain. It can recognize this reality and automatically reach liberation from rebirth if the soul has attained a high enough level of consciousness or spirituality during the life just past. The book instructs the soul to give up its self and ego and embrace this experience of the ultimate reality with love and compassion. The soul must recognize and open itself to this reality because it is one with the 'Clear Light', or it is itself the 'Clear Light' which it is experiencing. According to Buddhist doctrine, all humans have a Buddha-nature, so this 'Clear Light' is the Buddha-nature that has survived bodily death. It transcends time and the material reality of our common world. To recognize this at the moment of death is to reach a state of being called 'Dharmakaya', equivalent to Buddhist enlightenment.

However, it the soul does not recognize and completely accept this reality, it is pulled down to the 'Secondary Clear Light' by the weight f its own karma. Depending on how advanced the person's spiritual practices were during life, the soul would next be told to either (1) realize it was 'Dharmakaya', (2) meditate on whichever deity it had worshiped during life, or (3) meditate on the 'Great Compassionate Lord". Failing this, the soul would enter a second phase or Bardo at this point in its journey.

The soul spends seven days in this next Bardo and on each day the soul encounters a light from one of six different 'lokas' or worlds of reality. From each of these lights a Buddha soul emerges as a test for the soul. On the first day the supreme deity of the universe appears. If the soul of the person is good it will share the joy and happiness of the supreme deity's light and attain liberation, but if the soul was bad or evil in life then it will fear the light and continue on to the next day. On the second day the soul meets the second highest deity, but a smoky light also visits it from hell. This pattern repeats each day with lower deities appearing out of the light each succeeding day. If the soul has not yet been liberated from rebirth by accepting 'Dharmakaya', it will make it to the seventh day and face the 'knowledge-holding deities'. These deities are demonic looking. If the soul fails this test, then it is reincarnated as an animal.

Failing this test, the 'knowledge-holding deities' transform into the 'wrathful deities'. These are hideous demons that do great bodily harm to the soul and torture it. But the soul is instructed by the book to realize that these demons are being generated by its own bad karma. In a sense, these demons and the torment are the product of the soul tormenting itself and by recognizing this fact the soul is again given the chance for liberation. However, this liberation is to the second level of liberation, the best that the soul can hope for at this point. If the soul does not find liberation at this point, it moves onto the third Bardo.

In this third Bardo, or phase, the soul meets the 'Lord of Death" who passes judgment over the soul. If the soul protests and claims its innocence, it faces the 'Mirror of Karma" where its actions during life are reflected and out of which evil demons torture the soul for its past transgressions and evil deeds. But again, the book instructs the soul to recognize all of these demons and happenings as the projection of its own self and the product of its own making. If the soul recognizes or becomes aware that this is all of its own making, then it can still be liberated. If the soul once more fails to recognize this reality, then it is drawn into rebirth. The lights of the six 'lokas' return and the soul is reborn into the world of the 'loka' whose light shines the brightest. At this time, the soul is still suffering from the torture that it had just experienced, so it will seek shelter in what appears to be dark caves, but are actually wombs. Even at this turn of events, the book instructs the soul to reject what it sees and seek the Clear Light so it can be liberated from rebirth. However, the soul soon passes to a point of no return, beyond which the rebirth cannot be stopped or interrupted, so the book instructs the soul how to pick the best womb into which it will be reborn. The soul must concentrate on non-attachment and not be fooled by worldly pleasures or repelled by worldly ills. Only then can the soul be reborn in a higher level of spirituality, according to its karma and its level of spiritual awareness.

There is no other explanation of an afterlife, in heaven or hell as in the western tradition, or through reincarnation as is the eastern preference, which matches that given in the *Bardo Thodol*. The *Bardo Thodol* is unique in religious literature. Very few other religious works even come close to the complexity and precision of the explanation of reincarnation or death that the *Bardo Thodol* offers. In plain and simple fact, there are many correlations between the description of events in the *Bardo Thodol* and in the events described by people who have had Near Death Experiences in the past few decades. So it is at this point that science comes to look at death from its own perspective of our common natural reality.

CHAPTER 3
NATURAL PERSPECTIVES OF DEATH

First follow NATURE, and your Judgment frame
By her just Standard, which is still the same:
Unerring Nature, still divinely bright,
One clear, unchanged and universal light.

—Alexander Pope–1711

Through the ages, man's main concern was life and death. Today, for the first time, we find we must ask questions about whether there will be life before death.

—Albert Szent-Györgyi–1970

The road not so well traveled

That which is natural deals with nature and therefore lies within the realm of knowledge claimed by science. In fact, science was called Natural Philosophy before the 1840s when physics (the Greek word for nature) broke away as a separate scientific discipline and the other natural sciences of biology, chemistry, astronomy and geology as well as mathematics slowly emerged as independent academic disciplines. Natural is science and science is natural, or at least science defines what is natural in our common world. Natural Philosophy evolved in the west with the Greek philosophers in the last few centuries BCE, but proper science, as we know it today evolved in the seventeenth century as the first scientific revolution came to a head.

The first scientific revolution marks the end of one long evolutionary period of human thought, but it also marks the beginning of another long evolutionary period of human thought that ended with the second scientific revolution at the beginning of the twentieth century. Not only did science emerge from of the first scientific revolution, but the basic methods of conducting science also emerged from the work of Galileo, Isaac Newton and others in the seventeenth century. Through the efforts of these scientists and others, the scientific method that we know and use today was developed.

The scientific method has several distinct steps depending on how precisely it is defined by different scholars. First we observe nature, then we define quantities in nature that we can measure and we measure them. Then we look for patterns where events and phenomena occur again and again under the same conditions. We next develop a logical hypothesis to explain these events, phenomena and the patterns that they exhibit, followed by experiments that are designed to duplicate and test the hypotheses that we have developed. The hypothesis should predict the outcome of the experiment if it is correct. If the results of the experiment are negative, the prediction of the hypothesis is not confirmed, then we first check our experimental apparatus and design to make sure that it functioned correctly as planned. If there was no problem with the apparatus and design of the experiment, no flaw in the experimental procedure, then we return to our original hypothesis and either alter it or develop a new hypothesis to test and conduct a new experiment accordingly. However, when the experiment confirms the hypothesis with a positive outcome, such that the predictions of the hypothesis are verified, then we expand or generalize our hypothesis to test other conditions relevant to the event or phenomenon.

When the hypothesis is confirmed by experiment or further observation, it becomes a theory. If that theory is confirmed in all possible applications, under all possible related conditions, in all the different places that it is tested by different scientists, then that theory becomes a natural law and becomes part of the present scientific paradigm or worldview against by which we explain and analyze our material world. But the new theory must also be consistent with the accepted theories and laws that already make up the paradigm. If a new theory is verified but it does not conform to the old paradigm, theories and laws, then a crisis in science occurs that must be solved.

That part of the scientific method that includes the hypothesis, experiment and confirmation, is also called the experimental method. So the experimental method is itself a complete process within the overall scientific method. The scientific method, as here described, has literally propelled human thought to evolve and progress with respect to our knowledge of the natural world to the highest level that it has ever achieved and will continue to move the collective knowledge to a still higher level in the future. However, the experimental method is incomplete when it is used alone and does not help us to develop the knowledge to explain everything in our universe. Science does not like to be told so, but the experimental method is not everything and it is not absolute. It is a logical method and therefore does not explain intuitive leaps in knowledge, nor does it explain quantities or qualities that cannot be so easily defined and measured. During the second scientific revolution, a different way of conducting science was instituted. This method of further developing science was certainly not new, but it had never really been defined and categorized within science, yet it quickly became an integral and accepted part of science during the second revolution.

That method is called the 'gedanken' or thought experiment. Once something is given a name in science or our world, it can be considered real and a valid part of our world. So once named, the 'gedanken' experiment and its variations became a new method and accepted method in science. The notion of forming a hypothesis, conducting an experiment and developing a theory, in other words the experimental method, was abstracted to become a concept for the development of theories itself. The 'gedanken' experiment is an abstraction of an actual physical experiment in which the material apparatus has been replaced by thought only. The 'gedanken' experiment has allowed scientists to develop new theories, based upon logical deductions from previous theories and laws, leading to still higher level theories, or secondary theories, which could then be used to conduct real material experiments to confirm the higher level theories. However, in any and every case, nature is still the final arbiter of the validity of any theory and science must conduct real physical experiments and observations in the end to confirm the veracity, validity and falsifiability of all theories or science will succumb to supporting unjustified philosophical heresies instead of nature.

Albert Einstein formally developed the 'gedanken' experiment when he questioned the concepts of space and time upon which all previous theories of nature were based. Einstein redefined our fundamental concepts of space and time and in doing so set new standards and suggested new unobserved phenomena in nature that were later confirmed by observation and experiment. Yet, the confirmed theories and natural laws that were based on the older concepts of space and time were not abandoned or invalidated because they had already been verified. In the 'gedanken' experiment, the older verified theories must be incorporated into the hypothesis and the new theory, thereby extending the older theories into realms that are not available for experimental testing. The new theories based on the 'gedanken' experiments must be able to explain the successes of older theories within the range of conditions and possibilities under which they were originally tested and verified. But Einstein and others showed that we can change our basic assumptions, questioning our common perceptions of reality and redefining the physical quantities by which we define our physical and material reality and thus gain a more general and truer knowledge of a deeper and more fundamental nature. When we take a scientific look at death and the afterlife, that is exactly what we are doing: We are questioning our normal perceptions of the world and defining new quantities because we have no perspective or common perceptions of death by which to establish a science of death and afterlife.

All that we have at present to establish a science of death and an afterlife is concepts such as the soul, the spirit, ghosts and apparitions, but none of these concepts are subject to the direct and careful definitions that are required in the scientific method. So science has ignored death and the afterlife for two millennia, during which science itself has been evolving in other directions. So then how do we conduct a scientific investigation of death and afterlife? At this point, there are two scientific tactics for looking at death and afterlife. To begin with, science has an unwritten rule that is cannot use unknown quantities to describe unknowns. Hypotheses based upon unknown quantities and concepts are useless in science because they do not relate to science as it is already established. Since death is an unknown quantity, we have no physical perspective against which we can define and measure anything relating to the afterlife. We cannot scientifically or legitimately make up unfounded and unknown quantities by which we can define the afterlife, so science is forced to look for another method to deal with death and afterlife. In more simplified terms, we have no handle by which science can grab hold of the concept of an afterlife.

So science, as it is presently constructed and constricted, simply cannot deal with the concept of an afterlife. Science can develop neither a valid hypothesis nor experiments to either explain or test the concept of an afterlife and the scientific method seems to be failing on this account. The experimental method is therefore useless in exploring the scientific possibility of an afterlife at this stage of our conceptual development. So science can only fall, and has in the past so fallen, back on the second tactic or option for dealing with the possibility of an afterlife, and that tactic is outright denial of its existence. If the afterlife cannot be directly observed on momentary notice under known specific conditions, such that the rest of science whereby conditions of repeating the observations would be known does not imply an afterlife, then it does not exist. If something is not indicated by verifiable and repeatable observations, then it is not valid in the realm of science. Science does not technically deny the possibility of an afterlife, nor does it technically deny the possibility of concepts of death, it just says that they are not scientifically valid and so it will not consider them as scientific. In other words, phenomena and events dealing with the survival of something after death and an afterlife are neither indicated as possible by current or normal scientific theories and laws of nature, nor are they universally experienced by all people, equally, so they are not natural, i.e., occurring in nature or physical/material reality. Therefore, these events, like mind and life themselves, are relegated to either the super-natural or the para-normal.

It is rather strange that life and mind fit in this same category. But life, as something extra, has always been in this category. The split between matter and mind that marks the difference between what is scientific and non-scientific, dates back to the Greek philosophers in some rudimentary form, but was not institutionalized until the scientific revolution by Rene Descartes. However, the successes of Newtonian mechanics in explaining our material world and the development of Darwinian evolution in the 1860s again opened Pandora's Box and forced science to reconsider its decisions to keep life and mind out of normal science. The result was the development of a science of the mind called psychology, which was an unexpected and unplanned stopgap development to take the heat off of the rest of science, allowing it to deal with the material world alone. Psychology immediately settled into a nice profitable and safe existence whereby it could survive and grow within the scientific paradigm of its time and forgot about abstract concepts such as mind and consciousness. Under these circumstances, psychology did not want to rock the boat of science in spite of the historical impetus upon which it was founded. In other words, by not asking the really big ultimate questions about mind and consciousness that psychology was developed to answer, it quickly aligned itself with the materialistic worldview out of which it had developed in order to answer those questions.

It is no coincidence that the 'gedanken' experiment developed at this same time because the same historical forces that pushed the development of psychology also precipitated the 'thought' experiment. Although psychology was founded upon the questions of consciousness and the interactions of consciousness with our material/physical reality, the new science quite literally 'materialized' itself and studied behavior, how the mind interprets its reactions with other people and physical events in the material world, rather than concentrating on consciousness. Consciousness was therefore pushed to the back of psychological investigations and thought, all but completely lost for nearly six decades. The existence of life as something extra beyond the normal mechanisms investigated by science as well as mind and consciousness does not fit the material paradigm by which science views the world and nothing in science required or made necessary the explanation of those concepts at the time, so science had no need or use for them. But mind, life and consciousness still exist in nature as anomalies relative our scientific paradigms, so they have continued to pop up again and again to 'haunt' science. Quite simply, if life, mind and consciousness exist outside of the material world as explained by the material paradigm that is presently accepted by science, and then they should have some bearing on and play a significant role in the possibility of an afterlife. All that science says about dying, in fact all that science can say about dying under the present worldview and paradigms that it subscribes to, is that the material body dies and thus life ends. But if life, mind and consciousness exist beyond or 'outside' of the material body, then science is unable to claim that they end upon death and nothing of the person exists after death. So the mere existence of life, mind and consciousness poses a serious paradox for the paradigms of science and thereby necessitates a new look at the concept of death and afterlife.

During the 1960 and 70s, shifts in basic scientific attitude began to occur. To determine the exact cause would be impossible since the changes depended on the interaction between several developing trends in science and culture. In physics, general relativity had been ignored as an esoteric theory with no practical applications for the previous four decades, but new advances in technology and the space program gave general relativity a new relevance. The newly won relevance of relativity theory began to erode the basis of materialism inherent in modern physics, which had been previously dominated by the quantum worldview alone. At the same time, new physical interpretations of quantum theory were developed and popularized. Eugene Wigner and other physicists emphasized the role of consciousness in the ongoing evolution of physical reality according to the quantum mechanical worldview. Parapsychology began to emerge from its fringe science status and became a legitimate science in 1969 when the Parapsychological Association was admitted to membership in the American Association for the Advancement of Science. Normal science was beginning to soften just a bit toward radical ideas and develop an expanded and more tolerant worldview.

Western cultural attitudes were also undergoing a slow parallel change in perspective and orientation. Eastern philosophies and religions as well as martial arts became popular in the United States and Europe while scientists and scholars began to compare eastern ideas and concepts to modern physics and science. These ideas were popularized in books such as Fritjof Capra's *The Tao of Physics* while similar ideas were expressed in Lawrence LeShan's *The Medium, the Mystic, and the Physicist*. Psychologists began to look seriously at the concept of consciousness, which had been neglected since the founding of psychology seven decades earlier. Another popular book, Robert Ornstein's *The Psychology of Consciousness*, heralded the re-emergence of consciousness as a viable academic and scholarly subject. Physicists began to look into the paranormal in small but still unprecedented numbers while new phenomena, such as near death experiences, were 'discovered' and popularized within the general culture. Taken together, these changes amounted to a new awakening to the concept of death, inspired in part by the discovery of the near death experiences, although that discovery would not have made such an impact without the other changes in scientific and cultural attitudes that occurred at nearly the same time.

From Thanatology to NDEs

There is even a full-fledged science of death within the discipline of Psychology. It is called Thanatology after the Greek personification of death, 'Thanatos'. And, like other changes, Thanatology developed in the 1960s and 70s. The place of death in psychology, the science of mind, was in large part ignored until after the middle of the twentieth century, reflecting the positivistic attitudes and behaviorist mind set of the psychological community. However, in 1959 Herman Feifel published the book The Meaning of Death and caused a small crack in the wall of complacency surrounding the science of mind. Then in 1972 Robert Kastenbaum and Ruth Aisenberg published The Psychology of Death, widening that crack still further. But even before their book was finished for publication Elisabeth Kübler-Ross had already published On Death and Dying (1969). Kübler-Ross' book made the most dramatic breakthrough. She was interested in the dying person and through her research with them identified five states or psychological changes which characterize the dying: (1) Denial and isolation, (2) Anger, (3) Bargaining, (4) Depression, and (5) Acceptance. Not everyone experiences all of these states or even in the order given. These different states are not absolute, but relative to the person experiencing the NDE, yet they do mark the first awareness of the psychological problems confronting psychologists working with the dying. Kübler-Ross' book was a vast success, both in science and with a more popular audience. AS such, it opened up the field of Thanatology and became a standard text for health and other professionals who work with the dying.

Even while Kübler-Ross conducted her research, presented seminars on dying and wrote her book, her work was frowned upon by her superiors and her colleagues, reflecting the strict hands-off policy of science toward death and dying at that time. However, her book and subsequent publications have opened the field and forced the scientific and health communities to come to grips with the psychological problems faced by the dying and their loved ones who must carry on after the death. Yet Kübler-Ross' popularizations did not say much about the actual concept of death. It was not until the 1990s and the publication of *On Life and Death* that Kübler-Ross finally admitted in public that death was not an end but is in fact a transition or entrance into a new phase of being.

Even today, Thanatology deals with the study of death and dying only in the sense of the psychological mechanisms of coping with death. It also deals with our attitudes toward death, the behavior displayed after death known as grief, the moral and ethical problems associated of euthanasia and other related practical aspects of dying. Thanatology does not deal directly with either the concept of death or what happens to the person after death. Any considerations given to these aspects of death only come with how our beliefs in the afterlife affect our behavior and attitudes toward the act of dying itself. Thanatology is only a science of dying, not of death and like the rest of psychology, ignores the really big significant questions about the meaning and concept of death. The same is true for the medical community, which deals with death all of the time, only in a different capacity.

The medical community's main concern is the maintenance and well-being of the human body, just as the psychological community's main concern is with the maintenance and well-being of the human mind. So, to the medical community falls the task of dealing deciding when a person is dead. The medical community must decide that moment beyond which medical treatment and resuscitation attempts of the victim are no longer of any benefit. For centuries, death was thought to occur when either the soul left the body (a rather ambiguous moment) or blood stopped circulating, i.e., when the heart stopped beating. However, on some rare occasions people would be buried alive so these earlier concepts of the moment of death have lost all of their meaning. As health care improved, new criteria for determining death were developed. IN 1968, a committee at Harvard Medical School cited four criteria for determining medical death: non-response to external stimuli; cessation of spontaneous breathing; failure of motor reflexes except for primary spinal column reflexes, accompanied by fixed dilated pupils; and, a flat-line reading of an electroencephalogram (EEG). If these responses did not return after a twenty-four hour period during which the victim had no drugs in his or her system and was not suffering from hypothermia (cooling of the body), then the person was judged to be dead.

With the increasing number of political, ethical and legal issues concerning the time of death of a person, these criteria have been further refined as follows.

The official signs of death are:
1. No pupil reaction to light
2. No response of the eyes to caloric stimulation
3. No jaw reflex (The jaw will react like the knee if hit with a reflex hammer)
4. No gag reflex (Touching the back of the throat induces vomiting)
5. No response to pain
6. No breathing
7. A body temperature above 86°F (30°C), which eliminates the possibility of resuscitation following cold-water drowning
8. No other cause for the above, such as a head injury
9. No drugs present in the body that could cause apparent death
10. All of the above for 12 hours
11. All of the above for six hours and a flat-line electroencephalograph (brain wave study)
12. No blood circulating to the brain, as demonstrated by angiography

(The Gale Encyclopedia of Medicine)

Since people have been resuscitated after half an hour underwater if the water temperature was below 50°F/10°C, number seven must be carefully observed in cases where the body temperature at death was quite low. In more common medical terms, a person is not dead until that person is warm and dead.

The time of death is quite precisely defined for the medical sciences, but what happens at the moment of death is not. Medical science has traditionally accepted the fact, whether it is true or not, that dead is dead. Anything beyond the time of death, as determined by the above criteria, is not the concern of the medical community. And it is certainly not part of biology, which, by definition, is the science of life and living systems. The only science of death is Thanatology, but Thanatology has nothing to do with the meaning of death. So it seems that there is not even any room for death within the very structure of science. Whether the attitude of science and especially for medicine toward death is correct or not, only time will tell, but it does indicate two things about medicine: (1) the medical field is completely mechanistic and materialistic in this sense. If 'something' survives death intact, then that 'something' should have the ability to affect the health of the individual before the death occurred. By ignoring this simple fact, the medical field is ignoring a possible method or tool for bettering the health of the people it treats. (2) If 'something' of a person survives death, in that it still maintains a physical existence after the moment of death, it may be able to resuscitate patients after even longer periods of time after they have died, under special conditions. If we can discover what survives, then we can determine those conditions under which people could be resuscitated or we would better know when it is time to LET PEOPLE DIE and not even try to resuscitate them. The medical field is thus ignoring the possibility of 'raising the dead' in the moments just after death (before necrosis sets in) as well as tempting fate by keeping a body alive mechanically when the person has already departed the room. Similar problems exist for any science that ignores the possibility that 'something' of the person continues to exist after death, especially in the cases of both biology and Thanatology, so medicine is not alone in this criticism of its practices.

While these medical issues may not seem relevant with respect to ordinary deaths, they could have a more significant bearing on concepts such as suspended animation and reviving people who have been cryogenically frozen at the time of their deaths. These ideas could pose interesting ethical and moral questions for doctors and workers in the health care industry, but they are not pertinent to the concept of an afterlife. The primary obstacle to such speculations rests in the fact that brain necrosis and liquefaction begins fifteen minutes after blood flow to the brain ceases. After that point, life cannot be restored to a person under normal conditions.

The moment of death has gained a totally different meaning and relevance since the 'discovery' of near death experiences (NDEs) in the 1970s. Raymond researched the subject and published his findings in 1975 in his book *Life After Life*. Moody's book was a vast success and initiated a whole new field of research into the phenomena of NDEs, a term which he coined. In turn, research in NDEs has had a very strong influence on interest in what is now called the survival of consciousness, a scientifically neutral phrase for that part of the person that survives after the death of the body and the end of life. The phenomenon that Moody called the NDE was not, however, new and unknown at the time that he started his research.

NDEs actually have a very long history and had been studied in the past. This fact tells us a lot about the changes in scientific attitudes that occurred in the 1960s and 70s. Quite simply, what was so different about the 1970s and/or Moody's work that caused such a stir and initiated studies in the NDE phenomenon when past work on the phenomenon had not been so enthusiastically pursued? Others have made similar speculations and come to similar conclusions regarding this question. Perhaps NDEs were occurring more frequently by the 1970s due to recent advances in medicine rendering them a more compelling subject for science to review and study. This may be the case, but it cannot completely account for the rapid increase in studies of NDEs. Or perhaps, as some have suggested, the name NDE itself was a primary factor in their instant success and popularity. The name just caught on. It was the right name at the right time. But it is still easier to argue that attitudes in the scientific community had changed so much over the previous decades that the scientific study of death and subjects related to death were literally wide open and ripe for the idea of NDEs to take hold at just that time. Other trends in science and scholarship supported and reinforced the ideas and implications of NDEs at the time that Moody's book was published, so the concept developed rather rapidly. Together, all of these trends in thought, all arriving in full force during the 1970s, are now pushing the concepts of death and the afterlife to the forefront of science and academia. In other words, the time was right for Moody's book and concepts of the NDE to become popular.

An NDE can be defined as an event or phenomenon that occurs when a person comes close to medical death, or perhaps even dies, but is resuscitated or returns to life. The victim in this case has specific memories of events that transpired during that period of death or near death that are otherwise impossible to have had. The memories that a person has of this special experience would seem to defy material physical laws and cannot be explained in accordance with our common laws of nature. NDEs occur, especially in those cases where a person would have been declared legally dead except for the fact that he or she returned to life, during a period in which the normal five senses would have very nearly or completely collapsed. In other words, the person who had the NDE could not have gained the knowledge acquired during the NDE through normal sensory interaction with our common material world. This particular definition of the NDE is neutral in that it does not pre-suppose the possibility of the NDE being a preview of the afterlife.

Not all people who come close to death and not all people who die and are revived recall having an NDE. Only about one-third of the people whom we would expect to have an NDE (because they fit the medical criteria) actually recall having had an NDE. This does not mean that the other two-thirds of the people who could have had an NDE did not. Others could have had NDEs but their memories of the experience could have been repressed for some psychological reason or perhaps they had been given drugs that suppressed either the experience or the memories of the experience.

When working with the human brain or mind, ambiguity in such factors render the reality of the situation impossible to distinguish with any absolute certainty.

In reality, the NDE is defined by the characteristics of the experience rather than the proximity to death because an NDE can occur or not occur under so many different circumstances and conditions. When Moody wrote that first book about NDEs, he carefully noted the characteristics that he discovered. During an NDE, the person could experience moving through a dark tunnel, profound sensations of peace and tranquility, a life review that seems to occur in just a single moment, meeting spiritual beings which may or may not include loved ones that have already died, unusual noises, a sense of being out of your own body or floating above your body, seeing and hearing events while out of your body, experiencing a tunnel of light or just a bright white or brilliant light, meeting beings of light, a point or border beyond which you cannot pass and still return to your body, a choice whether or not to return to your body or being told that you must return, and then of course returning to your body. There seems to be a general sequence to these events, but they need not occur in the exact order in which they are here stated for all people. Nor do all of these events occur in all cases of NDEs. Different combinations of the events can occur for different people and no correlation between personality and the exact events experienced has ever been found. However, other researchers have discovered that these events are cross-cultural, occur in people of all ages, do not depend on religious convictions, do not depend on the level of education or lack of education, and occur with people who have never previously heard of NDEs as well as people who have had prior knowledge of the phenomenon.

Since Moody first publicized his findings, other researchers have added vast numbers of cases to the files on the phenomenon as well as claimed that some of Moody's characteristics do not occur with any frequency. It has been claimed that Moody's model of the NDE is not completely accurate. Although the details of his model may not be completely accurate, the general features of his model are accepted by science today. The most important additions to the list of characteristics comes form those whose NDEs were not pleasant experiences, or were not good and peaceful as Moody's list would indicate. About fifteen to twenty percent of NDErs have had bad experiences. Their NDEs could just be unpleasant, but they could also be oppressive, dark and even hellish. Some people have even reported that they were tortured during their NDEs. The truth is that NDEs can cover a broad range of types and events and no two of them are alike. The NDEs are as varied as the personalities of the people who report them.

Some researchers have tried to keep this in mind and instead of listing the characteristics and different aspects of the many different NDEs, have tried to categorize the major aspects of the experiences. Even before Moody wrote his book, Russell Noyes (1972) had described three different phases of the experience: resistance, the past life review and transcendence. Kenneth Ring (1980) later limited the phenomena to a "core experience" of basic elements: feelings of peace, the Out of Body Experience (OBE), entering darkness, seeing the light and entering the light or a supernal realm of existence. On the other hand, Bruce Greyson (1985) classified NDEs by other criteria. They could be cognitive and thus dominated by alternative thought processes, affective such that they are dominated by changes in emotional states, paranormal and thus involving various psychic elements, or transcendental and displaying mystical and otherworldly features. Michael Sabom (1982) proposed a simpler model of the NDEs based on the OBE portion of the experience. They could either be autoscopic or just consist of a localized OBE or transcendental whereby the consciousness seems to pass into another dimension, or a combination of the two. The idea behind categorizing the NDEs in these ways is to establish some basic criteria for explaining and understanding the experiences, which is the goal of all science.

Hypotheses explaining the NDEs are, however, rather sparse. Some skeptical scientists think them no more than hallucinations.

Another explanation sees the NDEs as products of the imagination that occur to protect the mind from facing the threat and reality of death. Still others see the NDEs a normal brain functioning at or near the moment of death or chemical reactions related to the shut down of the brain. However, there are those who consider the possibility that the NDE is just a preview of the new state of being into which the mind and/or consciousness is thrust upon its disconnection with the material body at death. While there is no quantifiable scientific evidence for this afterlife preview hypothesis, there is some circumstantial evidence supporting the possibility of an afterlife, but that evidence exists in events that occur after the NDE is completed and the experiencer returns to a more normal life.

There is a great deal of evidence that the NDE changes or alters the personalities of the people who have them. NDErs undergo personal transformations and in many cases that transformation is quite profound. Some of the typical aftereffects, as reported by many independent researchers, include spiritual transformations, a greater concern and compassion for others, a new appreciation of life accompanied by a loss of fear of death, decreases in materialism and competitiveness as well as a tendency to view oneself as just part of a far greater universe. These alterations in personality affect how the NDErs interact with other people, so they amount to behavioral changes as well as internal personality changes. NDErs tend to be more empathic toward other people and try to view things as other people would. Quite simply, the priorities of the NDErs are altered completely, but NDErs also claim that they have increased paranormal abilities after their brushes with death. This particular alteration could possibly be verified as science learns more about paranormal states and consciousness. While NDErs tend to be more spiritual, that spirituality is not confined to any particular religious view, but seems to be more universal in its scope. The increase in compassion is often accompanied by an increase in environmental awareness as if the compassion is not just for humans, and not even for animals, but for all living things. The proponents of the afterlife hypothesis claim that these changes are the result of having touched the light or accepted the light during the NDE, which is tantamount to being in contact with the afterworld or even becoming 'one' with the universe. Only something that drastic could cause such significant and long-lasting changes in the human psyche.

However, not everyone accepts the afterlife hypothesis so easily. Some of the aspects of NDEs can be duplicated by various abnormal brain functions and neurochemical reactions in the brain, so the skeptics believe that NDEs are just the collective effects of cardiac arrest or the cessation of brain functioning for short periods of time. For example, skeptics claim that neural noise and retinal-cortical mapping can easily account for the sensation of moving down a long tunnel. The sensation of peace and calm that seem to accompany the NDE can also be explained by the release of endorphins in the brain to counteract the stress of dying. Forms of anoxia in the brain may also account for the buzzing noise and perhaps even the bright lights of the NDE. Many of the 'symptoms' of the NDE can be chemically duplicated. A hallucinogenic drug called ketamine can induce the sensation of traveling through a tunnel and into light, the feeling that one is dead, OBEs and other aspects of the NDEs. However, the fact that the NDE can be in part duplicated by chemical, electronic or mechanical stimulations of the brain does not disprove the possibility that NDEs might actually be previews of the afterlife. They only 'prove' that the characteristics can be duplicated by artificial chemical processes. It is the overall experience which includes all of the various aspects that an individual has when the person is near or past the moment of death that defines the experience, not the possibility that the experience can be artificially induced or duplicated. By analogy, it is more than likely possible to induce a runny nose using drugs or chemicals, but that does not mean the common cold does not exist or that the common cold is a hallucination.

Probably the most serious challenge to the afterlife hypothesis comes from Susan Blackmore. Dr. Blackmore's opinions are skeptical, but they are born of good science rather than outright prejudice. In fact, Dr. Blackmore's credentials in the matter place her in a unique position to judge the validity and interpretations of the NDEs. Dr. Blackmore is an NDEr herself. She trained and entered the field of parapsychology to try and understand her experience, but after years of attempts to understand her own personal experience, she gave up and came to the conclusion that there was nothing to it. She also denies the existence of a separate human consciousness. Dr. Blackmore has noted four different arguments in support of the afterlife hypothesis and has offered alternate explanations in each case based on her own 'dying-brain hypothesis'.

1. The 'consistency argument' – According to the afterlife hypothesis, the consistency of the NDEs around the world implies the reality of the experiences and thus reality of an afterlife. No other interpretation can explain why different of people from vastly different backgrounds can have essentially the same experiences. Dr. Blackmore argues that the 'dying-brain hypothesis' better fits the consistency argument because chemical reactions in any dying brain can account for all of the same characteristics

2. The 'reality argument' – Since the NDEs feel so real to those who have them, they must represent reality in the next world. However, Dr. Blackmore claims that the 'dying-brain hypothesis' better accounts for this because it is the brain that determines what is real for us and what is not real.

3. The 'paranormal argument' – Those who have NDEs claim that they often involve paranormal events so they must be evidence that the NDE involves another dimension, another world, or the existence of a non-material spirit or soul. The evidence in this case is rather weak and there is always the possibility that the claims are fabricated. Even if the evidence were compelling it still might not indicate an afterlife. However, Dr. Blackmore claims that the 'dying-brain hypothesis' explains that NDErs seek paranormal evidence to bolster their impression of the realness of the experience while the stories of paranormal abilities and events are passed on and elaborated, but she also admits that

if more convincing evidence appears in the future she will have to back down from her claims. Even then, if new and compelling evidence does become available, a good amount of our present knowledge of physics, biology and psychology will also fall.

4 The 'transformation argument'– NDErs attribute their transformative changes to having had a spiritual experience in another world. On the other hand, Dr. Blackmore argues that there is no reason why the afterworld should be better or why it should have such an effect on people. It is merely assumed that experiencing the afterlife during the NDE would cause extensive alterations in the NDEr's personality. So the 'dying-brain hypothesis' is better because the experience merely makes people think how dying affects them and coming close to death can provoke the insight that the self is merely a mental construction initiating the transformation process. So the 'dying-brain hypothesis' can easily account for the mystical insights associated with the NDEs, according to Dr. Blackmore.

In all four cases, Dr. Blackmore argues that the 'dying-brain hypothesis' can better account for the characteristics of the NDEs. However, there are also some basic problems with her arguments. No one doubts that memories, thoughts and emotions evoke chemical responses in the brain. The real question is whether thoughts, emotions and memories are 'only' chemical responses in the brain. Dr. Blackmore's 'dying-brain hypothesis' assumes that this is the case, but in reality this is a chicken/egg argument. Which evolved first, the chicken of the egg? Are the chemical reactions the actual memories of the NDE, the cause of the NDE or the result of the NDE? There are two possibilities, or so we are led to believe. The memories could be real memories of independent events that occurred during death. In that case, there is an afterlife and something survives the material death of the body. A person near death, but not actually dying, begins to experience these physically real events as the brain begins to shut down. But the person does not die, so the person never fully experiences the afterlife or the afterworld, whatever it is. In those cases where the person actually dies and is either resuscitated or returns to life spontaneously by some unknown process, the person actually experiences or touches the afterworld by entering the light. Research by Owens, Cook and Stephenson in 1990 seems to indicate that the closer a person comes to death the more profound the NDE and the fuller the experience, so it can be assumed that people who actually die have a far greater probability of having a complete NDE.

On the other hand, the other hypothesis, as defined by Dr. Blackmore, might represent the reality of the situation. There is a scientific possibility that the NDE is not a real physical event but is chemically constructed or fabricated in the brain and/or the mind, at or near the point of brain death. However, in the case where the victim actually dies and is revived, the 'dying-brain' and similar hypotheses run into rather intransigent problems. There is no guarantee, nor is there any evidence, that a dying brain and a dead brain which is revived would undergo the exact same chemical reactions and thus fabricate the same set of memories. This point can be verified quite easily. If a person died by some very rapid means, without ever having suspected or expected the impending death, and was then resuscitated and still had memories of the NDE, then the NDE would more probably be memories of real physical events that occurred while the person was brain-dead. The brain could not have fabricated the memories simply because it was totally[7] non-functional while it was dead.

There would certainly be no memories of the stress or expectation of dying because the person died without warning, the death was neither expected nor suspected before the fact. In this case, the 'dying-brain hypothesis' might still be applied by arguing that the memories of the NDE were chemical reactions to reanimating the brain and the reanimation process caused the false formation of memories of an NDE, but then there would be no reason for people being near death without actually dying to have the same memories. So the issue between these two hypotheses could be decided quite easily by studying special cases of the type described.

If only one person died as described and had an NDE, then that person would either have to be lying or the 'dying-brain hypothesis' would be refuted. So it is possible to know which came first in this case, the chicken, representing the 'dying-brain hypothesis', or the egg, representing the afterlife hypothesis. This is the stuff of science. This is how science and human knowledge progress. Competing hypotheses are developed and tested to determine which hypothesis better explains the phenomena with respect to the overall worldview.

Seldom is science limited to only two courses of action and there is still another way of determining between the two basic hypotheses that seem to explain the NDE. Dr. Blackmore either implies or claims that such compelling evidence as would support the afterlife hypothesis would also necessitate fundamental changes in physics, biology and psychology and thus hints at this other way. Independent work in one of these fields could either imply or directly discover the truth of the matter. Physicists could discover another reality that somehow parallels our common reality or an extension of our common reality that we do not normally sense that could be identified with the afterworld. Or perhaps biologists might discover the reality of life, that 'something extra' whose existence many scientists have suspected and sought for several centuries. Some scientists have even come to give this 'something extra' a name, the biofield, and describe its probable characteristics. And perhaps, just perhaps, psychologists might discover the existence of the human 'soul'. Any of these possibilities could easily verify the afterlife without reference to either of the above hypotheses for explaining NDEs. The afterlife is not about NDEs, it is about literally everything that is physical reality itself, and so its existence cannot be boiled down to a simple 'yea' or 'nay' vote on the various interpretations of NDEs.

The 'dying-brain hypothesis' is just the latest installment of the same mind-matter or mind-body controversy that has plagued science for more than two millennia. Where do thoughts and memories reside, in a mind that is separate from and transcends the brain or in the brain itself? Does life transcend the material body or is life just the sum total of the interacting chemicals and mechanical actions that constitute the material body? Which is closer to the reality, the afterlife hypotheses or the 'dying-brain hypothesis'? All of these questions are the same, just asked at different levels of the same body of human knowledge and experience. Recent trends in science seem to be moving toward the possibility that life, mind and consciousness are not just electro-chemical mechanisms, but rather 'something extra' that transcends the mechanistic device or body with which life is associated.

These same trends in science have reintroduced the study of consciousness into the scientific mainstream. The correct question is not whether there is an afterlife, but 'if' there is an afterlife, what survives? The answer to this question will not come from the study of NDEs, nor will it come from deciding between those two hypotheses or two classes of hypotheses that attempt to explain NDEs. The questions raised by the reality of NDEs are far too fundamental to be answered at the level of NDEs. NDEs only exist at the observational level of the scientific method, not at the theory building point of the method. If anything survives that is subject to scientific scrutiny, then it is either consciousness or somehow associated with consciousness, so determining what consciousness is and how it is related to life will determine what survives the end of life and whether there is indeed an afterlife, and these questions are more basic to physics, biology and psychology.

The nature of consciousness

We could say that consciousness found science (again) in the 1970s or we could say that science found consciousness (again) in the 1970s. Either is correct because our science is the product of our human consciousness. Our science, in the form of theories and natural laws, amounts to how our consciousness interprets our physical/material environment. Since the 1970s, the science of consciousness has been slowly but steadily growing. The science of consciousness is now a valid interdisciplinary field within the scientific, academic and scholarly communities. By necessity, it is interdisciplinary because all of science, all of the various disciplines within science, indeed within culture itself, are either products of human consciousness or interact with human consciousness at some level.

The growth of the science of consciousness became evident to everyone in the 1990s when it reached a new level of sophistication. During that decade a system of international conferences based on the theme "Toward a Science of Consciousness" became a reality. In odd numbered years these conferences are held outside of the United States and in even numbered years they are held in Tucson, Arizona. The first of these conferences was only lightly attended, but these conferences have grown substantially over the last decade and are now very heavily attended by representatives from all of the academic fields. International organizations have also been formed to support this fledgling science. In particular, the Association for the Scientific Study of Consciousness (ASSC) was formed in the 1990s. The ASSC also holds yearly conferences. Several professional peer-reviewed journals are also being published, including the *Journal of Consciousness Studies* and the journal of *Consciousness and Cognition*. All of these milestones indicate a healthy, vigorous and progressive field of scientific endeavor. However, this endeavor is not without its own unique set of problems and its initial momentum of growth may be lost if some important new breakthrough does not come soon.

The interdisciplinary nature of the science of consciousness is both its strength and its weakness. Every discipline explains consciousness relative to its own relationship with consciousness, so there is no single explanation and no single direction toward which the science of consciousness can progress. The field may soon suffer from its own lack of a precise definition of consciousness. So the most fundamental problem is the problem of definition. It seems that everyone has a certain feeling for what consciousness is, but that feeling does not especially translate into an adequate definition of consciousness, even for an astute scientist or philosopher. Consciousnesses is not a thing like a proton or a concept like mass, so it cannot be quantified in the same manner as the other subjects treated by science. That is precisely why it has taken so long to begin developing a science of consciousness. It is like trying to grab a cloud. Maybe science can isolate it to some extent and even measure some of the quantitative results of its actions, but consciousness cannot be defined, quantified or measured directly. We only know consciousness exists by its results, and those results are interpreted by human consciousness itself. The philosophical implications of this circular process alone are tremendous and would be extremely difficult to overcome

First of all, we have philosophical debates whether consciousness exists or not. In the early twentieth century, behaviorism became the backbone of the new science of psychology, as based on the work of John B. Watson. Before that time, the existence of consciousness had been a purely philosophical question. However, the development of psychology as a science of the mind did not end the debate. Since introspection could not reveal consistent laws of thought that governed the various mental states, internal mental states of consciousness were just rejected as meaningless by the behaviorists. Consciousness itself was considered irrelevant to the objective study of mind, even though consciousness had been philosophically studied and thought about for hundreds of years. Consciousness was irrelevant because it could not be objectively investigated, so it was ignored in all psychological research. Yet science has now reached a point in its own progress where consciousness can no longer be ignored. All of the roads followed by several sciences lead to the necessity of dealing with consciousness at the scientific level. Although no single compelling scientific reason to develop a science of consciousness has emerged, the collective effect from all of the different sciences forms as compelling a situation as science should need. So science has developed or evolved a more accepting attitude toward consciousness as research on the brain grew more sophisticated and new questions were formulated based upon new neurophysiological findings that implied the existence of consciousness.

The brain had long been associated with consciousness, so it had been assumed that the brain was the seat of consciousness. In a sense, we think in our brains, so consciousness seems to exist in our brains. In fact, through use of modern technological devices such as the electroencephalograph (EEG) as well as NMR (nuclear magnetic resonance) and MRI scanning, specific areas of the brain could now be associated with specific conscious functions even if the actual processes of thought could not to be found or defined. But these new findings only raised new questions without answering the old questions about consciousness: Is the function of the brain consciousness itself or is consciousness more than just the functioning of the brain? This question parallels questions whether life is more than the sum of the individual parts of the body or organism, so it would seem that life and mind are the same kinds of 'things'.

The main question that needs to be answered still boils down to the concept of consciousness recognizing and realizing itself. First of all, humans can 'observe' their own mental states, an act called introspection. In other words, we are aware that we are thinking, as if we can somehow place ourselves outside of the thoughts themselves and think about the thoughts as they develop in our minds. I am thinking about writing this sentence at the same time that I am writing this sentence and you are thinking about reading this sentence at the same time that you are reading this sentence. Everyone is familiar with this type of thought process, which has been popularized by the Cartesian statement "I think therefore I am". In turn, this statement is related to the Cartesian dualism that has dominated science for more than three centuries, the separation of mind and matter. In the Cartesian sense, consciousness was part of the soul, so any study of consciousness would have been a religious undertaking, a matter of mind, rather than a scientific undertaking, a matter of physical matter. While the Cartesian philosophical imperative has long outlived its usefulness in science, there are those scholars and scientists who still think in similar terms, just disguised in modern garb. Some still wish to keep mind (consciousness) and matter separate.

We can react to external stimuli from our environment or act according to our internal needs (like eating) without thinking about the acts themselves. Some things we do without thinking, like breathing or pumping blood through our veins, but other things we do only by conscious thought and choice. Choice introduces the concept of Intentionality, which is another fundamental characteristic of consciousness. Intentionality specifies that consciousness is about something or that it has contents, its intentions. William James allotted consciousness five different characteristics: Intentionality, subjectivity, change and continuity, and selectivity. But more commonly, we know that when we think, we are aware that we are thinking, which implies that consciousness is separate from the thinking process. The fact that we can think about a subject even when we are thinking about thinking about the subject implies that consciousness is a 'thing' that is separate from our brain and the actual electro-chemical processes in our brain which are the material correlates of the thought process. There must be something that causes the thoughts to manifest in the brain; the electro-chemical processes did not initiate themselves, so that something is consciousness.

Since we know that we are thinking and we accept and utilize the thought processes themselves as tools then we also display the property of self-knowledge. It is this knowledge of self that distinguishes consciousness as a mental property or abstract 'thing' which we can try to define. Although there is no absolute test for determining consciousness, a primary test for determining which living beings could have 'consciousness' has been developed although it is controversial. It is called the 'mirror test' and it is based upon the idea of self-knowledge. The 'mirror test' was developed by Gordon Gallup in the 1970s and simply consists of showing an animal its image in a mirror and seeing if the animal recognizes its own image. As an indicator of consciousness, the mirror test has confirmed that humans, most great apes (except gorillas) and bottlenose dolphins are conscious beings.

Given the fact that we are conscious beings, such that we have consciousness, the next step is to define consciousness, and therein lays the rub. Science cannot settle on a single cross-disciplinary definition of consciousness that works for everyone in all of the situations where consciousness is known to play a role. In fact, the problem is that consciousness means different things to different people. Some scholars argue that consciousness simply cannot be defined at all, which just sounds like a Cartesian excuse for not dealing with the problems scientifically. For example, when a psychologist talks about conscious, sub-conscious or unconscious states of the mind, he, in all likelihood, is not addressing the same idea of 'consciousness' that a physicist invokes to account for the 'collapse' of the wave function in quantum theory. This distinction is all the more important because physicists are probing what seems to be a fundamental quantity within physical reality, rather than just a temporary state of mind, when they speak of consciousness. Even when a psychologist says that unconsciousness is a state of consciousness, he or she is implying the existence of a grander and greater 'consciousness' that is not defined by its states alone. A human can lose consciousness in the sense of sleeping or being unconscious, but cannot exist and still be human without a consciousness. Consciousness in this sense defines the person in the material body not the state of the mind. These are not silly word games, but represent the dilemma faced by scientists and philosophers when trying to deal with the concept of consciousness. So, depending upon how one defines it, consciousness can be an integral part of physical reality, just a state of mind, or anything in between. This dual notion of the concept of consciousness pervades and unnecessarily complicates the modern study of consciousness. The problem resides in the fact that physicists, at one extreme, deal (or attempt to deal) with very precisely defined quantities, while consciousness seems to be an ambiguously defined quality for all intents and purposes, rather than an easily definable and thus measurable quantity.

The basic dichotomy represented by these two extreme views is not new in the science of consciousness. Robert Ornstein proposed that there are two major modes of consciousness, one analytic and the other holistic. The two modes are complementary, each having its own functional area. They can be likened to the "rational" and "intuitive" activities of the mind. (Ornstein, 1972) Ornstein's "rational" could be thought of as representing the quantitative and logical aspects of consciousness that is addressed in physics, while his "intuitive" activities of the mind sound more like the psychologists' concept of consciousness as a state of mind. Other scientists are far less certain about 'consciousness.' Thomas Natsoulas narrowed his list to seven different uses of the term consciousness (Natsoulas, 1978) and there is no reason to believe that his choices exhaust the possibilities. Anthony Marcel and Edoardo Bisiach claim that the proliferation of different uses of the word 'consciousness' stems from the "domain or level of discourse" for the term. Sometimes consciousness is used as a functional term and other times it refers to phenomenological concepts. (Marcel and Bisiach, 1988) Whichever the case may be, the interpretation of the problem by these scientists is not all that different from Ornstein's two major modes, or, for that matter, from the differences represented by the physicist and the psychologist. These scientists have discovered essentially the same basic dichotomy in the use of the term, each expressing that dichotomy after their own fashion.

However, there are some distinct properties of consciousness that all can agree upon, so these form the basis of our ideas about a definition of consciousness. John Locke, the English philosopher, probably gave the earliest definition of consciousness in his book *An Essay Concerning Human Understanding*. Locke thought of consciousness as "the perception of what passes in a man's own mind". (Locke, II, 19) Since then there have been many different takes on the same idea. In 1899 G.F. Stout introduced the idea of introspection: "To introspect is to attend to the workings of one's own mind". (Stout, 14) The notion of introspection is rejected as impossible by some philosophers, behaviorists and materialists because it implies that mind is a non-material (some say non-physical) 'thing' that is capable of experiencing itself and the material world. They reject that possibility because such a 'thing' has no material basis in our physical world. In other words, introspection implies a higher order 'thing' that is separate from body and brain. This concept is especially hard for some to swallow since all evidence links both the mind and consciousness to neurophysiological constructs in the brain and the brain stem.

The concept of consciousness was complicated further when William James considered the notion of a 'stream of consciousness' in his 1890 book *The Principles of Psychology*. The 'stream of consciousness' differentiates between the whole of consciousness as a stream of visual, auditory, physical, associative and subliminal impressions that occur continually over time as opposed to a single thought or conscious act that occurs in a single moment in time. In essence, James' notion added the sense of time and duration to the mental concept of consciousness, which further clouded the concept. While scientists would have to deal with explaining how consciousness can be aware of itself in a rational manner, they must realize that consciousness is at the same time accepting new data that constantly changes it. Have you ever tried to catch a moving ball that is standing still? It seems an impossible if not a ludicrous undertaking because the ball is a material object that cannot be moving and stationary simultaneously, which implies that consciousness is not a material object because it can be changing and remain constant at the same time. Perhaps the concepts of change and constancy do not even apply to consciousness as they do to our concepts of material reality. So consciousness must be far more fundamental to our concept of physical reality, not just the material part of reality, than just the momentary electro-chemical interactions in the brain and brain stem, which are material constructs.

Physics entered the debate largely through the philosophical ideas expressed by Eugene Wigner in 1961. Wigner thought that quantum mechanics, as it was understood at the time, could not account for everything that exists and should therefore be modified to include consciousness and other 'self-reproducing states' that quantum mechanics deems unnecessary when dealing with purely physical problems. Quite simply, if consciousness is necessary to 'collapse the wave packet' and determine which way any physical system will proceed at any given moment in time, a process called measurement, then consciousness was introduced from outside of the quantum mechanical system to conduct the measurement process. Although this explanation oversimplifies Wigner's opinion on the matter, it still represents a valid view of Wigner's ideas regarding the role of consciousness in physical reality that has strongly influenced later work on the concept of consciousness. Wigner further concluded that some people wrongly reject the idea of consciousness even though we have direct evidence of its existence, so these people must not have the same sense of 'reality' as everyone else. By even making this statement, Wigner associated the existence of consciousness with the existence of physical reality. The concepts of consciousness and physical reality are intimately linked together, which means that consciousness is as much a part of physics, which studies physical reality, as it is part of biology or psychology if not more so.

Even though the door had been opened for physicists to play a significant role in the science of consciousness, the next really big step in the philosophical debate that could eventually lead to a physical theory of consciousness came from David Chalmers, a philosopher, in the 1990s. To clarify the issues in defining consciousness, Chalmers distinguished between the easy problems and the hard problem. The easy problems all concern 'objective mechanisms of the cognitive system' so they will eventually be solved by the cognitive psychologists and neuro-physiologists. Although they are very important to the overall concept of consciousness, the easy problems do not address the conceptual basis of consciousness itself. The hard problem does address this fundamental aspect. The hard problem is to determine how physical processes in the brain induce our subjective experiences. The real mysteries of consciousness are explaining what we think and feel about things, our internal interpretations of the raw data representing external stimuli gathered through our five senses. These are called 'qualia'; a term introduced by C.I. Lewis in his 1929 book *Mind and the World Order*. 'Qualia' are now understood to be knowledge that can only be gained through experience. For example, we know what the color red is and we can precisely define it, but we could never explain the feeling that it invokes to a person who was blind from birth and could therefore have never experienced the color red. The subjective experiences or 'qualia' associated with consciousness seem to emerge from the physical processes of interpreting sensory input, but cannot be explained by those same physical processes. That is the hard problem.

The existence of 'qualia' and the difference between the 'hard problem' and 'easy problems' seem to imply that consciousness is far more fundamental than what can be accounted for in our material worldview. While it is very probable that the easy problems will eventually be resolved through normal science by reduction of the whole brain and its parts to electro-chemical interactions, that same confidence cannot be placed in what we regard as 'normal' science since normal science seems no closer to answering the hard problem. The difference between the hard problem and the easy problems is just the modern version of the difference between mind and matter, mind and body or life and matter. The hard problem would seem to indicate that consciousness is 'something extra', beyond simple scientifically quantified processes and mechanisms. The hard problem is the same as that posed by the existence of NDEs. The 'hard problem' corresponds to the 'afterlife hypothesis' while the 'easy problems' represent the same point-of-view as the 'dying-brain hypothesis'.

The matter is quite simple to understand. If 'qualia' are something extra whose existence only the 'hard problem' can address, then true consciousness, although it interacts at the material level with the material brain, a process which can and will probably be explained by normal physics and biochemistry, is beyond normal physics and biochemistry's abilities of explanation. In that case, there is no reason for science to claim that nothing survives the death of the human or any other body because death is only the cessation of the biochemical and physical reactions and interactions by which the life process and mechanisms are explained. So, nothing in science prohibits 'something' from surviving death. Science can prohibit 'something' material from surviving because the material body dies, but a physical consciousness that is nonetheless non-material can survive. What is material in our reality is physical by definition, but not everything that is physical is also material. In other words, there are physical 'things' that are not material, such as the magnetic, electrical and gravitational fields. Explaining how consciousness can be physical while non-material is a truly 'hard problem', but at least it is now a better-defined problem.

Needless to say, there have been attempts to develop physical theories of consciousness. The Hameroff-Penrose model of consciousness is the best known of these attempts. In his book *The Emperor's New Mind*, the physicist Roger Penrose suggested that non-local quantum effects that occur within sub-neural structures in the brain could account for consciousness. He further suggested that changes in physics were necessary to develop a complete theory and understanding of consciousness. Normal physics, as it is presently understood, was just not enough to handle a model or theory of consciousness. However, suggesting a link between quantum mechanics and consciousness is one thing, but developing a workable theory is an entirely different matter. Enter Stuart Hameroff, an anesthesiologist by training and trade. Through his own researches on unconscious states during the application of anesthetics to a patient, Hameroff concluded that the anesthetic directly affects the microtubules that form the cytoskeleton of neurons to render the patient unconscious. Hameroff approached Penrose about the subject of microtubules and consciousness and the Hameroff-Penrose model of consciousness was born.

Microtubules are small nano-scale cylinders whose surface is constructed from tubulin proteins that connect in a specific spiraling pattern. The individual proteins can only exist in one of two different quantum states. When the tubulin proteins 'collapse' quantum mechanically into one state or the other, they form an electronic pattern over the surface of the microtubulin cylinders in the neuron. Hameroff claims that these patterns are the basis of consciousness in the brain. The individual microtubules form a quantum-entangled state within a neuron as well as with the microtubule cytoskeletons of other neurons to form a single coherent thought. However, the mechanism of this complex entangled state has not yet been explained. Hameroff rejects all suggestions that the entanglement could be electromagnetic in nature and claims that the entanglement is a completely quantum mechanical process. The theory runs into severe problems over this issue of entanglement. The brain is 'warm and wet', conditions that would tend to decohere (scramble) any large-scale quantum effects such as the type of entanglement proposed by Hameroff. Other critics argue that consciousness must be the result of much larger coherent patterns of activity that involve extremely large numbers of neurons and it is just that level of coherence that the Hameroff-Penrose model cannot explain.

On the other hand, many scientists believe that classical electromagnetic theory is adequate to explain consciousness. Various systems of thousands of neurons in the brain, called neural nets, could account for the complexities of thought associated with human consciousness. The neural nets would interact electromagnetically not quantum mechanically, thereby circumventing the issue of quantum decoherence in the brain. However, there is again a large difference between a few suspicions and expectations that electromagnetic interactions in neural nets can explain consciousness and a testable theoretical model of neural net consciousness. Still other scientists believe that nonlinear dynamics (in physics) or chaos theory (as it is called in mathematics) can supply the mathematical basis for a theory of consciousness.

If we could look at the individual motions of water molecules bouncing around in the atmosphere, we would see no specific patterns of action between the individual molecules and assume that their motions were completely random. Randomness implies that there are no forces coupling them together and their motions would have no effect on each other. But if we look at the same water molecules at a different level, a pattern called a cloud emerges from the random motions. At a still higher level, the pattern could be called a storm front or some other weather system. This entire ordered pattern develops out of the completely random motions of water molecules at a microscopic level. Chaos theory and non-linear dynamics explain how order can emerge from chaos and form the cloud patterns at a higher level of reality. Many scientists propose that the random firings of neurons at one level in the brain may allow the emergence of coherent thought patterns at a higher level, perhaps at the level of the neural net. Out of seeming chaos a complexity (the pattern) emerges that can be modeled by a mathematical formula called an attractor. The complexity that we detect as individual thoughts that emerge in this manner may then form a stable complexity at a still higher level called consciousness, just as the random motion of water molecules at one level emerges as clouds and weather systems at higher levels of reality. These complexities may or may not be related to the neural nets and electromagnetic coupling between individual neurons, but could eventually prove the existence of a new physical mechanism. However, once again good intentions and promising suggestions do not make a theory. So no verifiable and testable physical theories of consciousness have yet emerged from this direction of research.

Should a theory of consciousness be developed that explains consciousness as 'something extra' beyond the individual neurophysiological structure within the brain, the possibility that consciousness could survive death would have to be taken seriously by science. Science only has evidence that the brain dies at death. Otherwise, normal science is not yet ready to take on the fight to verify or establish the existence of an afterlife. That does not mean that normal science, science that is based upon accepted theories and laws, still wants to neglect the concept of death. Simply put, normal science does not know how to start a dialogue on the question from within its own established network of theoretical concepts. Yet this is not true out in the fringe areas of science, the foremost battle line of science. Scientists working in the field of the paranormal, in either parapsychology or paraphysics, are willing to look at what evidence is available to determine if anything survives death. In the parasciences, it is normal to think outside of the box, which is outside of the present paradigms of science, and consider the possibilities that normal science neglects *a priori*, the possibility of an afterlife.

CHAPTER 4
PARANORMAL PERCEPTIONS OF DEATH

I shall no doubt be blamed by certain scientists, and, I am afraid, by some philosophers, for having taken serious account of the alleged facts which are investigated by Psychical Researchers. I am wholly impenitent about this. The scientists in question seem to me to confuse the Author of Nature [God] with the editor of Nature [scientific magazine]; or at any rate to suppose that there can be no productions of the former which would not be accepted for publication by the latter. And I see no reason to believe this.

—C.D. Broad – 1925

Science beyond the normal

The goal of science is nothing less than the accurate and complete understanding of the world in which we live, whereas the world in which we live, in the general sense, should be considered as the whole physical and material universe. In other words, scientists believe the theories and laws that they develop locally from their observations of nature must apply to the whole universe. Within this context, science defines itself and sets the limits of what it can study as well as what is scientifically valid for study according to this goal. The world that science investigates would include both the outer world of material reality and the inner world that we call life, mind and consciousness as well as whatever else we may find when we probe the depths of our innermost existence and the furthest reaches of the universe. There is no part of our being that should not be included in the phrase 'our world' and therefore amenable to scientific inquiry and science assumes as much.

Yet there are phenomena and events that seem real on a personal level, but exist outside of science, so science actually works according to a double standard that it seems unaware of. In other words, it only seems that science describes an ideal world where no subject falls outside of the scientific perspective at any level of our observations of the world around us. The goal of science has been so deeply instilled in scientific thought that it now seems as though it is more of a moral purpose than a simple goal allowing some scientists to push forward into areas that have normally been off-limits to science. Some scientists sometimes act without even a second thought as to the consequences of their actions, believing they are fulfilling a false subjective purpose of science rather than an objective goal of science, even though they claim complete objectivity in their endeavors, while other scientists ignore some areas of observed phenomena for purely subjective reasons instead of objectively studying them. So science is not so perfect as it might seem to some scientists, scholars and academics.

At times, science unnecessarily limits the scope of its endeavors by diminishing the role played in nature by some types of events and phenomena that do not fit nicely into the prescribed scientific worldview and the paradigms (theories and laws) from which that worldview is constructed. Some scientists and scholars take their own private worldviews so seriously and absolutely that they deem certain reported phenomena impossible according to their *a priori* interpretations of what is or what is not scientific 'fact'. In some cases, it is admittedly difficult to determine the reality of phenomena, but science should make it a practice to err on the side of investigating questionable phenomena because questionable phenomena will always disappear under scientific scrutiny if they do not represent reality. However, scientists do not always follow this unwritten rule.

In general, real problems that occur in nature do not go away by ignoring them and reports of such phenomena do not go away just because some scientists and scholars believe that they are not worthy of scientific investigation. The persistence of some questionable phenomena that science ignores, such as the possibility of an afterlife as well as certain paranormal phenomena, continually cause problems for science precisely because science ignores them when in fact they will not go away. In the ideal sense that the realm of science is unlimited in its quest to understand nature, neither paranormal phenomena nor the possibility of survival would be ignored, but in reality they are ignored in most cases as *a priori* unreal. In the past, science has also ignored the direct study of mind and consciousness only to have them thrust once again against the scientific paradigms because mind and consciousness are part of nature as studied by science. To have a complete science, concepts such as mind and consciousness cannot be ignored, nor can they be sidestepped, while the independent existence of mind and consciousness implies the possible reality of some forms of the paranormal as well as the afterlife.

The method of conducting science by which reported phenomena dealing directly with mind and consciousness, including the survival of consciousness, are either accepted or rejected as real was originally based on sound rational thinking at a time when science was battling for its survival against superstition and the supernatural. When science first developed as Natural Philosophy, religion held an unwarranted stranglehold over education and human thought. Religion directly influenced every facet of human existence and the early development of science. It took several millennia to separate religion from science, resulting in Descartes' separation of MIND (the religious realm of nature) and MATTER (the scientific realm of nature). A scientific bias in the form of an absolute skepticism against some forms of reported phenomena then developed in western thought. The present attitude of science toward some parts of nature that resulted from this early bias has very nearly become just that which science sought to dispel in its early adversaries and competitors, such as absolutist religion, no more than an excuse to limit science by prior control of what phenomena can be studied by science. In the ideal philosophical reality espoused by science, anything that we observe should be open to scientific investigation and this includes various phenomena associated with the many forms of paranormal phenomena that we experience as well as the possibility of the survival of consciousness in some manner.

There is an old story dating from the middle ages of European history that illustrates this point. The minute details of this story as repeated here might not be completely true to the original story, but it is still a real story that survives in the historical record of that time. Two Roman Catholic monks were debating the number of teeth in the mouth of a jackass. One monk claimed that the jackass had thirty-eight teeth according to his patron saint, while the other claimed that the jackass had thirty-five teeth according to the testimony of his own patron saint. The debate raged on with no one getting the better of the other in the argument, until the two were interrupted by a novice who innocently asked "Why don't we just count the number of teeth in the mouth of the jackass and see who is correct?" Both monks berated the poor young novice for his impertinent suggestion that the saints could be wrong and then punished him for his lack of faith in the saints. This story is told to illustrate the philosophical problems inherent with the cult of scholasticism that ruled the thought and attitudes of many generations of scholars in Europe before the advent of science. Early science had to fight against scholasticism in order to grow and survive. Knowledge at that time was built upon the ruins of previous knowledge, whether or not that previous knowledge accurately described physical reality. If a past scholar or authority said that something was true then it was believed religiously, no matter what nature actually revealed. While the monks exhibited the worst of the scholastic principles of knowledge, the novice expressed the new scientific view that the scholars and philosophers should stop debating how nature 'should' work according to their philosophical whims and fancies and instead start observing how nature actually worked.

Similar stories abound throughout the history of thought, culture and science. In science we sometimes hold on to old outmoded theories and ideas long past the times when they should be laid to rest as footnotes in the history of human thought. To some extent that is to be expected since older scientists learned their craft and developed their personal worldviews under the influence of older theories that may not be as accurate as they were once thought to be, given the development of newer concepts and more accurate observations of nature. Older scientists do not easily adapt as the limits of science expand into new areas of research and application. Science decides what is scientific and what is not scientific according to its paradigms, but those paradigms undergo constant minor changes as they redefine themselves. When such a redefinition occurs, science can shift its focus and some scientists are unable to adapt to the new ideas and concepts by their own shift of focus.

By its own definition, science is conservative and sometimes overcautious to new ideas because human thought has been led astray too often in the past by whimsical ideas and speculations instead of sound theories. Science is terribly self-protective, that is why all new concepts, ideas and theories are subject to such strict and rigorous procedures of verification. However, scientific skepticism has become so powerful in some instances that a few scientists have become the scholastic monks arguing over the number of teeth in the mouth of the jackass. They just refuse to observe the evidence that is before their very eyes and accept the possibility of things that they cannot see or refuse to see. Some obstinate and biased scientists argue that science is always open to change at one level, but then refuse to even consider change because they assume that their theories and beliefs are absolutely true at another level. The phenomenon of death and the possibility of an afterlife are matters of this type, so some scientists may believe in the afterlife as a religious matter, but then dismiss that same possibility as a scientific matter.

A healthy skepticism is a necessary part of science. It serves as a form of internal regulation to rid the discipline of bad science and 'pseudo-science', but that does not mean that skepticism should be allowed to dominate science and tell scientists what is possible and what is not possible, nor what is real and what is not real. Surrendering to such absolute skepticism would just reduce science to a new form of scholasticism. There have long been tales of apparitions and ghosts, which were dismissed as superstition or worse, outright fraud, but the stories have been so persistent throughout history that the case for their validity has grown. When people who are not superstitious and not susceptible to fraud or false witness see ghosts or apparitions, their statements should be taken seriously. Science has dispelled many other superstitions as unfounded, and for good reason in nearly all cases, while science has also taken some superstitions and determined what truth there is in them and adapted what is left over to the scientific worldview. This has not yet happened with the concepts of death and the afterlife because science ignores them, but there is no reason to believe that this cannot or will not happen with some form of an afterlife. A very few objective scientific speculations on the afterlife have persisted in spite of all attempts of science to dispel the concept, to ignore the possibility or to relegate it to religion. The circumstantial 'evidence' of an afterlife has grown stronger over the decades and centuries and not withered away as science advanced.

No one can deny that death is real; it is observed all the time and phenomena associated with our continued existence after death are becoming more common. No one doubts that all things that live eventually die, but 'normal science' still ignores death as a concept and refuses to ask if there is more to death then just an end of life, even though life has never been completely or precisely defined and measured. Science is not conceptually equipped to deal with death and the possibility of an afterlife in any form whatsoever. Yet some scientists have bravely pushed to the forward fringes of science and dared to ask controversial questions about death. These scientists listen to those who have experienced or otherwise witnessed what seems to be evidence of some form of an afterlife. However, these scientists form only a very small portion of the overall scientific community and they are strongly criticized by their peers even though they are conducting scientific investigations to the best of scientific traditions under the worst of possible conditions. The longer that science allows the concept of death and the afterlife to languish, in spite of thousands of years of observational evidence that some part of each an every one of us could possibly survive death, the closer science drifts to either outright scholasticism or a dictatorship of the absolute skeptics. There is however one saving grace to this disgrace: The development of the paranormal sciences. These new sciences are barely in their infancy, but they still allow a scientific look at the possibility of an afterlife and at least enjoy some minimal protection by the larger part of the scientific community.

The modern paranormal sciences of parapsychology and paraphysics are far better suited to explain the concepts of death and afterlife than the normal sciences. They do not need to conform to the same limits and boundaries as normal science by their very nature. So parapsychology and paraphysics are actually at the forefront of the scientific exploration of those subjects that science cannot ignore, yet cannot accept. In fact, the science that we now call parapsychology has a long history of dealing with issues regarding the afterlife, especially in its most recent precursor, the psychic sciences that emerged during the era of Modern Spiritualism. The modern sciences of the paranormal can trace their beginnings to the possibility of communicating with the dead that was popularized in the later nineteenth century with the evolution of Modern Spiritualism. However, if some part or portion of us survives the material death of the body, then our present paradigm or worldview of physical reality is too limited and would need to be revised. Such a revision would be a matter for physics and paraphysics, not psychology and parapsychology.

The spirit of science

Many people have dated the rise of 'Modern' Spiritualism from 1848. During that year, mysterious rappings occurred in a cabin occupied by the Fox family in Hydesville, New York. These rappings were eventually attributed to the spirit of a murdered salesman and news of this discovery traveled far and wide. From this humble beginning, various methods of communicating with spirits of the dead were popularized and developed during their rapid spread from America to England and Europe as new mediums emerged from all corners of western society. But the rapid spread of Modern Spiritualism cannot be accounted for by the Fox episode alone, to do so would be to perpetuate a simple legend. In other words, a single historical incident cannot explain why a large number of people in western society underwent a radical change of attitude and began to accept the reality of communication with spirits within just a few years. Such a rapid spread of any particular idea, including communication with the dead, could not have occurred in an intellectual vacuum, but could only have occurred if fertile ground already existed for the idea to take root. By the time that the rappings occurred in the Fox cabin, the road to Modern Spiritualism had already been paved by earlier movements in the occult, including forms of common witchcraft, phrenology, Mesmerism, and earlier types of spiritualism as well as other related disciplines and the scientifically respectable speculations of natural philosophy that were part and parcel to the European intellectual heritage in the decades preceding the 1840s. However, the rise of Modern Spiritualism also depended to a large degree on the recent successes of science and out of the combination of the two a new psychic science began to emerge.

Science had been beating at the door of religion for quite some time, making forays and expeditions into areas of thought that were delegated to the realm of religion as well as encroaching on areas of study that were once the sole concern of religion and superstitions, while industrialization was rapidly changing the social norms of human culture in the early nineteenth century. Faith in science and technology was growing at a far more rapid pace than religious faith, at least at the level of established religious doctrines. For nearly a century, geological discoveries were challenging the priority of the Christian view of creation. According to a large number of scientists, the geological features of the earth had evolved over millions of years by a slow physical process called 'uniformitarianism', challenging the Christian religious myth that the earth had been created in its present form only six thousand or more years earlier, over a period of six days. Then came the evidence of human and animal evolution. Charles Darwin published *The Origin of Species* in 1859, but the ideas surrounding evolution had been developing for some time and a populace that was becoming more educated than ever before, due to industrialization, was quite familiar with the newest scientific advances.

The Darwinian theory of evolution, a completely Newtonian mechanistic theory, postulated that humans had evolved from lower forms of animals and life, but could say nothing about the origins of the one aspect of humanity that distinguished all humans from other species of animals, the human mind. The human mind was unique and its existence offered as much of a challenge for the established religions as it offered for science, although religions could just state that the mind was created without any further explanations as to the plan, purpose or means of creation. Science had incorporated the Cartesian distinction between the realms of MIND and MATTER (or mind and body) over the centuries and had capitulated the concept of the mind to religious speculation by refusing to directly with concepts such as the meaning of life and the ultimate nature of life. However, the old Cartesian barrier was slowing eroding away as the successes of Newtonian science to explain matter and our common world grew. No science of mind was even possible before the 1840s as the other branches of science grew and matured. As far as science was concerned before the 1840s, the human mind was solely the province of religion and the metaphysical speculations of a few philosophers. This separation of powers in one sense rendered the concepts of death and afterlife solely within the realm of religion and in another sense protected the objectivity for science in reducing and measuring nature. Mind could be neither reduced nor measured by any standards that had yet been developed, so death and afterlife were completely beyond any possibility of definition or measurement and thus were not part of the realm of science.

But Darwin's theory of evolution changed the status quo and forced science to look at the concepts of life, mind, consciousness and death again, if only very grudgingly. Science could no longer afford to look the other way and ignore these issues since Darwinism thrust them into the scientific domain. Even for the general educated population, the theory of evolution implied a possibility that something could survive the death of the human body and that implication placed the subjects of death and afterlife in the sole possession of individuals rather than either the established religions or the scientific establishment. So when the story of the rappings in a small cabin in New York was heard the world, it found an eager and enthusiastic audience in the educated masses who were on the verge of reclaiming their own spirituality and destiny from the Church and developing their own scientific worldviews based upon their limited knowledge of science. Thus it would seem that the new spiritualism was not religious in the same manner as established religions, but still had the philosophical trappings of a religion while enjoying an air of scientific confidence, at least at a laymen's level of the understanding of science. The level and quality of science included in Modern Spiritualism depended upon each person's individual definition of science and personal worldview, but it was absolutely not a part of established science. So Modern Spiritualism differed from its earlier predecessors not only by its overtly personal religious components, but also by its character as a popular semi-scientific movement, if not a pseudo-scientific movement..

Once a large enough portion of society accepted the notion that mediums could communicate with the spirits of the dead, the challenge was then put to the scientific community to react to the new 'scientific' trappings of these modern spiritualists. The scientific community did not answer the challenges of Modern Spiritualism in anything that resembled a common front. Some open-minded scientists accepted the possibility that the phenomena reported by spiritualists were real and began to investigate the phenomena, but this approach represented only a small undertaking within the scientific and scholarly communities as a whole. The majority of members of the scientific and scholarly communities either reacted very critically and skeptically or just remained neutral in the debates over the scientific validity of the newly reported phenomena, reflecting the older worldview of a separation between science and matters concerning religion as well as the conservatism that normally marked science. Still other scientists and scholars saw in spiritual phenomena a new class of physical and mental phenomena that they thought could be explained by science without hypothesizing the spirits of once-living humans, while another small group of scientists and scholars accepted the reality of survival and the possibilities of communication with the dead. These scientists speculated upon the physics and psychology of what we presently call paranormal phenomena and can be seen as the true forerunners of modern parapsychologists and paraphysicists. The scientific reactions to modern spiritualism certainly took a variety of forms, but the most common form was outright rejection.

At one extreme, J.K.F. Zöllner, a German astrophysicist, claimed that the spiritualist phenomena were proof of the existence of a fourth dimension of our common space where spirits dwelled. He thought that the various phenomena experienced by spiritualists during séances and sittings were empirical data supporting his hypothesis of a fourth spatial dimension. Zöllner's reputation as a scientist was ruined when his primary test subject, an American medium named Henry Slade, was caught faking his communications with spirits. Mathematicians had only recently discovered the scientific possibility that space could indeed be four-dimensional and not just three-dimensional, as we commonly perceive. The hyper-dimensionality of space was a popular academic theme during the same period of time that Modern Spiritualism became popular, so it would have seemed to some that the two different ideas could be joined even though that was never the intention of the majority of mathematicians and scientists. However, the belief by a few that the fourth dimension of space could be a home to spirits and ghosts did more damage to science than necessary and discredited not only scientists but the very notion that the four-dimensionality of space could actually represent physical reality.

At still another level of the reaction to the changing role of mind in science, Sigmund Freud turned to psychiatry to explain the human personality as the common face of the human mind. Gustav Fechner developed psychophysics while Wilhelm Wundt took some of the first steps toward experimental psychology. Ernst Mach, another German physicist, turned instead to the philosophy of science as Newtonian physics began to encroach upon areas that were earlier outside of its domain, such as life and mind. Rather than condoning the older separation between MIND and MATTER, Mach postponed the need to answer any questions about the role of mind in science through philosophical arguments. He circumvented the whole paradox of the nature of MIND and ultimate nature of MATTER by claiming that our theories and laws of nature were based on our human sensations of physical reality, not on physical reality directly. Therefore, we could never really know the physical reality (MATTER) that science studied and speculations on the ultimate nature of reality were useless. We could only base our science on what we could actually see or sense and then develop theories about material reality only as filtered through our sensations. Natural laws were merely convenient and efficient ways of expressing our common human sensations of reality. Mach's work was fundamental to the developing 'positivist' school of thought and would play a significant role in the evolution of scientific attitude and thought in the first half of the twentieth century.

On the other hand, a leading American scientist and philosopher by the name of William James developed new ideas about human consciousness and became one of the principle founders of the newly developing science of the mind, psychology. Fechner, Wundt and James were important figures in the development of the general concepts of psychology as well as experimental psychology. All three thought that the mind was separate from the body (or matter), but that the human mind was still reducible in a manner by which it could be 'measured' and investigated by science. Fechner sought to study the functional relationship between physical and mental phenomena. He found differences between an inner and an outer psychophysics, his name for the relationship between mind and matter. Fechner believed in the simple truth that mind and matter were just different manners by which we sense a single reality. For his part, Wundt relied on experimental introspection to determine the limits and characteristics of the human mind. He concluded that every mental element paralleled a real physical element in our material reality. He then developed his method of experimental introspection to probe the mind, based on a belief that sensations occurred in the mind, which was not material itself, even though the sensations originated from the material world outside of the mind.

On the other hand, James' approach was more mystical and more closely related to religious and psychical views of the mind. His publication of books like the *Varieties of Religious Experience* in 1902 did not exactly endear him to the hardcore part of the scientific community. On the other hand, James was also a pragmatist who emphasized cause and effect, prediction and control as well as behavior rather than introspection as a direct look at the mind. James also developed the notion of 'habit' in reflecting the interaction of mind and physical environment and what is habitual in nature is quantitatively measurable and thus open to objective scientific analysis even though 'habit' originates in the subjective mind. Unfortunately, James' methods also led to the development of the behavioral school which adopted a positivistic and materialistic approach to psychology that denied the importance and significance of a separate mind and consciousness, thereby circumventing scientific interest in consciousness until the 1970s. In a very real sense, the newborn science of mind, psychology, lost consciousness for most of the twentieth century. But James was also interested in mysticism and the psychical aspects of consciousness. He was a founding member of the American Society for Psychical Research (ASPR) as well as an influential member of the Society for Psychical Research (SPR) in Great Britain.

The cultural background of society after the 1840s was so fertile for change that spiritualism took root before the new science of psychology could emerge, but the same historical forces and trends played out in the development of each of them. A whole new variety of psychical experiences and phenomena, based completely on the possibility of survival of the spirit, evolved to characterize the Modern Spiritualism movement. Quite simply, mediumship became a growth industry in America and Europe as well as a major performance art. Some mediums practiced showmanship to at least as great an extent as communication with the dead, which alone would have provoked concern by scientists. What had begun as the rappings of a dead salesman in a small house in New York evolved into full-blown séances and the public performances of mediums. The séances provided not only forums for communication with the dead, they also provided platforms for grandiose physical effects that included grand 'apports' (materializations) of flower bouquets and other objects as well as the materialization of rain showers inside the Victorian living rooms and studies of well known and respected scientists.

Spirits became visible during séances as 'ectoplasm' and other otherworldly forms of matter oozed from the mouths and noses of the different mediums to form recognizable apparitions. Some mediums used cabinets in which their spirit guides appeared. Photos of the spirits and apparitions were taken and many still exist, however, too many of the old photos of spirits of the dead bear remarkable resemblances to the mediums that called the spirits up out of their nether world habitat. In one famous incident, a well-known and well-studied medium by the name of D.D. Home reportedly floated out of a second story window in the home of an investigator and into the window of another room under the watchful eyes of a group of distinguished scientists. Acts of levitation were also reported in many other cases. Tables tilted and jumped around as if they were alive. Solid brass rings lying on tables somehow ended up wrapped around the legs of the tables as if they had passed right through the wooden legs and writing mysteriously appeared on chalkboards that were completely enclosed and inaccessible to the mediums. All of these phenomena were decidedly physical in character and quite flagrant and spectacular in practice, flying in the face of Newtonian science. The written reports of scientists and investigators on these matters are quite comprehensive and it seems that the scientists investigating the phenomena took adequate protective measures to guard against fraud. But fraud by the mediums did occur. Nearly all of the mediums were caught committing fraudulent acts at one time or another and the lists of fraud grew longer as time passed, slowly discrediting the whole Modern Spiritualism movement. So the scientists who investigated spiritualist phenomena were seen as being duped by the overall scientific and academic community who were more than happy to just ignore the phenomena.

The new form of spiritualism hit England much harder than other countries and English scientists reacted differently to the spiritualist movement. A few forward thinkers formed a professional scientific society to deal with the matter of the survival of spirit. Well known and respected scientists, such as the physicists Sir Oliver Lodge, William Crookes and William F. Barrett, supported and studied mediums such as Mrs. Piper and D.D. Home in spite of harsh criticism from the scientific community at large. These scientists never developed a proper theory to support their observations, the data they obtained or the fantastic claims of the mediums, so their scientific efforts were ultimately fruitless and are today considered an aberration within the overall progress of science, at least according to the standard accepted version of history.

The negative view of the contribution of these scientists to the overall progress of science is to be expected, although not absolutely true, since conservative science won the historical battle and thus gained the honor of presenting its own viewpoint as historical truth. The work of these scientists and their contributions to the development of science is still ignored as if it did not impact or affect the progress of conservative science. At least history owes them credit for the helping to lay the scientific foundations for the development of psychology as an independent science as well as the later development of parapsychology, both of which deal with questions surrounding human mind and consciousness at some level. Perhaps science owes more to the Modern Spiritualism movement than it would like to admit. The Modern Spiritualism movement was a popular reaction to the same historical forces and scientific progress that forced the first development of a science of mind, psychology, and the rise of Machian positivism as well as the Second Scientific Revolution.

In spite of the popularity of Modern Spiritualism within general society, the scientists studying psychical phenomena never constituted more than a small number within the scientific community although there was enough interest to warrant the development of professional organizations for the 'systematic investigation of this subject by science.' William F. Barrett attempted to present a paper before the annual meeting of the British Association for the Advancement of Science (BAAS), but its publication was rejected. So Barrett called for a meeting of like-minded scientists and scholars in London at the close of 1881, which led to the foundation of the SPR early in 1882. The SPR was the first professional organization of its kind. The founding members included Barrett, F.W.H. Myers, Henry Sidgwick, Lord Rayleigh, William Crookes, Sir Oliver Lodge and Alfred R. Wallace. The inclusion of Wallace in this group is especially significant since Wallace is the co-founder of the theory of evolution with Darwin. Still, the more general scientific establishment would not give the scientific investigation of the spiritualistic phenomena a fair hearing in spite of the scientific stature and reputations of these men.

Scholarly research and legitimate scientific papers on psychical matters were shunned by the larger part of the scientific and academic communities although they occasionally found their way into the smaller conservative scientific journals and popular periodicals. However, the SPR offered safe haven for like-minded scientists to meet, discuss the issues, study psychical phenomena and support the publication of their findings. It is a shame that some scientists appeared to control other segments of the scientific establishment to such a degree that they could *a priori* disregard all spiritualistic phenomena, which would seem an unscientific position to take, but that was the case. Many scientists today take the same anti-scientific position with regard to parapsychology. However, the negative attitude of the majority of scientists and scholars should not be interpreted as implying that they did not believe in the spiritual world. They could deny that the reported psychical phenomena had any scientific merit, while still holding to their personal beliefs of a separate spiritual (in a religious sense) world. The tendency to separate the spiritual world (of religion) and the material world (of science), just as had been done over the past centuries, was still quite strong. Academics and scholars were still influenced by the original Cartesian separation of MIND and MATTER, which seemed built into the art of science as practiced. Meanwhile, William James and other American scientists founded an American Society for Psychical Research (ASPR) in 1884 by on the other side of the Atlantic Ocean.

Still other scientists, influenced by the same questions and cultural factors that gave rise to modern spiritualism and the SPR, approached the question of survival from a purely physical and completely non-spiritualistic direction. In 1875, P.G. Tait and Balfour Stewart published *The Unseen Universe: or Physical Speculations on a Future State*. This particular work held a peculiar position with respect to the spiritualist movement. It was not a work on spiritualism and had nothing to do with spiritualism, yet it so closely paralleled the attitudes of some spiritualists that it could not be ignored. The book was surprisingly and unexpectedly popular among many scientists who were openly anti-spiritualistic. It must be remembered that the movement of Modern Spiritualism dealt to some degree with attempts to put older forms of spiritualism on a scientific basis. Any work which was spiritualistic or shared common fundamental characteristics with spiritualism, while being authored by so widely known a scientist as Tait, deserved special attention. Writing a book on the physics of death and afterlife, even during the height of the Modern Spiritualism movement, was an extremely brave thing to do. Tait and Stewart were so apprehensive about criticism from the scientific community, that the first printing of the book was anonymous. However, the book was so well received and even praised, indicating the immense popularity of the book throughout culture and science, that later printings carried the authors' names. Again, it is important to emphasize the fact that their book had nothing to do with spiritualism and merely offered 'physical speculations' about the possibility of something continuing beyond the death of the material body.

Tait and Stewart left their anti-spiritualist position to no one's imagination although they qualified their opposition to spiritualism. In *The Unseen Universe*, they argued for the survival of the human spirit after the death of the human body on strictly thermodynamical principles. They made no references to either religious or spiritual beliefs in their attempt to explain a possible form of afterlife. Yet they were clearly influenced by the same desire to find a place for afterlife in science that had inspired the spiritualists. Tait and Stewart specifically denied any belief in the phenomena reported by mediums and the spiritualists, calling them "pretended manifestations".

> We do not therefore hesitate to choose between the two alternative explanations, and to regard these pretended manifestations as having no objective reality.

> 49. But while we altogether deny the reality of these appearances, we think it likely that the spiritualists have enlarged our knowledge of the power which one mind has on influ-

encing another, and this is in itself a valuable subject of inquiry. We agree too in the position assumed by Swedenborg, and by the spiritualists, according to which they look upon the visible world not as something absolutely distinct from the visible universe, and absolutely unconnected with it, as is frequently thought to be the case, but rather as a universe which has some bond of union with the present. (Tait and Stewart, 70)

Tait and Stewart disavowed the physical manifestations displayed by the mediums although they adopted the same fundamental ideas of a unity between the material and spiritual worlds as the spiritualists. The unity of these worlds, as well as the mere existence of the spiritual world, was derived by Tait and Stewart in a logical manner based on the scientific theories of their time. Tait and Stewart did not need the psychic phenomena that were reported to support their work, which was purely physical speculation.

They claimed that the Second Law of Thermodynamics implied that the heat in the universe would be dissipated so that 'the final state of the present universe must be an aggregation (into one mass) of all the matter it contains' at a uniform temperature. Given the sheer volume of empty space, if all matter and heat were to evenly dissipate over space, the overall temperature would be quite cold. This notion was popularized independent of the spiritualism movement as the 'heat death of the universe'. They concluded that the present visible universe had begun and would end in time, so that 'Immortality is therefore impossible in such a universe'. The visible universe must therefore come to an end in both of its forms as transformable energy and as matter at some point in the distant future. Science could not escape from this conclusion! Nevertheless, the Principle of Continuity upon which all such arguments were based still demanded a continuance of the universe, forcing Tait and Stewart to conclude that there is something beyond that which is visible. This invisible realm, the 'Unseen Universe', existed independent of our commonly sensed visible universe, but was in contact with the visible universe at all times.

The Conservation of Energy also played an important role in this scheme. The aether that scientists of the era believe permeated both the visible and invisible realms acted as a medium of energy transfer so that the universe as a whole remained constant in energy. In this way "the Great Whole is infinite in energy, and will last from eternity to eternity". (Tait and Stewart, 172) Tait and Stewart tried to show that "immortality is strictly in accordance with the Principle of Continuity". (Tait and Stewart, xxii-xiii) With the immortality of man assumed through the auspices of the Unseen Universe, Tait and Stewart were "led by scientific logic to an unseen, and by scientific analogy to the spirituality of this unseen" and concluded, "the visible universe has been developed by an intelligence resident in the Unseen". (Tait and Stewart, 223) Tait and Stewart were clearly influenced by the Darwinian theory of evolution, but went far beyond Darwinism without directly invoking any religious notions such as God. In some ways their ideas were not all that different from those of Wallace, who accepted some spiritualistic phenomena as real. In spite of its initial popularity, *The Unseen Universe* seems to have disappeared from science after the turn of the twentieth century as spiritualism itself began to lose credibility. This loss of popularity paralleled the development of a new revolutionary period of science.

Perhaps the most influential if not the most enduring work to come out of the efforts of the scientists in the SPR was another book about survival and a possible afterlife. Frederic W.H. Myers summarized the scientific research on death conducted during the last three decades of the nineteenth century in his book *Human Personality and the Survival of Bodily Death*. The book was published posthumously in 1901, so it can be regarded as a compendium of the efforts of the nineteenth century scientists to deal with the concept of an afterlife. In his introduction, Myers aptly and accurately characterized the position of the scientific community on the afterlife concept.

IN THE long story of man's endeavours to understand his own environment and to govern his own fates, there is one gap or omission so singular that, however we may afterwards contrive to explain the fact, its simple statement has the air of a paradox. Yet it is strictly true to say that man has never yet applied to the problems which most profoundly concern him, those methods of inquiry which in attacking all other problems he has found the most efficacious.

The question for man most momentous of all is whether or not he has an Immortal soul; or – to avoid the word *immortal*, which belongs to the realm of infinities – whether or not his personality involves any element which can survive bodily death. In this direction have always lain the gravest fears, the farthest-reaching hopes, which could either oppress or stimulate mortal minds.

On the other hand, the method which our race has found most effective in acquiring knowledge is by this time familiar to all men. It is the method of modern Science – that process which consists in an interrogation of Nature entirely dispassionate, patient, systematic; such careful experiment and cumulative record as can often elicit from her slightest indications her deepest truths. That method is now dominant throughout the civilised world; and although in many directions experiments may be difficult and dubious, facts, rare and elusive, Science works slowly on and bides her time, – refusing to fall back upon tradition or to launch into speculation, merely because strait is the gate which leads to valid discovery, indisputable truth.

I say, then, that this method has never yet been applied to the all important problem of the existence, the powers, the destiny of the human soul. (Myers, 1)

Myers correctly noted the fact that science had previously ignored the question of an afterlife, which he termed 'immortality'. However, Myers clearly considered the soul to be that part of the person that survived death. His opinion followed a long prescribed religious tradition that the human soul survived death; the only difference being that Myers thought the soul could be studied, at least indirectly, by science. His purpose in writing the book was to break down the artificial wall that existed between science and the supernatural (Cartesian MATTER and MIND) and what he produced was probably the greatest analysis of mediumship, psychic abilities and the afterlife that has ever been published. *Human Personality and the Survival of Bodily Death* was certainly the most comprehensive book ever written on the subject. It was not a simple catalog of anecdotes regarding psychic phenomena. Myers was the first to describe the psychic experiences of people in different (altered) states of consciousness; asleep, awake and hypnotized. His book was also one of the first truly scientific studies to propose that psychic abilities were natural phenomena to which he applied the scientific methods of observation, definition and analysis.

Myers was a transitional figure between more modern concepts of consciousness and the older association of consciousness with the soul. He sat on the divide of human intellectual history between the strictly Newtonian worldview of a mechanical 'clockwork' universe and the inclusion of the strictly non-mechanical concept of mind in science resulting in the development of psychology. Myers developed the concept of the 'subliminal self' to explain the differences between the scientific concept of a soul and the older religious concepts of the human soul. He thought that the spirit or soul endured through death to an afterlife as did the spiritualists, but unlike the spiritualists he did not attribute normal psychical phenomena to the dead spirits. He instead attributed psychic phenomena, what we would today call psi or the paranormal, to the soul or spirit while it still inhabited the living body.

According to Myers, we live in a far greater universe than we 'normally' sense, but we still 'subliminally' sense that greater universe. The 'subliminal self' is a vast organism of which human consciousness is just an accidental part. The 'subliminal self' is the life of the soul, but not of the body, while the faculties that we call paranormal perceptions are the sensations that are directly perceived by the 'subliminal self' just as our normal perceptions of the material world are sensed by the body. Although Myers did not accept the spiritualists point-of-view on the survival of the spirit, his model of the 'subliminal self' and its relationship to the soul greatly enhanced the possibility of survival. The 'subliminal self' did not diminish during life and could therefore not diminish at death, so its purpose must be survival. He offered no further explanations of the data that he had gathered and offered no theory to explain the mechanisms of psychical phenomena, even though he better defined the problem of death and afterlife than had anyone before him.

So the nineteenth century ended with no real concept or theory of the afterlife, even though a general recognition of death by science itself did begin to develop. Recognition by a few scientists that both the proof and an explanation of the afterlife were well worth pursuing also emerged. At least the concept of an afterlife was no longer ignored, but it was still far from a distinct part of science. However, a science of mind did emerge in the form of psychology and the concept of a separate consciousness was recognized. Mind and consciousness were both considered separate from body and matter, as if there were dual worlds, the mental and the physical-material worlds, which interacted to create our universe. After all of the work and research on mediumship, no evidence was ever developed or provided that could verify the reality of communication with the spirits of the dead, or even if they existed, although the purely mental aspects of psychic phenomena had gained some credence in the small part of the scientific community represented by organizations such as the SPR and the ASPR. At least the scientific organizations that could lend support to psychical research were in place so scientists could finally find a forum for their ideas and publications on the subject. In spite of this progress, scientists still thought of death in terms of a soul that was somehow related to mind and consciousness, if they thought of the concept of an afterlife at all, rather than the survival of mind or consciousness. So it would seem that there were now three realities to contend with, the mental reality, physical reality, both of which were amenable to scientific investigation, and a spiritual reality which was still religious in nature, but not quite as completely religious as it had been a century earlier.

After 1900, spiritualism experienced a downturn in popularity within general society, as did the scientific investigation of spirits and the afterlife. After three decades of investigation, nothing of specific scientific value had been developed other than a lot of observational data and anecdotes. It was not until the 1930s that a new form of psychic investigation again opened a small window on the interactions of mind and matter. The new branch of science that evolved, called parapsychology, was based on behaviorist and materialistic methods, not on the concept a separate mind, consciousness, spirit or soul. Unlike previous psychical research, parapsychology did not emphasize the soul or spirit within a scientific context, both of which were finally purged from science. Parapsychology, as developed by J.B. Rhine, was a laboratory science and therefore not completely amenable to the concepts of death and an afterlife, although his work was ultimately related to those concepts. Just as science as a whole was finally beginning to look beyond the normal boundaries that it had set for itself and a few scientists were beginning to take psychical phenomena and the afterlife seriously, what was thought to be normal in science changed due to the Second Scientific Revolution that began in the period between 1900 and 1905. Scientific research in the psychical arena did not adjust to the changes in normal science until the 1930s with the work of Rhine.

The loss of soul and spirit

Fifty years earlier, statistics had been adopted by physics as a scientific method to determine the macroscopic effects (heat and temperature) of large numbers of randomly moving atoms and molecules in a gas. Statistics was first used because scientists could not measure or otherwise determine the individual motions of all of the atoms or molecules at the same time in a volume of a gaseous substance, so they worked instead with the statistical averages of the gas molecules. Statistics used in this manner complied fully with the Newtonian worldview. However, after 1900 the quantum theory institutionalized randomness and indeterminism into the very fabric of physical reality, at the sub-atomic level. The problem was not the technological inability to measure the positions and velocities (momentums) of sub-atomic particles simultaneously, instead it was deemed impossible to do so under any circumstances by the very indeterministic nature of physical reality. Nature itself acted and reacted probabilistically in its smallest and most fundamental regions of material existence, in spite of scientists attempts to explore and determine the nature of the reality thus explored. It took nearly three decades, from 1900 to 1927, for science to come to the complete realization of this subtle and fundamental change in our concept of physical reality and the basic tenets of the quantum theory. In the meantime, a social upheaval of worldwide proportions also occurred in the form of World War I, changing the cultural underpinnings of society.

During these three decades, Modern Spiritualism declined as a popular social movement. It seems as if the magnificence, in the form of the grand physical effects of mediumship, began to wither away. In actuality, the study of the psychical aspects of the human mind was slowly reacting to socio-political and scientific changes. Frank Podmore, member and chronicler of the SPR, had noted in 1902 that the emphasis of psychical research had changed to the mental aspects of telepathy, clairvoyance and precognition. Since no convincing or compelling objective evidence for survival and communication with the spirits of the dead ever materialized, other than subjective anecdotal evidence that is always considered suspect in science, psychical research began to concentrate instead on the mental aspects of telepathy. But research in this area also began to change radically during the 1930s as the new paradigms in physics began to affect other areas of scientific inquiry.

The next four decades of psychical research can be summarized in a single word, parapsychology, and associated with the name of one single scientist, J. B. Rhine. No other scientist or scholar was as influential as Rhine on the future course of the scientific study of psychical phenomena and no other scientist did as much to define the discipline of parapsychology. Rhine was invited to Duke University in North Carolina in the early 1930s. At Duke, he began a long and illustrious career in the field of parapsychology by conducting simple laboratory experiments in ESP and similarly testing other mental aspects of the paranormal. In this manner, Rhine quietly brought psychical studies out of the parlors of practitioners and supporting scientists and into the laboratory where they could be isolated and studied under controlled conditions, conforming to the norm for the rest of science. These changes in fundamental method were matched by the name change to parapsychology, marking the development of a new lab science. Under these new conditions, the emphasis of research in the paranormal changed radically from a search for the afterlife and the confirmation of phenomena associated with the afterlife to the study and statistical analysis of ESP. Spirit and soul were literally purged from science, even that small part of science that tried to legitimize the paranormal aspects of nature, just as the direct study of consciousness and mind were purged from psychology to study behaviorism after 1913. In both cases, the strong influence of Machian positivism is quite evident.

The first simple experiments were conducted with Zener cards, a deck of twenty-five cards with five different symbols: A star, circle, cross, square or wavy lines. A sender would turn a card over and another person would try to read or identify the card through the senders' thoughts. Probability would dictate a one in five chance of correctly reading the card from the sender each time a card was turned over, so these simple experiments were analyzed statistically. Statistics thus became a primary tool for the parapsychologist. Over hundreds and thousands of attempts to read the senders' thoughts, a very slight variation from the statistical norm would indicate the very subtle influence of extrasensory perception (ESP).

These and other experiments were thereby placed under very strict control to meet specific scientific standards and prevent fraud. Scientists no longer looked for the spectacular results of mediums and other psychics that were non-repeatable under laboratory conditions, but instead sought evidence for the existence of the very subtle influences that were statistically significant and hopefully repeatable when others conducted the same experiments under the exact same conditions.

Unfortunately, too many other unknown variables and factors affect the outcomes of the experiments and they were not always repeatable. When dealing with subjective quantities and the direct interactions of human mind and consciousness, the conditions are never completely controllable. Like psychology, the existence of mind and consciousness are assumed at some level, but neither was directly tested. Only the result of the direct conscious interactions between minds was studied. This change marked a certain maturing of the science of psychical research, but it also made obtaining proof of the phenomena that much more difficult and a far more tedious routine for obtaining data developed.

Skeptics and critics quickly identified problems with the experimental procedures. Over the next six decades there seemed to be a neck-to-neck race between the parapsychologists and the skeptics, during which the parapsychologists fine-tuned their experiments again and again as the skeptics made ever more stringent demands on the parapsychologists' experimental methods to forestall any possibility of fraud or deceit, whether intentional or accidental. Yet the skeptics would never accept even the possibility of positive experimental results from the parapsychologists because they assume the absolute position that the paranormal is impossible and would thus *a priori* assume the probability of fraud and deceit if positive results were obtained. The end product of this cat-and-mouse game was the development and application of far more comprehensive and exacting experimental methods and mathematical analysis than any other field in psychology. Many scholars and scientists have stated that if normal psychology were expected to follow as strict a methodology as parapsychology, many if not all of the theories of normal psychology would be discarded as unsupportable. Yet parapsychology is still troubled by undue bias and outright prejudice within the scientific and scholarly communities. The parasciences have been held up to standards that no other science has had to contend with and have actually gained strength by being forced to conform to such high standards. It almost seems as if debunking parapsychology and paranormal phenomena has been reduced from a scientific challenge to a political propaganda war.

The precise nature and specificity of the criticisms have forced parapsychology to carefully define its limits and the scope of the discipline. In 1946, R.H. Thouless and W.P. Weisner conducted a series of experiments from which they concluded that a single agent or mechanism was behind all of the known ESP and PK effects. They named this agent 'psi.' So, rather than speak of paranormal or psychical phenomena, it is now more appropriate to speak of psi phenomena or a psi effect. The various phenomena that parapsychologists and other scientists now recognize and investigate are commonly categorized into two separate groups: ESP (extrasensory perception) and PK (psychokinesis) phenomena. Classical parapsychology is purely a lab science and only covers ESP and PK, but as studies of ESP and PK have become more numerous and complete, parapsychology has again expanded to include more of the older psychical arena of phenomena. According to the SPR, the field of study of parapsychology includes more than just ESP and PK, or more than normal psi phenomena.

> The principle areas of study of psychical research concerns exchanges between minds, or between minds and the environment, which are not dealt with by current orthodox science. This is a large area, incorporating such topics as extrasensory perception (telepathy, clairvoyance, precognition and retrocognition), psychokinesis (paranormal effects on physical objects, including poltergeist phenomena), near-death and out-of-the-body experiences, apparitions, hauntings, hypnotic regression and paranormal healing. One of the

Society's aims has been to examine the question of whether we survive bodily death, by evaluating the evidence provided by mediumship, apparitions of the dead and reincarnation studies. (SPR, 1)

This list represents nearly the whole range and scope of the modern variety of psychical experiences, as exhibited or experienced in the laboratory and society in general. The new category of NDEs has been recently added. NDEs hold a special position because they are also studied (and sometimes dismissed) within normal science. Although parapsychology deals more with the ESP and PK phenomena that can be tested under laboratory conditions, parapsychologists are beginning to conduct more field investigations and collect anecdotal evidence for the other forms of psychical phenomena listed. Parapsychologists feel that ample evidence to support the reality of psi has been found since the 1930s, so they have expanded their studies to again include new data from spontaneous events that can be used to develop hypotheses and theories of psi.

The lack of repeatability of the experiments is the major concern that has haunted the science of parapsychology. Even the thoroughly controlled experiments in ESP and PK are not always repeatable. Whether evidence of psi is collected in the laboratory under stringent conditions or in the field with no control of the conditions, the repeatability of data gathering under the same circumstances, a requirement of all scientific studies, has been the hardest problem to overcome. Not knowing what psi is, and not yet having any theoretical basis for testing psi, experiments that limit and control the variables that influence psi effects cannot presently be developed with complete confidence and accuracy. Anytime that science is working directly with mind or consciousness, the task proves extremely difficult because experiments provide examples of the mind studying itself. In a sense, scientists are shooting in the dark without a theory or working hypothesis to act as a light and guide them through the darkness. Valid and crucial verifiable information is hard to obtain under these conditions, so parapsychology has experienced a very slow although steady growth during the middle decades of the twentieth century.

At the time, many parapsychologists believed and still believe to this day that psi is a purely mental phenomenon and physics has nothing to do with psi. This opinion fits well with the lack of complete objective control over the subjective nature of ESP experiments. Otherwise, the old habits of a separation between MIND and MATTER are hard to be discarded. If psi is purely mental, then it can never be 'objectively' defined and subjected to reduction and a model of mind and consciousness. Consequently, even some of the best and most qualified parapsychologists have ignored any possibility of a physical theory of psi, surrendering the development of a physical theory to an even smaller number of parapsychologists and scientists than even their own ranks could muster. On the other hand, physics has traditionally been considered inadequate to cope with psi phenomena, so developing hypothetical new quantities or entities to explain psi would seem both easy and necessary for any scientist interested in a physical theory of psi, let alone one who was not up to par on his knowledge of physics. Physicists were more involved in other matters during the period when parapsychology was developing and establishing itself as a legitimate science, from the 1930s through the 1960s, so they did not care about psi phenomena and surrendered the discipline to the parapsychologists. Physicists were busy applying quantum theory to physical systems, such as those studied in atomic and nuclear physics, which seemed by far the more pressing concern for science during the decades after the flowering of quantum mechanics in the 1920s.

Interest in developing a physical theory of psi did however grow, if only slowly, throughout the period from 1930 to1970. More physical theories of greater variety emerged in the 1960s than ever before, but those physical theories were largely developed by non-physicists. In 1969, the Parapsychological Association was admitted to the American Association for the Advancement of Science. Although some scientists fought the admission and still fight for the removal of the Parapsychological Association, the new elevated status of parapsychology gave the science of psi a new legitimacy that it had never before enjoyed. Admission to the AAAS also brought more non-parapsychologists into contact with the parasciences as well as providing an important new forum for established parapsychologists to present their evidence for the reality of psi.

After 1970, paraphysics finally began to come into its own as the word paraphysics began appearing in a few specialized books and scientific articles. The appearance of the word in literature marked a turning point in the evolution of this science and the beginning of a return of physicists to the study of psi phenomena. The factors influencing this return are numerous and mark a specific shift in attitude of the physics community as a whole. One aspect of this change can be found in the first development of 'observational' theories of psi, as others call them. The 'observational' theories of psi demonstrate the introduction of a new level of competence in physics into psi research. Quantum theory had been incorporated in some manner in several earlier theories of psi, but not at the same level of sophistication as became evident with the 'observational' quantum theories of psi. The earlier theories had used quantum effects to explain psi, but the new 'observational' theories were based directly upon the philosophical foundations of quantum theory, right up to the point where the observer and the observed quantum system interact.

When you take a closer look at the historical development of physical theories of psi and place them within the context of recent developments in theoretical physics, new and significant developmental trends appear in science that will eventually affect the continuing development of all areas of science. These are not trends in the science of psi alone, but trends in physics that directly impact the study of psi. Trends in other areas of science and culture parallel these trends and together they all make up what we believe to be the recent history of scientific development and form the starting point of new developments in science and culture. Tracking these trends gives us some information about where science is headed and the effect that science will have on culture and vice versa in the future.

One important trend is marked by the acceptance of consciousness as a legitimate area of research in physics over the past two decades, while another trend marks the march toward what physicists call a 'theory of everything', a TOE. The physics of consciousness is intimately bound with two other important trends. On the one hand, researchers in psi are now coming to a consensus that present day physics should be able to cope with and explain psi; if not now, then in the very near future. On the other hand, psi and consciousness have been bound intimately together by researchers in paraphysics and parapsychology as well as scientists who are not involved in the study of psi. Just as consciousness has been deemed something beyond the electrochemical and mechanical operations of the material brain, separate from both brain and body, psi seems to operate separate from, but still connected to, our material reality and thus unconfined by any material limits of action. So just as the survival of consciousness could account for an afterlife of the individual person, this type of afterlife would be intimately bound to psi while the individual was still living and all are bound directly to what we commonly call physical reality at its most fundamental level. These ideas are not that different from the concepts developed by Myers, but placed in modern terms and related to modern physical concepts.

A simple example of the relationship between psi and consciousness can be found in Robert Ornstein's 1972 book *The Psychology of Consciousness*. Although Ornstein is not a parapsychologist, nor does he specifically study psi phenomena, his study of the scientific concept of consciousness has naturally and logically led him to a short discussion of psi (Ornstein, 188-191, 220-224) because they are intimately bound together. This intimate connection means that psychology and the study of consciousness cannot be dissociated from recent trends in pure physics such as the evolution and proliferation of new interpretations of quantum theory and the functioning of quantum reality as well as the recent revival of unified field theories based upon the field theoretic view of general relativity. Nor are the individual trends just cited completely independent of one another, even though they are all distinguishable as unique in their own rights. Should all of these trends come to a head at or near the same point in time, in the form of a single broad-based theory, then a Third Scientific Revolution would surely emerge.

Even given these recent trends and other developments in the science of psi, we still cannot answer the question "Does psi exist?" with any precision at this time, although we will be able to answer that question in the near future. However, an alternate question could be considered now in its stead: "Is the pre-paradigmatic period of psi drawing to a close with the development of a new paradigm in physics that will be more amenable to psi?" The historical evidence indicates that the answer to this second question is a qualified 'yes' and when this question is answered we will also learn the truth about death and the afterlife. Because of the delicate emotional and cultural nature of these questions, no amount of anecdotal evidence will ever convince everyone of the reality of either psi or the afterlife, at least as long as our physical theories of reality do not specifically account for their existence. So a new physical theory of reality is called for and this call corresponds to recent work in pure physics to develop a TOE.

Any evidence that could be found would be found in the association of psi effects with the consciousness/matter interaction, the modern version of the MIND versus MATTER question. In this case, the flowering of the many different contending theories of psi developed by physicists characterizes a pre-paradigmatic or pre-revolutionary period in science. The pre-paradigmatic period is also evident within the context of the trends noted above, although they do not guarantee a change in paradigm. These trends only indicate a probable change of scientific attitude toward psi research, which could and most probably will be part of a larger paradigm shift in physics itself. The full content of the new, suspected paradigm would include the development of a TOE. By its very definition, a TOE must include an explanation of psi, if there is any evidence of a psi effect, and a model of consciousness, if not some definitive statement on what happens when we die. So scientific concepts of death and afterlife cannot be purged from normal science forever (as some would like) and keep coming back to haunt science at each point of progress in science, while paranormal science has been slowly progressing toward the solution of these problems, much to the chagrin of practitioners of normal science.

By the 1980s, scientists were willing to admit that the unification of quantum theory and relativity, as Einstein had attempted, was desirable and the search for a TOE was engaged by mainstream scientists. The first suspected TOE was called supergravity. It was developed in the 1980s, but the theory failed. The next attempted TOE came in the 1990s. It was called Superstring theory, but Superstring theory then evolved into Brane theory. At present, the Superstring and Brane theorists believe that their work is only a precursor to a final TOE that they call M-theory. No one seems to know what the 'M' in M-theory really stands for, but it has been suggested that it could stand for 'Messiah' (at least the Jewish word for Messiah) and the concept could be bound up in the Jewish mystical practices of the Kabbalah, a rather strange turn of events for a physical theory, but one that demonstrates the relationship between modern science and the expected limits of its future applications. It has now been suggested by the Superstring and Brane theorists that our universe is a four-dimensional Brane (membrane) curved in a separate five-dimensional bulk, (Hawking, 180-184) which may or may not be related to consciousness and therefore associated with psi and death by logical extension.

Unfortunately, the search for a TOE has followed along the lines of quantifying the continuous gravitational field, changing continuous space-time into small discrete lumps of space-time. In other words, scientists are trying to develop a unification theory based on the quantum assumption that there are fundamental particles of empty space as well as fundamental particles of time itself. Yet Einstein's approach to unification, the forerunner of the modern TOE theories, was conducted from the opposite direction. Einstein felt that the ultimate unification of physics would proceed from the concept of the continuous field, with the discrete quantum emerging from the continuous field rather than the other way around. Einstein may well have been correct, because the final M-theory, toward which Superstrings and Branes are only steps, is beginning to look more like Einstein's five-dimensional unified field theory. Whether modern scientists like the idea or not, science is approaching the very limits that define the paranormal and separate the paranormal from the normal in science with these advanced theoretical models of reality. Science in general and modern physics in particular have come to the point where they can no longer dismiss the reality and the direct physical interactions of consciousness that we now refer to as paranormal, whether scientists want to recognize that fact or not.

If nothing else can be concluded from an analysis of the development of physical theories of psi, it can at least be discerned that consciousness acting non-locally is intricately related to the physics of psi, if it is not psi itself. And consciousness acting non-locally could account for the paranormal phenomena that are normally associated with apparitions, ghosts and other 'things' related to the afterlife. So, physical theories of psi which account for the non-local action of consciousness at the quantum level are very likely candidates for a theory of psi within the coming paradigm shift, while the interaction of consciousness and matter haunts the very core of some TOE theories. The non-local interaction itself has not yet been defined or explained in physics, but it goes by the name of 'entanglement'. In the meantime, psi seems to be a field effect as opposed to an example of the discrete nature of matter as expressed by traditional quantum theory.

Entanglement seems more of a field effect itself, even though it is even though it is an essential part of the quantum theory. Therefore, a field theory of psi also seems to be a requirement for any future theory of psi as well as a TOE. Perhaps even the question of quantum versus field has reached a level indicating that it may soon become irrelevant, both in pure physics as well as paraphysics. The quantum is discrete and the field is continuous, so the quantum and field are mutually incompatible, even though unifying these two concepts is essential to the development of a TOE. The future paradigm in physics will be based upon the field or quantum as well as both and neither, all at the same time. This possibility seems illogical, impossible and ridiculous, as if the physical reality described by the suspected TOE were a Zen Buddhist 'koan' (what is the sound of one hand clapping?), but that is the reality of nature and the nature of reality. Quantum and field will coexist in a new synthesis leading to a new paradigm and part of that paradigm will deal with psi and death, just as it will deal with life, mind and consciousness and their role in physical reality.

Although many scientists will debate this conclusion, the new area of study called the 'physics of consciousness' is presently attracting many scientists who would otherwise not enter the fields of parapsychology or paraphysics, while the study of psi is a valid endeavor within the study of consciousness. Consciousness studies now offer a safe haven for parapsychologists and paraphysicists to talk about phenomena considered paranormal, as well as a safe haven for physicists to talk about paranormal subjects without really mentioning the paranormal and incurring the wrath of their colleagues and skeptics. If the study of the 'physics of consciousness' can show what consciousness is, isolate human consciousness within its overall physical model of reality and demonstrate that consciousness is a physical although a non-material 'thing', then it will be theoretically possible to demonstrate whether consciousness survives death or not. All roads that lead to developing a science of consciousness eventually lead to a definitive answer whether consciousness survives death or not, without any reference to the existence of either a soul or a spirit. If mediums actually communicate with the dead, the fact of the communication implies that the part of us that survives is conscious of that survival as well as its own consciousness or self, so what survives death must be either consciousness or related to consciousness in order for the communication to occur.

Another new type of paranormal phenomena also emerged during the decade of the 1970s: remote viewing. Remote viewing is a form of clairvoyance, but unlike normal clairvoyance, remote viewing is taught to individuals, so it is neither spontaneous nor does it require a 'psychic' or a medium. Remote viewing is a learned ability and thus the practice assumes that all conscious beings have the ability or are capable of developing it at some level. Remote viewing is a 'directed' paranormal response in that particular methods or conditions are set-up or prepared to enhance the experience of detecting a specifically chosen 'target'. Unique and specific targets are chosen to view remotely and then the viewer is asked to concentrate on the target and draw the results of his or her impressions. Simple geometric figures and impressions are drawn and the relationship of these simple figures in the drawing gives the overall sense of the target viewed. Then the drawings are compared to the target to decide whether the viewing session was successful or not.

The practice was developed at Stanford Research Institute (SRI) in California by a group of physicists under the direction of Harold Puthoff and Russell Targ. Their first subject of investigation, Ingo Swann, coined the term 'remote viewing'. Remote viewing was developed under the sponsorship of several U.S. government and military intelligence agencies as a method for spying during the height of the Cold War. It was developed because rumors came from behind the Iron Curtain that Soviet intelligence agencies were conducting similar programs to spy on the West. In fact, the U.S. government funded a secret program under the name 'Stargate Project' to the tune of several million dollars to collect intelligence data. The program was first made public in the middle 1990s when news organizations learned of it through the Freedom of Information Act. It was publicly concluded and announced that the program failed and it was ended, but individuals who participated privately confirmed individual successes within the program. If the results of the remote viewing program are judged by the overall percentages of its successful 'hits' on projected 'targets', then perhaps it was a failure, but in some very specific individual cases the program had significant successes. The occasional successes of a few remote-viewing attempts out of the many tried, as well as the high improbability that any remote-viewing attempts would ever be successful, give credence to the reality of the experience. Under the presently accepted scientific paradigms, remote viewing is considered a complete impossibility, but occasional attempts to view specific targets are so exact that they are statistically impossible to have been guesses. The occasional successes of this type are the best evidence for the reality of remote viewing.

Other purely technological advances have also introduced various new forms of communication and new varieties of psychical experience. Modern tape recording and other recording devices have been used to detect and record strange voices and noises. These are called Electronic Voice Phenomena (EVPs). The EVPs are usually associated with hauntings, but no conclusive evidence has been offered on the actual origins of the recorded noises. People only need to leave tape-recording devices, either digital or audiotape (analog), in quiet places where hauntings are expected to occur. When they replay their tapes they sometimes hear strange voices against the background static and noise. Some people are also studying the video equivalent of EVPs, where images seem to emerge from the background noise (snow) on television screens. These phenomena have not been confirmed and introduce a whole new group of problems that would render them especially difficult to verify.

New photographic techniques, including thermal imaging and infrared photography, are being widely used in field investigations of paranormal events. In many cases, amateur scientists rather than professional scientists are conducting the field investigations. Experimenters and investigators are gathering data, but that data is not being turned into sound hypotheses, theories or models of the activity that they purport to demonstrate. A solid background in science is usually needed to properly interpret the data obtained, at least if it is to have any scientific value. These new techniques and technologies are offering new data to the ever growing body of evidence that something strange is happening during haunting investigations, if not new evidence for survival itself. On the one hand, the techniques for recording the phenomena are simple and cheap, so everyone who even makes a simple attempt to gather data is doing so. Simple photos of ghosts, orbs and other phenomena can be made by anyone in a cemetery after dark, although there are some special methods to enhance the possibility of photographing some form of the anomalies. In some cases, nonscientists are leading the charge into these new areas of research, while hardcore scientists are ignoring the data, simply because it is so commonly found and thus open to fraud and deception. On the other hand, parapsychologists and investigators no longer have to rely on anecdotal evidence alone. The new technologies allow empirical data to be collected at haunting sites and other types of field investigation sites. So these new phenomena present a double-edged sword for science: They increase the overall database of information for science and maybe give new clues to the nature of the phenomena, but they also lead to nonscientific speculation that forces science to ignore them or look at them with far too great an amount of suspicion.

However, another new technological advance is far more significant for the study of psi: The development of computer technologies. PK events of the past had always been spectacular and highly suspect for fraud. PK events such as poltergeist activity had also been spontaneous, and thus difficult to study in a scientific manner. With the use of computers, a whole new variety of PK events that are measurable in the laboratory have been successfully documented and studied. Psi can be used to affect the outcome of random number generators (RNGs). This form of psi action is called micro-PK, as opposed to the traditional form of PK which is now called macro-PK. Experiments involving micro-PK and RNGs can be controlled under laboratory conditions, so micro-PK has been added to the variety of psychical experiences testable in the laboratory in the last few decades, supplying another source of data for the investigation of psi. Yet computers are also control and monitoring devices, so they can be used to gather information and data electronically and analyze that data. In this case, computers are used in conjunction with other types of electronic monitors to investigate hauntings as well as analyze photographic images and other forms of data to a degree of precision unimagined just a few decades ago.

There can be little doubt that significant changes in the scientific attitude toward psychical experiences were in the works by the 1990s. They were forced upon science by the many different advances that occurred in vastly different areas of science. Some scientists are finally becoming aware of this change, even if that awareness is reluctant. Otherwise, several significant experimental advances have also occurred in the 1990s that are beginning to change the overall equation of psi research and any such change in psi research could only bode well for determining the reality of the afterlife. In particular, the constant harassment of scientists investigating the paranormal by skeptics reached a new level when the skeptics challenged parapsychologists to conduct a statistical analysis of all experimental results related to ESP that were conducted over the past five decades. To answer their complaint that the positive results were fixed by the selective censorship of negative data, the skeptics required that all of the negative results for the existence of psi that had been neglected or discarded in the past studies now be added into the new analytical study. Jessica Utts, a professional statistician rather than a parapsychologist, undertook such a new study with a startling outcome. The new statistical meta-analysis has shown beyond a reasonable statistical shadow of doubt that some unknown process that cannot be accounted for by normal statistical variations has occurred in past ESP experiments.

Utts took all of the reported data for a period of over six decades and added in a negative fudge factor to compensate for any possibility that the scientists who conducted the research could have discarded their negative results and thus slanted their results. The skeptics themselves suggested this compensation factor, but the new analysis showed that a definite and measurable effect had taken place to a very large statistical probability. It would therefore seem, and some previously skeptical scientists now agree, that the statistical evidence either supports the action of psi in past ESP experiments or some other unknown force is at work. Science doe not like unknowns, so ESP becomes the most logical answer, no matter how unlikely, to explain the experimental data. These findings represent a significant victory for parapsychology over the skeptics. Yet this victory was also supplemented by other trends in science.

In the late 1970s, Robert Jahn and Brenda Dunne established the Princeton Engineering Anomalies Research (PEAR) laboratory to study the direct affect of mind on mechanisms. In other words, the PEAR group studied the effects of micro-PK. Jahn and Dunne have come to the conclusion, after two decades of research, that there is a small but significant effect of mind on mechanisms that cannot be accounted for as a statistical anomaly. So, it would now seem that there is substantial evidence for micro-PK. On the other hand, Dean Radin conducted a series of experiments in the early 1990s to test for an effect called 'pre-sentiment', which is a low level spontaneous and subconscious precognitive effect. Radin's experiments clearly indicate that the subjects he tested demonstrated physiological changes corresponding to fear and other emotions 'before' the test subjects were shown pictures meant to evoke those same emotional responses. Radin's experiments therefore indicate the existence of subconscious precognition or 'pre-sentiment' in the subjects he tested.

Dick Biermann, a professor in Holland, has repeated Radin's experiments and has obtained the same results. Biermann later conducted a search of similar psychological experiments that were conducted to find other commonly accepted psychological effects, having nothing to do with pre-sentiment or precognition, to see if the data from those independent experiments indicated the existence of the same pre-sentiment effect that Radin detected. In the cases that he reviewed, Biermann discovered that the data from other experimenters' research shows the exact same precognitive effect, even though the scientists in those cases were not looking for evidence of precognition or presentiment and had never noted the effect that was clearly indicated in their data. It would seem that this experimental evidence verifies the existence of psi. If it does not demonstrate the existence of psi for the case of pre-sentiment, then it poses a very serious challenge for any skeptic who wishes to challenge the results as false. The verified existence of psi would indicate an intimate non-material connection between material objects, other minds and an individual's consciousness. Such a connection would further indicate that consciousness is separate from the material body and brain although it interacts with the material body, so it would be safe to conclude that the death of the material body and brain would not necessarily guarantee the death of consciousness. Once again, the confirmation of psi in paranormal phenomena enhances the prospects for verifying the reality of some form of an afterlife.

And finally, a new breed of mediums has become popular in the last few years of the twentieth century. These mediums do not levitate themselves or other objects, make flower bouquets appear out of nowhere, cause rain to fall in closed rooms, cause apparitions to appear, rappings, or any of the spectacular physical paranormal effects that mediums discharged (and many times faked) during the era of Modern Spiritualism. These new mediums merely sense impressions from the dead relatives and friends of individuals and interpret those impressions in a meaningful way for the recipients of the messages. These new mediums rely on communications only. These mediums have gained a small amount of scientific credibility through the researches of Dr. Gary Schwartz in Tucson, Arizona. He has published the results of his researches in the popular book *The Afterlife Experiments*. The fact that the new mediums are only sensing impressions from the dead sounds vaguely similar to the descriptions of remote viewers who sense impressions of their targets, but do no completely see (or have complete visions of) their targets. In both cases, mistakes are made because the readers, whether mediums or remote viewers, are *interpreting* impressions rather than receiving direct and concise messages. However, in both cases there are sometimes very spectacular bits of information given that defy all odds and attempts at fraud. It is neither the small bits of information nor even the misinformation given that proves or disproves the veracity of mediumship or remote viewing. It is the occasional spectacular result and the frequency of the spectacular results that seem to indicate the reality of these psi effects.

All of these changes, both inside normal science and outside of normal science in the realm of the paranormal, have come together in the last decade. When all of the progressive trends in normal science are taken together with the individual advances in the study of psi, something wonderful seems to be emerging. Vastly different trends in vastly different areas of science are all leading to the same conclusion: A new revolution in science is just beyond the horizon. When different independent sources say the same thing, then the likelihood and probability of the truth of that thing is all the more assured, as is the case in the scientific studies of the paranormal. A newly published book by Dean Radin, *Entangled Minds*, summarizes and documents all of the recent advances in paranormal studies within a scientific context. Radin clearly shows that the paranormal in science is all but normal. The evidence is overwhelming and seems to fit with recent advances in purely normal science. More specialized books by the other scientists mentioned above have also been available for some time. Even though these books have not proven popular enough among members of the scientific community to shake the very foundations of science, they have introduced more than a few ripples and demonstrated more than a few cracks in the edifice of current scientific paradigms. The very fact that these books are being published and circulated demonstrates that issues regarding the paranormal are far from dead in science (as the critics and skeptics would like to, and often do, claim) and verifies the probable impact that such research will eventually have on hard science.

Science has failed to realize what has been happening over the past few decades. The meta-analysis that was conducted by Utts has verified ESP. The experimental work conducted by Jahn and Dunne has confirmed the action of PK on the micro-level of reality while Radin's experimental work has confirmed precognition. Remote viewing is real and has given exceptionally good results in specific cases even though, according to normal science, remote viewing should be impossible. However, none of these effects indicates the possibility that the afterlife is real. The reality of these paranormal phenomena indicates that consciousness and mind can act directly between other consciousnesses and minds as well as directly on matter without the intervention of material mechanisms. The fact that mind and consciousness can act independently of the material living body with which they are associated further implies that they are independent 'structures' or 'things-in-themselves' that are associated with but not wholly part of the living material body. Therefore, it is not guaranteed that the mind and consciousness must cease to exist when the material body dies. So, while paranormal phenomena such as ESP and PK do not directly support the possibility of an afterlife, their reality nonetheless bolsters the case for the reality of the afterlife. When the various paranormal phenomena are counted with the various phenomena that indicate the possibility of an afterlife directly, like mediumship, NDEs, DBVs and such, the scientific case for the physical reality of the afterlife becomes highly probable. In any case, these phenomena seem to be implying the existence of a much vaster physical reality that can be normally accounted for by modern scientists. What then is the basis of this new physical reality?

All of these effects are the result of psi, which for all intents and purposes seems, as does consciousness itself, to be a 'subtle energy' effect. In physics there are only two kinds of energy; kinetic energy, the energy of moving matter, and potential energy, the energy resident in the relative positions of material bodies. For example, heat energy is actually the additive effect of the kinetic energy of molecules and atoms while nuclear energy is the potential energy derived from the unstable relative positions of protons and neutrons in atomic nuclei. Potential energies are related to physical fields, so we associate physical fields with their potential (what comes before potential energy). We are all familiar with the potential in an electric field, which is measured in volts. By analogy, we can conclude, with a very high likelihood, that the 'subtle energies' associated with psi phenomena are also derived from field potentials. So we can say that psi is a field potential structure and mechanism and it is amenable to physical analysis through its physical influences on our material world. These influences would be extremely subtle and extremely difficult to observe and then only under perfect conditions, which is why paranormal phenomena are so elusive and difficult to detect and replicate on demand. In cases of interaction between a surviving consciousness and our material environment, an example that we would commonly call a ghost or apparition, the 'subtle energies' involved only allow a very low and extremely subtle level of material interaction, one that is almost imperceptible, immeasurable and only slightly detectable under the best and most favorable conditions, which could explain hauntings, DBVs, and other such phenomena.

Apparitions and Ghosts R Us

While psi phenomena represent extremely subtle field effects and thus seem to manifest in the material world as subtle energies in the form of micro-PK, presentiment and ESP, the surviving consciousness of humans must also be associated with physical fields. In other words, the surviving consciousness could also manifest its own effects in our living material world by the same mechanism of subtle energies. From the physics point-of-view, the conditions for such manifestations would therefore be quite specific and somewhat rare because of the extreme subtlety of the energies involved. But from the cultural point-of-view, we have a very long history of them. They go by the names of apparitions, hauntings, ghosts and visions, to mention only a few. Records of these phenomena are as old as written language itself, but no systematic studies of them were conducted until the late nineteenth century. Again, Newtonian physics had become so successful that it sought new frontiers to explore, but science was not yet prepared or ready to tackle the problem in the late 1900s and failed in this respect. Science was not quite mature enough even given the earlier founding of the SPR because it had to deal directly with mind and consciousness before it could adequately consider the possibility of an afterlife.

Yet the SPR and similar groups began to collect data, evaluate and interpret reported hauntings, visions and sightings of otherworldly beings as best they could with their limited scientific knowledge. They thought that their science was adequate to the task, but we now know that late nineteenth century physics and science were rather primitive and incomplete in some areas. Today we accept four types of direct observational evidence that at least seems to support the existence of an afterlife or the survival of consciousness in some form; NDEs, DBVs, reincarnation and ghosts (hauntings). But reincarnation is substantially different from the others. While both NDEs and DBVs are quite personal occurrences, involving only the individual experiencing the event, the appearance of apparitions and ghosts are not individual events or at least they need not be individually experienced, which means that the events can be physically measured and recorded by artificial means.

In particular, Death Bed Visions (DBVs) have been known, popularized and documented for some time. DBVs are actually far more common then one might guess. People near death, sometimes minutes away from death and sometimes hours away have reported visitations by a variety of apparitions although mostly by friends and relatives who have already passed over. There is, unfortunately, a very fundamental problem with DBVs: The only first hand witness to the vision dies, so if details of the vision were not shared with others before the death of the experiencer, no one would ever know that the DBV had taken place. This simple fact presents a serious problem for investigation of DBV phenomena. Other than this simple limitation, DBVs do not seem to be limited by cultural or social boundaries. They occur and appear to have the same characteristics in all cultures.

William Barrett of the SPR conducted one of the first serious studies of this phenomenon and published his findings in the 1926 book *Death Bed Visions*. Barrett concluded that it was more common for dying people to see friends and relatives who had already died in the moments preceding their own deaths than one would expect. In some special cases, verifiable evidence of the visions is supplied when the dying person sometimes seeing a friend or relative who the dying person could not have possibly known is already dead. It would only be discovered at a later time that the relative who had been seen was already dead, although the dying person had no way of knowing that fact. Such cases would seem to verify the survival of consciousness at some level. Otherwise, the DBVs could always be argued away as the illusions or hallucinations generated by the dying brain. In other cases, children have reported seeing angels without wings. Supporters of the phenomena argue that if the DBVs of these children were hallucinations, then the children would have given the angels wings because angels with wings more closely fit their preconceived notions of angels. This last issue barely provides circumstantial evidence for the reality of the phenomenon.

The next important study on DBVs did not come until the 1960s and 1970s and was not made public until 1977, when Doctors Karlis Osis and Erlendur Haraldsson published their book *At the Hour of Death*. The research for this book was funded in part by the Parapsychological Foundation, indicating the expected importance of the subject to the parapsychological community. Osis and Haraldsson conducted a vast survey of doctors, nurses and health care workers dealing with the dying. They first sent out questionnaires and followed up with interviews of those who responded to the questionnaires. As a matter of comparison and contrast, as well as other reasons, they worked with subjects in both the United States and India. All in all they conducted thousands of surveys over a period of more than a decade.

The conclusions reached by Osis and Haraldsson are quite shocking to the normal western cultural worldview. The vast majority of people who experience DBVs claim they are visited by people with whom they are familiar (departed friends and relatives) rather than religious, spiritual or mythical figures. Many times the visiting apparitions state their purpose to take them away and guide them into death. The DBV experience reassures the dying person, rendering the death an easier and less traumatic experience than it would otherwise have been. The dying express happiness with the experience and do not show any fear of the apparitions that appear in the vision as they might if they had seen a ghost under different circumstances. The dying person's moods improve and in some cases their health improves, at least before the actual death occurs. The DBVs occur while the dying person is cognizant and aware of his or her environment and surroundings, as if the apparitions visiting them are imposed on the normal physical reality of the situation. In other words, the DBVs are not hallucinations that would normally supplant or replace the actual physical reality experienced by the dying person with the whole vision. And finally, neither the dying person's belief in an afterlife or lack of belief in an afterlife makes any difference. Both the vision and the dying person's reaction to the vision follow the same overall pattern for believers and nonbelievers.

These characteristics, as documented by Osis and Haraldsson, are actually quite similar to the characteristics of NDEs as documented by top researchers over the past three decades. It would almost seem as if the consciousness of the dying person slowly (and consciously) slips into a state of death, experiencing different levels of death depending on the particular conditions of their dying and the unknown personality qualities that cannot be specified. In other words, DBVs are low-level NDEs that occur while the dying person is still conscious and aware of his or her material surroundings. In a sense, they are in-body experiences rather than out-of-body experiences as occur in full NDEs. The NDE just represents the next level of sensations before actual and final death.

While it could be argued, and surely is argued, that these 'symptoms' only characterize the cessation of brain activity, it would seem more logical that the consciousness is slipping into something (a state of being dead or an afterlife) instead of slipping out of something (death as a complete and total end of life followed by nothing). Individual DBVs make for good stories and interesting anecdotes among family and friends, but they do not offer any real compelling evidence of an afterlife. Yet collectively, the fact that DBVs are universally experienced and are quite similar across social boundaries, even in vastly different cultures with different cultural views of death, is itself rather compelling and strongly implies a logical and scientific case for the afterlife. At the very least, the research of Osis and Haraldsson warrants further studies of the DBV phenomenon.

But again, DBVs are personal experiences, so they are quite subjective to the individual who is dying. On the other hand, common hauntings are both collectively and individually experienced. One would think that collective observations of hauntings would give them more credence in the scientific community as well as a more solid and objective foundation for scientific study, but instead they are practically shunned, if not completely ignored, by science as innocent superstitious myths and worse. Skeptics will nearly always fall back on accusations of fraud and deceit if the hauntings are too realistic and cannot otherwise be dismissed or refuted. The problem facing science comes with modern technology, primarily in the form of photographs, that seem to verify at least some type of anomalous physical effect that is associated with the observations of ghosts and the phenomenon of haunting.

Hauntings occur when apparitions or ghosts are associated with a particular place or a particular person. Apparitions are different from ghosts in that apparitions can be appearances of the dead or they can be non-local appearances of the living. By definition, ghosts can only be apparitions of the dead. Hauntings are recurring phenomena, rather than one-time sightings and, as such, have been reported by different observers. Different observers can experience the same haunting simultaneously, seeing the same or similar events at the same time and location. On the other hand, observers can see the same or similar events at the same location but different times. So the phrase 'collective sightings' can have several different meanings. In any case, such sightings tend to decrease the possibility of subjective bias and increase the objective scientific value of the phenomena reported, although the subjective impressions of the people involved can never be completely eradicated. In the end, seeing ghosts or apparitions is a personal event that cannot be directly shared.

One basic class of hauntings deals with historical events and seems to represent some type of energetic imprint or impression on the local environment where the historical events originally occurred. Hauntings in this class can sometimes be accompanied by sounds and smells or they might include changes in temperature in the immediate vicinity of the haunting. Such characteristics are not peculiar to these haunting, but can also occur with other types of hauntings. People have reported seeing replays of portions of the D-Day invasion in Normandy, specific Roman battles along the moors of northern England and running gun battles between Union and Confederate soldiers at Gettysburg. These are not human reenactments by enthusiasts dressed in period costumes, but small windows into the actual battles. These hauntings have a holographic or holomovement quality to them, like real three-dimensional movies. They do not affect the surrounding natural environment in any lasting way that would leave physical or material evidence of their passing. They just seem to be three-dimensional images that are shared by more than one person at a time, rather like a recording of the actual events being replayed again and again for different audiences when specific conditions are just right.

Different observers see the same scenes at different times, as if they were plays being staged automatically by the ghosts of the people originally involved in the real battles. The people witnessing the haunting do not interact in any manner with the ghosts or apparitions and the observers' presence does not seem to have any affect on the haunting, just as if the apparitions were unthinking zombies carrying out some automatic ritual. That is why these particular hauntings are described as energy imprints in the physical environment rather than visitations of ghosts and apparitions. At the moment, the 'why' and 'how' of these hauntings completely escape any reasonable attempt to explain them. The best scientific guess, and it is only little more than a guess, is the energy imprint hypothesis, although a counter argument could be made that a time window to the past and the actual battle had been witnessed in the modern setting. This second hypothesis is nearly as radical as the energy imprint hypothesis, but radical events call for radical hypotheses and innovative explanations. Science has no way of knowing if these apparitions are real physical beings with a physical existence or whether they are just pure mental images (a form of ESP) in the mind or minds of the observers. In any case, either explanation is difficult to accept from either a cultural or a scientific point-of-view, even given the fact that this form of haunting could eventually say nothing about the possibility of an afterlife.

In other words, ghost hunting has become a popular pastime among a growing number of enthusiasts in the past two decades. Some people take ghost hunting quite seriously and the sport almost becomes an avocation instead of a hobby. Numerous ghost-hunting societies have popped up all over the United States and other countries and their efforts are supported by the recent technology revolution. A vast amount of literature on the subject of hauntings, apparitions and ghosts has also appeared in the popular press in the past few decades, as have numerous television documentaries and specials on the subject. It seems that the public viewing and reading audience cannot get enough of this fascinating subject. Ghost hunting has almost become the new 'modern spiritualism movement' of the twenty-first century and beyond. So there have been numerous reports of all forms of hauntings and the amount of literature on the subject is growing rapidly. Some of the existing literature is good, some is bogus and some is in between, but all of the literature depends on the truthfulness and the accuracy of the authors and observers are not good sources of data in science. At the very least, there are no scientific controls in the vast majority of sightings because sightings are spontaneous events.

Some literature is scientific in that it meets scientific standards of observation and analysis, while some is completely non-scientific and only meant to titillate the reader and viewer. Some of the stories and investigations are reported rather amateurishly, some are professionally documented and some are completely pseudo-scientific. So everyone who reads the literature or views programs on the subject of ghosts and haunting must pick and choose, evaluate and compare what is reported before reaching any conclusions on the authenticity of the events recorded, even though the conclusions reached will normally reflect each person's *a priori* opinion on the subject of ghosts.

With today's technologies the objectivity of ghost sightings and hauntings is not at question because we can now photograph and take other verifiable measurements at haunting sights. However, the question of 'what do these measurements and observations mean?' is the real problem for science and culture. In other words, science no longer has to depend on subjective anecdotal evidence alone in the case of hauntings. Ghost hunting is no longer about the stories that can be told, but about the measurements that are made and the pictures that are taken. The problem for science is not with the evidence or even any lack of evidence for hauntings, it is with the interpretation of the data and the possibility that data could be contaminated or even faked by overly enthusiastic amateurs and professionals. Anyone can take pictures in a cemetery or haunted house, especially since good digital cameras have become cheap and affordable.

The following pictures were taken with an inexpensive 1.3 megapixel digital camera. They were taken with a flash just after midnight in May of 2004 in nearly total darkness. The flash in the camera was weak and only had a range of about five or six feet. The photos were taken at the Borland Springs Hotel near Parkersburg, West Virginia. The remains of the hotel are located on the side of a hill in the woods above a stream that once sported a mineral spring. The mineral spring dried up many decades ago, but during the late 1800s the hotel was very popular with tourists before the spring dried up. Only the rock foundations of the hotel still exist, with trees growing in the center where the hotel lobby and rooms once stood. The wooden hotel structure burned to the ground about a century ago. This site is quite famous as one of the most haunted locations in West Virginia. Its history is well documented and includes duels to the death, suicides and other stories of tragic deaths.

The first picture in the series shown below was taken shortly after midnight, the witching hour, and nothing unusual can be seen. Photos were being taken at random around the ruins when a group of three students claimed to see the figure of a man in the shadows leaning against the tree shown in the pictures. So the photographer returned and took new pictures of the area around the tree. When the images were reviewed on the camera and later printed, the latter pictures showed groups of large luminous 'globes', more commonly called 'orbs', around the site (See figures 1 and 2). These orbs are quite amorphous seem to be about one and one-half feet in diameter and overlap each other. The presence of the orbs cannot be accounted for as a late night mist because there was no mist in the air that evening. Nor could a mist account for the spherical shape of the orbs that appear in the pictures.

The orbs seemed to move around relative to the fixed frame of the picture (See figure 3), so they could not be due to microscopic water droplets or smudges on the lens of the camera. Nor do they seem to result from any type of diffraction pattern, say from a dust filament or hair that could have momentarily lighted on the lens. Strangely enough, when the photographer turned around to snap a photo of the students who originally saw the man leaning against the tree, the orbs seemed to disappear and then reappeared when the photographer turned back away from the students and continued taking pictures of the trees in the dark background. In some printed pictures the orbs do not disappear and seem to be blue tinted, but in other prints made the orbs appear to be white and do disappear. The ultimate tint or color of the globes depends on the printing process and the color spectrum used to view the images. The orbs also seem more distinct and show up better when the pictures are viewed on a brighter monitor. These orbs do seem to have a definite blue-white tint to them, which could eventually be a clue as to their origin. Only after these photos were taken did orbs appear in numerous photos taken in all directions around the hotel site.

A simple scientific analysis of the photos can be conducted by anyone with a basic knowledge of science and technological procedures. The orbs could either be the sources of the light or they could be reflecting the flash. It seems more likely that they are reflecting the camera flash since the orbs only appear when the flash is used. However, the trees behind the globes also reflect the flash so the light is both reflected by the orbs and passes through the orbs in both directions. So, whatever the orbs are, they could be both reflecting and refracting (transmitting) the light waves from the flash, which means that the source of the orbs must be very small. It would thus seem that the globes just dust particles, insects or water droplets in the air that are out of focus, but this fact needs to be established: It cannot be taken for granted in each and every case where orbs appear.

Under these conditions, there is no concrete evidence that these orbs are ghosts or apparitions. Such claims need extraordinary evidence that is irrefutable before orbs could be attributed to hauntings. They are just anomalous phenomena that seem to defy simple explanation by the ordinary laws of optics, at least without further testing. However, within the context of the history of the hotel the fact that the orbs only appeared after a group of students claimed to have seen a shadowy figure leaning against the tree, suggests, but only suggests the very slim possibility that the appearance of the orbs is indeed associated with a haunting. It is too easy to let one's imagination run away with snap conclusions under such circumstances. Other pictures were taken earlier by another student showing orbs of a more common variety in a nearby area in full daylight, which eliminates the possibility of flashback from the camera. Since no flash was used in this instance, the orbs must have been their own light source or somehow have the ability to distort the ambient light in the background. These pictures are not conclusive, but none-the-less suggestive and thus 'prove' nothing. These photos just raise new issues and questions that may or may not prove easy to solve.

At a much later date, a series of pictures was taken in a three hundred year old cemetery along the Ohio River. In this case, a 3.2 megapixel Kodak camera was used. The higher resolution of the camera makes focusing errors more unlikely, but did not eliminate small particles that are out-of-focus as a possible source of orbs. The most common source for orbs is dust particles, raindrops, snow, insects, water mist, and similar natural phenomena that reflect the camera's flash. All of these possibilities must be eliminated before a claim that the orbs are due to hauntings can be taken seriously. The problem is that it is nearly impossible to eliminate all of these normal possibilities. The following pictures (See figures 4 and 5) were taken about one o'clock in the morning during the middle of a ten-day heat spell. The temperature was about eighty degrees Fahrenheit, even in the early morning. So there was no rain nor snow and no mist of any form. There seemed to be no dust nor insects in the air, at least no white flashes from either dust or insects were evident when flashlights were used just before the pictures were taken, but that does not rule out the possibility that dust and insects were the source of the orbs. It only makes dust and insects slightly less likely than if they had been seen in the area prior to taking the photos. All possibilities must be taken into account. During this experiment, only two photos showed distinct orbs. Several photos taken just before and after the photos showing orbs had neither orbs nor evidence of a source for the orbs.

On a separate occasion two weeks later under similar conditions, a dozen or more photos were taken in the same location with the same camera. In this case, the camera was mounted on a tripod and a series of photos were taken as fast as the camera's flash could recycle. An attempt was being made to catch orbs from either insects in flight or dust particles moving in the breeze to demonstrate a source of the orbs. Photos were taken at about ten second intervals for two or three minutes. All of a sudden, one photo (See figure 6) shows more than twenty orbs when no other photos, of the two dozen taken, showed any orbs. There was no evidence of either insects or dust at all. The fact that orbs showed up in only one photo is strange, but still inconclusive. Speaking probabilistically, if that either insects or dust particles caused many orbs, then the insects or dust particles should have been present in more than just one photo. But these results are still inconclusive and the orbs could still be coincidental cases of camera flashback from small particles that are out-of-focus.

The most significant photos were taken during a trip to Europe in July of 2007. The following photos (See figures 7, 8 and 9) were taken in Saint Stephan's Cathedral in the Altstadt (Old City) district of Vienna, Austria, using the same 3.2 Megapixel Kodak camera. This church is several hundred years old, so it has seen a significant amount of Viennese history. The pictures were taken at about eleven in the morning on a very bright and sunny day. There were no sources of light inside the cathedral and camera flashes were strictly forbidden. Hundreds of tourists were milling about the cathedral, admiring the works of art and the architecture, but the only light source for everyone to see the artwork came from sunlight streaming through the large windows. Under these conditions, orbs were quite apparent in the photos taken.

Several of these orbs appear to look just like the orbs photographed in cemeteries. They have a sphere-like appearance with some form of internal structure. However, a new form of orbs has appeared in these photos. These new orbs are bright spots of pure light. These light spot orbs are not the product of reflections from flashes or any other source, since there were neither flashes nor bright sources of light inside the cathedral. The ambient light in the cathedral was not very bright except when looking directly into the windows. The indirect light from the windows that allowed pictures to be taken at all was quite dim. The normal orbs might still be a product of dust particles in the air, diffracting light from the windows rather than reflecting light as is the probable case with cemetery orbs, but that possibility is shrinking, while the light spot orbs cannot be so easily explained and remain anomalous. The light spot orbs are just too bright to be from reflecting or diffracting light within the cathedral hall.

Figure 1—Borland Springs Hotel

PARANORMAL PERCEPTIONS OF DEATH 177

Figure 2—Borland Springs Hotel

Figure 3—Borland Springs Hotel
The orbs have moved relative to the camera lens, so it cannot be an object on the lens.

Figure 4—Orbs in a 300 year old cemetery

Figure 5—More orbs in the cemetery

Figure 6—Orbs appear in this photo, after not appearing in other photos.

Figure 7—Bright orbs St. Stephan Cathedral in Vienna

Figure 8—Bright orbs and cemetery type orbs in St. Stephan Cathedral

Figure 9—Bright orbs as well as cemetery orbs in St. Stephan Cathedral

Figure 10—An orb in a passage at the Hofburg in Vienna

The final photo shown is even more problematic for the hypothesis that all orbs come from the reflected light of flashes and out-of-focus dust particles or insects. This picture was taken in the early afternoon hours in a passageway exiting from the main courtyard of the Hofburg (Royal Palace) in Vienna. This photo (See figure 10) was taken with the same Kodak camera. The passageway runs from the central courtyard to the residence areas of the palace. The picture clearly shows an orb just inside the passageway, at a height of about three meters at the left side of the photo. No flash was used because the sunlight coming through the arches of the passage is extremely bright. The photos from Saint Stephan's Cathedral and this passageway clearly indicate that orbs are not just a product or optical artifact of a camera's flash. However, these photos do not indicate or give any further evidence for what the orbs might be. These orbs are clearly anomalous phenomena that need to be identified.

Pictures like these are not all that unusual for haunted sites. Other photographers have photographed globes, orbs, strange mists, bright spots and other strange luminous objects. Sometimes the orbs are moving and sometimes they seem stationary, at least for the moment during which the camera flashes. Both independent observers and admitted ghost hunters have taken still pictures and others have taken videos, using both film and digital cameras. On some occasions the photos taken more closely resemble people, a fact that further defies logical explanation. Some photos are less spectacular and some are very spectacular, but whichever the case may be they do offer a good example of the growing body of evidence that 'something strange' that cannot be easily explained by normal science is occurring at haunting sites. This matter can only be resolved by a comprehensive scientific study of the characteristics of different cameras with different electronic elements, programs, focusing characteristics and flash characteristics. Such a study would need to be conducted with as large a number of different conditions as possible so that it could be used to determine the true physical origins of the whole variety of orbs that are commonly photographed. Only then could it be determined if orbs are true depictions of apparitions and truly represent hauntings.

During another investigation of a haunting, this time in a house occupied by a college student and her husband, the pictures that were taken revealed nothing, but other physical effects were detected. In one particular closet in the master bedroom, where strange drafts of wind and cold 'spots' had been reported, an electric field meter was used to test the inside of the closet entrance. When the electric field meter was pushed into the closet, the needle pulsed so strongly and rapidly that it banged against the retaining peg at the end of the red zone of the meter and caused a clicking sound somewhat like a Geiger counter. Even though a strongly pulsating electric field was detected, the hair on the back of the investigator's arm was not affected and did not raise as it normally would if near an electrostatic charge. This effect was highly unusual.

Upon returning home, the investigator tested the electric field meter by placing it near the picture tube of a twenty-two inch television screen while the television was warm and operating. Enough electrostatic charge had collected on the television screen to raise the hair on the back of the investigator's arm, but that was only enough to move the electric field meter needle into the red zone, let alone peg the meter as the closet had done. So whatever affected the meter needle in the closet was not an ordinary static electric charge. Again, this was an anomalous phenomenon that seems to defy logical scientific explanation.

The strong electric field in the closet could not have been generated by the presence of electrical charge or ions in the air or on the clothes in the closet. Yet it was an extremely strong pulsating electric field. According to the commonly accepted interpretations of the electromagnetic theory, an electric field can only be generated by either electrostatic charges or a varying magnetic field. A pulsating magnetic field would surely generate a pulsating electric field at the same frequency and this is normally called an electromagnetic wave. The frequency of the wave in the closet would have been a few Hertz or cycles per second judging by the pulsating rate of the needle in the electric field detector, which would have meant that the wavelength of the generated wave would have been on the order of one-hundred million meters, which is ridiculous. Paradoxically, an electromagnetic wave of that length would have been so weak that it could never have moved the needle of the detector. Something very strange occurred in this instance, but this does not mean that ghosts or apparitions were present. It just means that something very strange and unexplained, something anomalous, was happening. These effects were clearly anomalous according to the present scientific worldview, but do not seem related to a specific haunting, at least no ghosts presented themselves at the time.

These are all firsthand accounts of particular anomalous events, so this author is completely confident of their complete authenticity. However, no ghosts or apparitions were evident, nor were they sighted and confirmed in either case. Yet ghosts and apparitions or ghost-like figures have been photographed in many other cases on record and similar physical effects have been detected by electronic equipment at or near the time of other sightings. What is needed is a proper analysis of the data by physicists, engineers and expert physical scientists to try to determine the true causes of the anomalous effects detected in these and similar cases.

Only when correct and comprehensive physical descriptions of the anomalous events can be made, can scientists begin to look for patterns and develop hypotheses to explain the reported anomalous events. But first, scientists need to be aware of these anomalous events and learn not to ignore them. Whether or not such data would constitute compelling evidence for the afterlife is not an issue at present. Science must first determine the reality of the phenomena and dismiss them a priori. The evidence presented here for ghosts and hauntings just barely rates as circumstantial and no more regarding the reality of an afterlife. However, the total weight of all of the sightings throughout history together with the recently collected physical evidence from pictures, electronic devices and measurements is beginning to look a bit more compelling than past individual sightings and personal evidence alone, although to what end is not presently known. At the moment, truly compelling evidence of an afterlife needs to come from other quarters.

Small, Mediums and Large

The practice of mediumship, in one form or another, is as old as human history itself. From the legends and philosophies of ancient Greece and other early civilizations to the modern sensitives and practitioners of the art of communication with the dead, the human race as a whole has shown a keen interest in the prospect of communicating with its dead ancestors, relatives and friends. However, mediumship has been at odds with science and the scientific worldview since science first began to eradicate the hold of superstition and the supernatural on human thought, so the human interest in communication with the dead poses a paradox for our modern society since it cherishes and favors its scientific ideals. Unfortunately, mediumship has become synonymous with showmanship and spectacular physical tricks during the only era in which science took an interest in the practice, the late nineteenth century. Even if communication with the dead were possible, it has been unnecessarily hampered by its connection to mysterious and sometimes outlandish physical phenomena that go against all accepted scientific principles. Even today, most scientists consider any of their number who approaches mediumship as a legitimate subject for scientific investigation as perhaps eccentric if not an outright unscientific fool, and that appraisal of the situation is being kind. No matter how liberal the scientific community as a whole has become, its position on such controversial issues is unlikely to change without substantial positive 'proof' that we survive material death, let alone that we can communicate with the survivors who have died.

Why then does a large portion of the general populace continue to believe in communication with the dead in the face of open skepticism, scientific criticism and common sense? It would seem that we humans have an innate intuitive sense that 'life' and 'mind' are more than just the sum of the physical components of the living body and brain and, therefore, some part of the person must survive the death (cessation of electro/chemical action) of the material body and brain. This intuitive notion, seemingly bereft of either scientific or logical support, persists even in the scientific community, albeit at a level of personal belief rather than theoretical or practical imperative. Many scientists still attribute the possibility of survival after death to religious faith rather than scientific reason, although the religious aspects are slowly disappearing in the ever-increasing secularization of our modern societies. Although some scientists may believe that the soul survives death, they also believe that the 'soul' is not a scientifically definable, and thus reducible, quantity or entity. Even if this 'soul' survives, it could have no contact with the material and thus scientifically defined world or reality. It is therefore logical to conclude that the scientific community will never accept mediumship until it is proven that some physical portion of 'mind' or 'consciousness', at the very least, survives the death of the living body and brain.

Technically, nothing in science can be 'proven' with total and absolute certainty. Absolute certainty in our material world is a philosophical impossibility. So the term 'proof' is used here rather loosely. It would be far more accurate to say that science 'verifies' theories, but sometimes theories become so strongly believed that verification is taken for granted, which is tantamount to a proof. We only need to cite the theory that the earth rotates on its axis and follows an elliptical path around the sun to demonstrate a particular theory that no longer needs to be verified and has thus become accepted as truth. So a 'proof' of survival can only come in two different forms. First of all, some form of survival must be established as an experimental or observational fact. However, there is very little basis, except for anecdotal evidence and the independent verification of facts gleaned by mediums from communication with the dead, to experimentally 'prove' the reality of survival.

All experimental 'proofs' of survival seem overly subjective rather than objective, and if they are not then the skeptics will still claim that they are too subjective for scientific purposes. And, of course, it is always difficult to use evidence from mediumship to 'prove' the reality of the mediums' claims, just as it is meaningless to use a word to define itself. The very nature of mediumship negates any independent verification that it really constitutes survival of mind, consciousness, soul or spirit at death. Mediumship provides no insight into what exactly survives, even if mediumship were demonstrated to provide a real form of communication with the dead. It is very possible that what we call communication with the dead is just a mixture of clairvoyance, ESP and retrocognition, which are collectively called GESP (general ESP) or super-ESP. Even if science accepts survival in some manner, mediumship may not even involve survival at all. The best that could be hoped for would be some spectacular event such as the return of someone from death, and then the skeptics still might not accept such a momentous event because the skeptics could always come up with an alternative explanation or just claim fraud. This last instance is well demonstrated by skeptics who refuse to accept the NDEs reported by people whose deaths have been clinically documented, which was as close as anyone could come to actually returning from the dead in a scientific sense.

On the other hand, the possibility of survival could be implied by the development of a physical theory of death, or better yet a physical theory of reality from which some form of an explanation of death and survival would evolve logically from the theory. In other words, a valid scientific theory of physical reality that predicts or requires survival from fundamental principles would be acceptable to science, but as of yesterday nothing comes close to this possibility. Science presently will not consider survival because its worldview includes nothing by which it can explain or even attempt to explain survival. Science has no meaningful language to describe the afterlife, so it cannot even talk intelligently about the subject. Ideally, all experimental evidence of survival and any viable theory that is developed would eventually work hand-in-hand and complement each other, but in the absence of either one of them, the scientific community would always view the other with suspicion. Without theoretical support, any physical observation or other form of evidence for survival will always be open to counter-interpretations (and/or misinterpretations) that do not involve survival. So, in the absence of a viable and testable theory, all experimental work meets with over-restrictive conditions due to the objections of the numerous skeptics who refuse to believe in survival a priori, if only for the sake of their refusing, and demand an extraordinary and even impossible amount of evidence. If the skeptics who fight against anything having to do with survival are not enough to contend with, those scientists who wish to investigate survival and the possibility of an afterlife also have to contend with a very conservative scientific community that will never settle the issue of survival until the theoretical basis of survival is at least made plausible, if not fully established outright.

In his 1974 foreword to Edgar Mitchell's book, Psychic Explorations, Gerald Feinberg, an open-minded physicist rather than a physicist working in parapsychology, stated that parapsychologists and other researchers should start looking for the properties of psi rather than seeking even more statistical evidence of its existence, implying that the time had come to develop a theory of psi. All of the above-mentioned scientists who have conducted experiments to verify the paranormal over the last half century have stated that they are in need of such a model to further their own research. Without a theory of psi, definitive experiments to 'prove' or rather verify the existence of psi cannot be developed. A theory, or at least a good working hypothesis, is necessary to identify and eliminate the unknown variables that are now thought to be corrupting some experiments, so that individual variables that are associated with psi can be tested.

All the major paranormal researchers recognize the need for a theory of psi, even if it is only a working hypothesis rather than a final theory. In many cases, they have taken their experiments, methods and procedures as far as they can without a theoretical model to help them shape and guide the next phases of their experimental research. Although Feinberg was talking about psi phenomena in his comment rather than the survival issue, ESP and survival are related, so Feinberg's statement also holds true for death and survival. A physical theory of death and survival is absolutely necessary before physicists and other scientists will accept the possibility of communicating with the dead. In the absence of such a theory, the concept of communication is completely baseless and thus useless in science.

Any theory developed for the sole purpose of explaining survival, would be based upon experimentally determined properties of survival, which is really not a good idea. There are no criteria, scientific or otherwise, for determining such properties. Even if we could eliminate all of the many problems associated with collecting physical evidence of survival that have been cited by the skeptics, we would still be left with seemingly insurmountable problems related to the collecting of valid experimental data, which is a necessary first step in constructing a theory of survival from scratch. However, a few brave scientists have still investigated the mediums in their attempt to isolate and validate the phenomena. Gary Schwartz has made a name for himself through the scientific investigation of various aspects of the survival issue.

Schwartz has investigated several of the best-known modern mediums during the past few years in his laboratory at the University of Arizona. In particular, he has investigated the claims of John Edward and James van Pragh, among others. In his tests, Schwartz has calculated that some of the mediums are correct nearly 80% of the time. He has taken extraordinary precautions to protect against fraud and deceit in his experimental procedures, so his calculations should be accurate. But even then, Schwartz has noted that it is not the high percentages of hits that are conclusive. It is the occasional spectacular insights made by the mediums that are the most convincing, and such insights are only convincing to the people that they affect. These spectacular successes cannot be measured in the laboratory, but they do indicate that the mediums are experiencing some form of extrasensory communication, most probably with the dead. However, these results beg the question; "Does this prove survival?" And that question cannot yet be answered.

In his analysis of the controversy, Schwartz and his colleagues have pointed out that

Throughout recorded history, claims have been made that sensitive individuals (for example, mediums and persons who have experienced the death of a loved one) can sometimes receive and share meaningful information with persons who were deceased. In those instances where three possible non-controversial hypotheses of (1) deliberate fraud, (2) faulty perception (including misperception and self-deception), and (3) statistical coincidence, can be ruled out, three possible controversial hypotheses remain: (4) the psi hypothesis (in its extreme form, sometimes termed super-*psi*), including telepathy, remote viewing, and clairvoyance, (5) the systemic memory hypothesis (6) and the "survival of consciousness" (SOC) hypothesis.

If we exclude the first three of Schwartz' hypotheses as irrelevant by meeting the criticisms of the skeptics and correcting any faults found in the experimental procedures and methods, we can better define the problems associated with mediumship and survival. A similar situation has already occurred in normal psi research where experimenters have honed and perfected their procedures and methods according to the criticisms of the skeptics.

We can also eliminate most of the real problems introduced by the concept of super-psi (or GESP) of (4) by carefully designing all future experiments to negate the possibilities of clairvoyance, remote viewing, telepathy or retrocognition. This may be difficult to accomplish in practice, but it is not impossible. Doing so would leave the systemic memory hypothesis of (5) as the only alternative to explain the experimental results demonstrating survival. However, systemic memory can be eliminated as a possibility by careful design of experiments and a strict choice of test subjects. Scientists could then, and only then, after all other logical possibilities for alternative explanations have been eliminated, consider using their experimental results to test the "survival of consciousness" hypothesis. Even then, one important point would still haunt scientific efforts to 'prove' survival and the existence of some form of an afterlife: There is no theoretical basis for such an assumption, even given today's most advanced theories of physical reality.

Once the explanation of experimental results has been limited to the survival hypothesis alone, physical quantities associated with survival could conceivably be isolated, defined and measurements taken, laying a foundation for the development of a physical theory of survival. This is the point, thanks to the research of Schwartz and his colleagues, which paranormal science is now approaching. This procedure would reflect the normal manner by which science progresses, according to the scientific method. However, this possibility seems extremely unlikely given the present temperament of the scientific establishment at large. Even if all other possibilities for explaining mediumship could be eliminated leaving only the survival hypothesis, there is no guarantee, and as yet no real indication, that science could isolate and define any physical quantities associated with survival. In fact, it seems more likely that science does not even know how to proceed toward this end. You must have at least some idea of what you are looking for before you can define and measure it, and as yet, there is no physical basis or framework for understanding the physical nature of death in the context of a definition of life, let alone the conscious survival of an undefined 'something' after death. So science is left in the lurch.

To solve this impasse, science needs to recognize what is merely physical in this world as opposed to what is both physical and material. Physical and material do not mean the same thing. Everything that is material is physical, but not everything that is physical is material. The material body and the material brain die, but that is no guarantee that the complete physical being of the person ends at death. After accepting this simple truth, science needs to question what kind of physical (non-material) quantities are associated with life and could possibly survive death? What type of physical but non-material structures could account for life, mind and consciousness, such that the mind and/or consciousness are physical structures that could survive the material death of the body as well as the end of the physical structure that is life? Until science seriously attempts to answer these questions or questions like them, science will never be able to attempt an explanation of death, the afterlife and the possibility of communication.

In normal science, anomalies discovered in the laboratory or observed in nature would be made to fit into, or be interpreted within the context of, the present scientific paradigms and worldviews. It is only when 'accepted' anomalies cannot be interpreted within the known theories that make up our paradigms that interpretations are sought from outside of our accepted theories and paradigms. There is absolutely no place for an explanation or theoretical model for any form of survival within the present paradigms of science, so it is safe and even desirable to look toward new and previously unexplored areas of science outside of the box that is normal science to explain communication and survival. Since we have no theoretical basis or constructs to guide us in explaining survival, then we can only base the survival hypothesis on intuitive leaps of knowledge or experimental findings. If a hypothetical or theoretical model of the afterlife were to be based on no more than an abstraction from intuition, then science would have nothing better than religion, if not religion itself, because science must rely on experiment and factual observations. Only experimentation and factual observation can determine between religious ideas and abstractions based on faith and scientific concepts based on verification.

Our present view of physical reality, limited as it is by materialism, is neither broad nor complete enough to embrace concepts such as life, mind, consciousness, psi and survival, all of which imply yet another way to proceed toward a model of survival and the afterlife: develop a more complete model of reality. Life, mind and consciousness already exist in the present scientific paradigm of reality, but the understanding of them is totally incomplete within the present scientific paradigm that guides scientific research and study. Since we have no real context for looking at experimental findings and simple observations, we can instead change the theoretical model of reality upon which we base our present worldview. If we assume that consciousness survives material death or that whatever survives death must at least have an element of consciousness to it, if it is not consciousness itself, then science must develop a physical model of consciousness before it can even discuss the survival of consciousness in any logical manner.

The possibility of the survival of consciousness alone does not warrant such a basic change, but our present worldview is already changing due to other trends in science and these changes are already affecting the possibility of incorporating survival into the newly evolving worldview. In other words, confirming the reports of mediums is not nearly enough to warrant changing all of science to fit the facts that are confirmed. Confirming and verifying the reports of mediums, no matter how scientific the methods used, will not change science without an accompanying theoretical model that fits the present scientific paradigm. The reports of mediums are not just compelling enough, no matter how spectacular they are. Mediumship alone, no matter how well documented it is, does not present a 'crisis' for accepted scientific paradigms that would warrant altering our scientific attitudes toward the afterlife. So it is already becoming likely that science will soon have matured to the point of explaining the afterlife as theories of death and survival logically emerge from a newly evolving worldview of physical reality.

In many ways, a physical theory of reality that implies or better yet requires a continuance of consciousness after life ends would be preferable to a separate theory of an afterlife that could only be linked to either our present or a future theory of physical reality. If consciousness or some other independent 'something' survives death, then it would mean that the present scientific concept of physical reality is woefully incomplete because that 'something' would necessarily be connected to the material body during life. The logical necessity of survival, emerging from an independent theory of physical reality, would surely influence the scientific community to a far greater degree than developing a simple model that could be pasted to our current collection of theories of reality. This alternative is the only acceptable possibility for establishing survival as a scientific reality in such a manner that the scientific community could accept survival, and survival is a necessary precursor to any possibility of communicating with the dead. Whichever the case may be, it certainly seems as if science will not accept the reality of afterlife and probably not even the continuance of consciousness past the moment of death, at least from the evidence so far described, even though a very impressive case has thus far been built for the afterlife. On the other hand, the scientific evidence for reincarnation presents an even stronger case for the possibility that consciousness survives.

There and back again

While DBVs, ghosts, apparitions and hauntings are certainly pertinent to the question of an afterlife, they are not persuasive enough to change fundamental and time-tested attitudes within the scientific and scholarly communities. NDEs are relatively persuasive, but not quite compelling enough unless the 'dying brain hypothesis' can be eliminated as a counter explanation to the 'afterlife hypothesis'. The same 'dying brain hypothesis' can also be used to explain away DBVs 'if' they were ever to become an issue in science. However, the 'dying brain hypothesis' has nothing to do with hauntings and ghosts, so these phenomena could break the deadlock, 'IF AND ONLY IF' hauntings could be directly linked to the concept of an afterlife and survival. However, that possibility is not on the agenda for scientific research anytime in the foreseeable future. Yet there is still one other form of evidence for an afterlife that has to be considered, and that is reincarnation. The 'dying brain hypothesis' is totally irrelevant to the concept of reincarnation, so any scientific evidence that could support reincarnation could completely undermine the case for the 'dying brain hypothesis'.

Reincarnation is an extremely old concept. The ancient Greek philosophers talked about rebirth and a few even claimed they were reincarnated from earlier lives. Reincarnation is mentioned in all major religions, although it is not commonly accepted in the western religions that are based upon the Judaic/Christian/Islamic creeds. Yet reincarnation is a fundamental tenet in the eastern religions and therefore a very large proportion of the world's population and cultures accepts the reality of rebirth of the 'soul'. In spite of the official position of the western religious establishments that reincarnation does not exist, many mystical traditions that are associated with the western religions include a belief in reincarnation. In the religious perspective, reincarnation is called 'transmigration of the soul', which, of course, presupposes the existence of an immortal, indestructible, immutable and ineffable soul that can survive material and physical death. The concept of a soul is no longer part of science although there is nothing in science that would deny the possibility of reincarnation if 'something' survives death. In other words, if a physical consciousness survives death in any form, then reincarnation would be just as feasible as other forms of an afterlife. While the possibility of reincarnation would seem far less likely than afterlife alone within the context of normal science, since reincarnation presupposes survival, which is yet an open question, and an afterlife, the possibility of reincarnation actually provides an evidentiary window on the afterlife that is missing with other survival phenomena such as ghosts, hauntings, NDEs and DBVs. Quite simply, if a person has memories of a past-life, those memories could provide verifiable evidence that the person had indeed lived before!

The concept of reincarnation was first popularized in modern science in 1952 with the case of Bridey Murphy. Bridey was a nineteenth century Irish woman from Cork who apparently reincarnated to a young Colorado woman by the name of Virginia Tighe. An amateur hypnotist placed Virginia in trances to reveal her past as Bridey. The story of Bridey Murphy was popularized by magazine articles, newspaper accounts, a best selling book and a movie, and thereby initiated modern interest in past-lives and reincarnation within western Christian societies. Since then, a small cottage industry in Past-Life Regression (PLR) through hypnosis has evolved. In PLR, a person is hypnotized and ostensibly taken back to a time before they were born (into their present life). If successful, the person in the hypnotic trance 'remembers' events of a previous life or lives. It is believed that some psychological and medical problems that people have in their present lives are holdovers from previous lives, so the complete treatment of a patient must include knowledge of how past-life events influence present-life psychological and medical conditions. For example, some people believe that phobias in their present lives are results of events in previous lives. A person who previously died from drowning could have a deathly fear of water in the next life. Skeptics claim simply that we do not have past-lives, so the practice of PLR is completely bogus. Otherwise, they criticize the method of PLG and point out that hypnosis is bogus. Both of these arguments are personal biases and no more, whether or not any particular regression reflects the truth or not. There is nothing in science or logic that warrants a *priori* the impossibility of reincarnation and our knowledge of past-lives, just as there is nothing in science that supports them.

Recent scientific evidence implies that hypnosis is real and that it can be used beneficially in psychological and medical therapies. Using modern MRI scanning techniques, science is presently learning how hypnosis works and providing evidence that it does work. However, this evidence neither supports nor disproves the possibility of past-life memories or regressions, so the validity of hypnosis as a therapeutic procedure has no bearing on the question of the afterlife at this point in time. Skeptics also claim that the present knowledge of how memory and consciousness work, literally 'everything that we know about them' at present, does not support reincarnation. But this argument is an empty fool's paradise because science presently knows very little about how memory works and has no model for explaining consciousness. If the skeptics have some special knowledge or information about the memory and consciousness and their inner workings, then they should share this knowledge with the scientists who are working to better understand memory and consciousness. In other words, science cannot say that present theories of memory, mind and consciousness negate the possibility of reincarnation or past-life memories because there are no such theories, only speculations and assumptions whose details have not yet been worked out.

PLR is currently used as a therapeutic tool and is supported by a professional organization, the Association for Past-Life Research and Therapies that was founded in 1980 by a small group of psychiatrists and psychologists. In 2000, the group became the International Association for Regression Research and Therapies. While membership is not limited to practicing scientists, the organization is based upon scientific principles. Unfortunately, PLR has been incorporated into some pseudo-scientific belief systems, which prejudices the scientific case for using hypnotic regression to verify past-life experiences. While therapists are interested in using past-life memories as a tool to heal, other scientists are interested in using past-life memories as verification of a greater physical reality that includes an afterlife. The therapists are interested in personal solutions, while research scientists that take PLRs seriously are interested in the more global aspects of PLRs and the greater philosophical questions the regressions can help to answer.

Within this context, Dr. Ian Stevenson, a psychiatrist at the University of Virginia, has laid down the fundamental groundwork for the scientific investigation of reincarnation and past-lives. Dr. Stevenson began his work in 1960 after hearing a child in Sri Lanka tell of a past-life. Since then, Dr. Stevenson has written scholarly books on the subject and published numerous articles in scientific journals and periodicals. After three decades of research, Dr. Stevenson has concluded that reincarnation is a fact and that the 'transmigration' of consciousness does occur. His work offers the 'best available proof' of an afterlife and verifies that consciousness or some part of consciousness survives death. Yet his work is still relatively unknown within the general scientific community. Dr. Stevenson does not rely on hypnosis, thus circumventing the problems associated with that practice. He investigates children who have spontaneous past-life memories.

In some cases, young children remember their past-lives spontaneously. In general, as the child grows older the memories of their past-life fades away, although in some rare cases the memories persist into adulthood. The memories are extremely vivid and detailed for some children. In those cases where memories are very detailed, the memories can be checked and verified for accuracy and this is just what Dr. Stevenson and his research teams have done. They have studied hundreds of cases where the memories of past-lives have been fully documented. Such evidence would be very difficult to fake; yet Dr. Stevenson and his colleagues have been scrupulously careful in their work to protect the integrity of their findings against fraud and deceit. In some of the most startling cases, the evidence is not subject to the falibility of the human mind and memories, but is instead purely physical and objective.

In cases where a child has past-life memories of a violent and sudden death, the child sometimes bears birthmarks that coincide with the physical wounds that caused the death in the previous life. Of all the children who are born with birthmarks and deformities, a small percentage simply cannot be explained by normal scientific means. When these children have past-life memories, the birthmarks can be traced back to events in the previous life. In his 1997 book *Reincarnation and Biology*, Dr. Stevenson presents his studies of such birthmarks. He focuses mainly on deformities, anomalies and birthmarks that cannot be attributed to either heredity or occurrences at birth (both prenatal and perinatal). The objective physical evidence that Dr. Stevenson presents for reincarnation in these cases is nothing if not spectacular. In one particular case, a young boy was born with what looked like a mottled scar on his chest. The boy remembered dying from a shotgun blast to the chest as well as his name in the previous life. Dr. Stevenson not only found the man who had been killed by the shotgun blast, but also recovered the official autopsy report for the death. He then proceeded to match the shotgun pellet entry wounds from the coroner's report with the individual markings on the child's chest. This evidence would seem incontrovertible and yet it is only one instance of evidence supporting the reality of reincarnation among many similar instances, albeit the most startling confirmation of physical reincarnation.

Dr. Stevenson has also found that children have phobias in cases where their previous deaths were caused under special circumstances (fear of water in the case of drowning), likes and dislikes associated with previous lives (which he calls filias), a continuity of habits (alcoholism, drugs and so on) that manifest at an early age, as well as special abilities and talents that are carried over from the previous life. In some cases, the knowledge of a foreign language (xenoglossy) 'transmigrated' to the new life with the conscious memories of the past-life. Dr. Stevenson's and similar studies offer the best scientific evidence that consciousness survives, evidence that is both objective and verifiable. Yet if Dr. Stevenson's work were all that was available to support the afterlife hypothesis, then it would still not be enough to support the reality of the survival of consciousness, as startling and persuasive as it is. Alone, it is neither persuasive enough nor overly compelling even though it presents a real challenge to the present scientific worldview. But Dr. Stevenson's evidence is not alone. It just completes a body of evidence, ranging from circumstantial to quite substantial, that includes ghosts, hauntings, modern mediumship that has been scientifically measured, NDEs and DBVs.

Taken together, this scientific evidence should be both compelling and persuasive, but it is still insufficient to sway the scientific and scholarly communities. It may be compelling and persuasive, but it is not yet overwhelming and only an overwhelming amount of evidence will convince the scientific community that survival is a reality. This evidence is simply not enough, nor will experimental and observational evidence ever be enough, under any circumstances, until it is supported by a theoretical framework or model that can at least relate the evidence to the rest of science. Even when all of the evidence for psi and paranormal phenomena are added to the weight of the evidence for survival, the scientific community is still not swayed to accept either possibility. Science can only work with evidence and phenomena that it can interpret relative to what it already knows about nature and physical reality as presented in its worldviews. Science simply cannot incorporate reincarnation and an afterlife into its present paradigms and worldview. The best that science can hope for, once it accepts the possibility of survival, is to relate these concepts to its present worldview by expanding that worldview, and that possibility seems highly unlikely. In order to expand its worldview and redefine what science accepts as physical reality, science must redefine the fundamental assumptions upon which its worldview is based. Yet doing so would constitute the development of a whole new paradigm to replace the present paradigm. In other words, it would initiate a new scientific revolution. Quite simply, the basic quantities upon which the scientific worldview is based are not precise enough to distinguish between living and dead matter. In fact, the whole question of matter is at stake in this pursuit.

This course of action actually suits physics quite well, since physics is the science that deals directly with physical reality at its most fundamental level. For all of its complex theories, assumptions, attitudes and instruments, physics is merely a logical study of the natural world which is conducted by reducing all events and phenomena to their most characteristic and fundamental natural components. The fundamental components utilized by physicists in this quest are 'matter' and 'matter in motion' against the background of relative space-time. Nothing seems simpler than this methodology, but physics does not yet distinguish between living and dead matter, so physics cannot yet develop a theory of death and an afterlife. In physics, matter is just matter, whether it is living or dead. So what, in physics, could distinguish between living matter and non-living matter? Living matter can move itself about in space, but living matter is not really a kind of matter. Life is a relationship between different materials or forms of matter, a specific material structure. Pure physical matter is not important in life; it is the structure if not how that matter is put together to form a body that is important. Without the structure, the relationship between the parts or portions of matter, living matter is not different from dead matter. Life is a special physical structure that is more fundamental than the matter from which the living body is constituted; so it is necessary to develop a physical model of life before science can even begin to understand the physical nature of the afterlife.

CHAPTER 5
THE SOFT LIFE

> *For science is, I verily believe, like virtue, its own exceedingly great reward. I can conceive few human states more enviable than that of a man to whom, panting in the foul laboratory, or watching for his life under the tropic forest, Isis shall lift her sacred veil, and show him, once and for ever, the thing he dreamed not of; some law, or even mere hint of a law, explaining one fact; but explaining with it a thousand more, connecting them all with each other and with the mighty whole, till order and meaning shoots through some old Chaos of scattered observations.*
>
> —Charles Kingsley – 1874

That's Life

In many respects, our universe is a vast system of dualities. For every up there is a down, for every right there is a left. We have females and males, young and old, sick and healthy. The notion of duality extends beyond the world of things and quantities to qualities such as love and hate, intuition and reason, emotion and logic, happiness and sadness, and good and evil. These dualities are called symmetries in physics and the idea is a basic principle in some areas of physics and science. For example, every particle has an anti-particle. The electron has a positron and the proton has an anti-proton. Even the photon, a particle of light that has no mass, has an anti-particle: The photon is its own anti-particle. Anti-matter also forms a duality with matter, but anti-matter is not mind or life even though mind and life also form dualities with matter. Life and death also form another system of dualities. However, life forms another duality: From mind and matter, we get life and matter. This particular duality has influenced the course that science has followed for more than three centuries. Although this duality is not wrong, as dualities go, it is not the best possible duality of life when it comes to finding a theory of life or defining the concept of death.

No one seems to have realized that life, depending on how it is regarded, forms a much grander set of dualities. First there is life and matter, or more precisely animate and inanimate matter. In this context the duality is immaterial to physics. It makes no difference whether a person with a mass of 75 kilograms (animate matter) or a box of 75 kilograms (inanimate matter) is falling off of a cliff. They both follow the same gravity gradient downward at the same rate of acceleration. Physics is indifferent to animate life and normal matter in this regard as in many other cases. But there is also the duality of life and entropy, also a matter of concern for physics. Life is an organizing principle, and thus living organisms are highly structured and ordered, while entropy is a measure of the disorder of a system.

According to the second law of thermodynamics, entropy naturally increases. This characteristic of entropy renders entropy 'time's arrow' because entropy is thereby the only quantity or 'thing' in physics that determines the forward flow of time. If a system, such as a living organism, displays a decrease in disorder or a tendency to move toward a more complex and more ordered state, then a larger system that includes all of the air, water, food, and energy that is taken into the living organism must be considered to save the second law. Within this larger system, it can be shown that the overall entropy increases in accordance with the second law of thermodynamics. Therefore, living organisms could be considered small orderly sinks that are a natural part of physical reality within a larger ocean of entropic disorder. When the second law is expanded in this manner to ever larger systems to save the second law, the second law eventually comes to incorporate the whole universe and then it an be stated that the total disorder in the universe naturally increases.

This extrapolation leads to a concept called the 'heat death of the universe', whereby temperature in the universe will eventually even out with all points in the universe coming to have a single common temperature. Because the universe is so vast in size, this temperature would be quite cold compared to the relative warmth of the planets, stars and living organisms that now inhabit the universe and all life would die. However, if true, this possibility would only reach a conclusion so far in the future as to seem unimaginable. Also, the overall effect of other 'things' in nature, such as attractive forces of gravity and the existence of life, are not taken into account when the 'heat death' is considered.

Nor has any mechanism for the principle of entropy ever been discovered. It could well be that the entropy is related to the condition of the universe as a whole, such that entropy presently dominates the physical being of the universe because the universe is expanding. If this conjecture were true, then life, which is the opposite of entropy (life would be a neg-entropic mechanism), could also be spread out evenly across the universe as a natural consequence of the laws and symmetries of nature. However, life would not be quite as plentiful as entropy while the universe continues to expand and life itself might somehow offer a solution to the continued expansion of the universe and offer a solution to the problem of the 'heat death of the universe', a possibility never before considered in science.

But the really fundamental duality that concerns life would be life and death, where death is a state of non-life. Normal inanimate matter is also a state of non-life in the sense of animate and inanimate matter, but the distinction between life and death is different. The duality of life and matter deals with normal matter that has never been alive as opposed to death, which represents the matter in a body that was once alive. So the very idea of a duality of life and death, as opposed to life and matter, would be that there is 'something extra' in a living body that is not matter or material, since the matter in a living body would be the same as the matter in a dead (once living) body at the moment of death and immediately afterwards, before the dead body decomposes.

The very fact that a dead body naturally decomposes, such that entropy rules over the matter in the body only after life has departed the body, would seem to indicate that the living body had 'something extra' before death that distinguished it from the normal matter in the body and protected it from entropy and the normal laws of the universe before death. This argument implies that 'something extra' existed in accordance with yet separate from the material body only while the body was living, which further implies the possibility that 'some-thing' might have survived the death of the body. At the very least, it requires that science try to define that 'something extra' before scientists can make a definitive statement that nothing survives when a body or organism dies. Science can neither confirm nor deny that some part of a person survives death intact since science does not completely understand what life is. Although not openly admitted by scientists, scientists practice this fact every time they express their 'opinions' that 'nothing' survives or 'some-thing' survives. Some other 'thing' is associated with life, which science has not yet identified or studied.

It should therefore be clear that to understand death, you must have some understanding of its opposite, life, and that is where science has failed. First physics failed to understand life, then biology failed and finally chemistry has failed to define life in any meaningful manner. Certainly, religion has not defined life, nor has any other scholarly or academic discipline defined life. So no individual or group has a specific, exclusionary or absolute right to claim what survives (or doesn't survive) death or the conditions of its survival, at least not until we have a better understanding of life. So where does the answer to this dilemma lie?

The first feeble attempts to explain life came from Natural Philosophy, by applying the principles of physics. Scientists and scholars considered only the overall functioning of the body, its mechanics, a simple and rudimentary reduction of the living organism. They concluded that life was 'something extra', more than the sum of the individual parts from which the body is constructed, as exemplified by the Frankenstein monster. Then biology looked at the organs and how they functioned individually and worked together to keep the body alive. Life was reduced to anatomy and physiology. Then the reduction went further and science looked at the chemical processes involved in life, in disciplines now called organic chemistry and biochemistry.

For a long time, this last reduction seemed to be enough, or at least it kept science busy and little was said about life as 'something extra'. But now, as progress in chemistry is pushing toward a point where an end to the quest of life should be in sight, or can at least be imagined, 'life' has not been found or detected. Nor have the basic properties of the general concept of 'life' been identified. The chemists can only say how life operates within living organisms, not what life is. This last failure implies that the reduction must be continued to a still deeper level of nature than that normally explored by chemistry and this deeper level brings us back, once again, to physics because physics is the science that is basic to our concept of physical reality. It is again beginning to appear that we will have to determine what physical reality is to determine what life is, and thereby understand death. There are now strong indications to the effect that, and many scholars and scientists believe that, life is a general principle associated with physical reality itself.

Physical Reality

The structure, function and form of physical reality is the biggest question of all time and the answer has been sought for as long as humans have been able to think in abstract terms. Answers have been sought from the perspectives of religion, philosophy, science and general human thought, each looking for a confirmation of its own version of reality. Life, mind and everything else are just small parts of our total reality. Religion seeks to answer the 'why' of that reality by utilizing intuition and emotion based on faith while science only looks at the 'how' of that reality using logic and reason based on observation. Understanding our reality is the single goal of all science, with each discipline within science working to understand its own part of that reality in conjunction with the other scientific disciplines. The separations between physics, astronomy, biology, psychology, medicine, geology, chemistry and all of the disciplines of science are artificial human constructs and do not really reflect either the ordering or nature of reality.

Science has come a long way since the days of the Greek philosophers, who took the first recognizable steps toward science. In physics, the first and oldest of the sciences, we have two basic views of the fundamental structure of that reality and those views both originated with the Greek philosophers. Reality seems to be either continuous or discrete. At present physicists work with two basic theories, each of which represents one of these views. Relativity theory represents the continuous view of reality and quantum theory represents the discrete or non-continuous view of reality. Relativity theory is based on the concept of the continuous field and quantum theory on the discrete particulate nature of our world. Although these theories are quite modern, dating back only a century, the concepts upon which the theories are based have identifiable precursors dating back thousands of years, to at least the ancient Greek philosophers.

Aristotle thought our world was continuous, or rather matter was continuous. He called that continuity the plenum. There was no such thing as a vacuum or void in Aristotelian physics, what we call space without matter. So Aristotle had no concept of space as a construct of relative positions of objects, he only talked about place. On the other hand, Democritus developed the first idea of atoms. The atom was the smallest discrete bit of matter and a void existed between atoms. These were early precursors of our own modern concepts of field and matter, from which our physical reality can be theoretically constructed.

When we speak of empty space, a void or a vacuum, we are only referring to a region or volume lacking the presence of any form of matter. A lack of any matter is a void or vacuum. However, a void or vacuum is not totally empty because all of space is filled with one type of field or another. The common fields that we experience in every day life are the electric, magnetic and gravitational fields, while electric and magnetic fields form light when they are coupled and varying at the same rate or frequency. All of space is filled with these individual fields and electromagnetic waves. Electric fields originate in either positive or negative charges, and magnetic fields can be either north or south directed, so they can be blocked by their opposites, while electromagnetic waves can just be blocked, so a void can be made without the presence of these fields. But there still remains the gravitational field. As far as we know, gravity cannot be blocked in any way, so there is no true void or empty space which does not have at least a gravity field passing through it. In other words, gravity fields fill all of known space to the extent that there is no point in our universe that is devoid of gravity, no matter how weak the gravity field might be at distant points. Gravity is therefore the basis of our physical universe.

The problem in modern physics, which attempts to describe physical reality such that it forms our most fundamental understanding of physical reality, is that the physical quantities of particles and fields, or the physical concepts of discrete and continuous which they represent, are mutually incompatible. If the physical universe is discrete or particulate at its most fundamental level, then the discrete parts of the universe are separate and have nothing, literally 'no-thing', between them, not even void (which is a thing) or empty space (which is a relative thing). We cannot even comprehend this possibility. However, this possibility is exactly what quantum theory postulates. In fact, the most advanced forms of quantum theory assume that there are even particles of space and time themselves, and these smallest parts of the universe have assumed sizes of about a Planck length and a Planck unit of time. The overwhelming majority of scientists in the scientific community believe that the quantum theory is the correct and proper approach to understanding physical reality, or that physical reality is discrete at its most fundamental level of existence.

But pure field theorists who work with relativity theory think that the underlying basis of physical reality is not discrete as the quantum theorists contend. To them, the most fundamental level of reality is the continuous field. No matter how far we look, how deeply we seek, at smaller and smaller scales, we could never find a void or place where the field is discontinuous or absent. In this case, all of reality would be just a single field out of which the electric, magnetic and gravitational fields evolve under the specific conditions inherent in our four-dimensional space-time. Even matter is a specific construct or manifestation within the single field. The main proponent of this point of view was Albert Einstein, who first formulated the special and general theories of relativity. Einstein spent the last three decades of his life seeking a 'unified field theory' that he hoped would explain the physical reality underlying the quantum.

The problem that science now faces, from both philosophical and practical considerations, are that these two views are mutually incompatible. So the continued progress of theoretical physics seems to have come to a roadblock. Our four-dimensional physical reality cannot be both continuous and discontinuous (discrete). Yet the relativity and quantum theories that represent these two approaches seem absolutely correct within their own domain of application. Since quantum theory has been the most practical theory in its numerous applications to our world, most scientists just assume that continuity will eventually be shown to be an illusion and the final reality will be discrete in the smallest possible regions of our world. The key for the quantum theorists is to explain gravity, the last of the natural forces to fall under the quantum umbrella of theories. And toward this end the quantum theorists are trying to develop a theory of quantum gravity.

Three competing quantum based theories are presently being developed as the solution to reality; the supergravity, superstring and quantum loop theories. However, some scientists working on these theories believe that these theories might not represent the final theory of reality, although they do believe that these theories will eventually lead to the final TOE or 'theory of everything', which some call 'M theory'. In its present state, superstring theory is not complete and is presently evolving into Brane theory. Branes are membranes of various dimensions that exist in higher-dimensional bulks. So superstring theories are already evolving toward the suspected 'M theory'. In the meantime, each of the various quantum based theories is beset by fundamental problems, which might eventually render these theories useless for science, while none of them is even close to an attempt to account for life, let alone death.

Scientists who believe that the continuous field is the most fundamental 'thing' or reality are conducting very little theoretical research on physical reality in spite of the fact that this was Einstein's approach to the problem. Yet the philosophical arguments supporting continuity are much stronger than those supporting the discrete. Quite simply, if physical reality is discrete at its smallest levels, then what lies between the discrete 'points' at the smallest level of reality? It is logically impossible to build a continuous field, infinitesimally small point by point, out of larger discrete point quantities. You just can't build smaller things out of larger things. At best, all you can hope to accomplish is to mimic continuity using a purely mathematical model based on discrete units, but that still would not be reality, it would only be an approximate picture of reality. That is all that the quantum approach can ever hope to accomplish and that is exactly what the quantum theory does at present, build a very good and accurate picture of reality.

On the other hand, it is philosophically possible to build larger units or discrete parts out of smaller points in a continuous field. So the continuous approach to describing reality is philosophically superior. However, this reduction cannot be completed within either a four-dimensional space-time framework or a three-dimensional space framework with a separate time dimension. A higher dimension is needed to accomplish the reduction and explain the quantum. A theory of this type has been developed, but largely ignored by science until a few years ago. In 1921, Theodor Kaluza added a fifth mathematical dimension to Einstein's four-dimensional space-time continuum in order to unify gravity and electromagnetism as different aspects of a single combined field. Scientists have never refuted Kaluza's theory, but it has not been verified either because it only mimics electromagnetism and offers nothing new beyond classical electromagnetic theory by which it could be tested.

Oskar Klein developed a combined quantum-gravity model based on Kaluza's five-dimensional extension five years later. The extension of our common four-dimensional space-time into five dimensions needed to be extremely small to guarantee that the fifth dimension could not be perceived or detected in any manner, a fact that allowed Klein to incorporate the quantum into the single five-dimensional field. However, Klein was never able to finalize his theory and eventually rejected it. After a long hiatus during which no progress was made and the theory was all but forgotten, the Kaluza-Klein theory was conveniently rediscovered in the 1980s and adopted as the basis of both the eleven-dimensional supergravity and ten-dimensional superstring theories and finally leading to Brane theories. Perhaps some of the problems that these new theories are experiencing are due to the faults in the Kaluza-Klein theory upon which they are based.

On the other hand, Kaluza's original extension of general relativity spawned another lesser-known line of development. In 1938, Einstein and Peter Bergmann demonstrated that the extension in the fifth direction does not need to be extremely minute, as Klein had reasoned, but could be macroscopic in size. Einstein abandoned this avenue for developing a unified field theory in the early 1940s because he could not justify the fact that the macroscopically extended fifth dimension that was indicated by the mathematical model could be neither perceived nor detected. However, this model does lead to a theory that can explain the quantum and unify all of physics by making a few minor changes that Einstein did not try. The newer theory is called the 'single (operational) field theory' or SOFT. According to this theory, the single continuous field fills or coexists with all of five-dimensional space-time while our material four-dimensional world is just a 'sheet' consisting of a very dense portion of the field within the five-dimensional single field. Yet the 'sheet' is completely continuous with the single field, not something separate. The quantum is defined by the 'effective width' of the 'sheet' in the fifth direction, which is extremely small. This single field is the precursor to all other physical fields. It yields matter, gravity, electricity, magnetism and the other natural fields within our relativistic four-dimensional continuum. Each common physical field is a particular construction of the five-dimensional single field as experienced within our four-dimensional space-time.

Under these circumstances, it is a safe bet that any future theory that attempts to unify all of physics will depend upon on at least a five-dimensional space-time if not a strictly five-dimensional structure. All of theoretical physics is presently headed in that direction while the Einstein-Kaluza model is the only successful theory to unify gravity and electromagnetism. So the Einstein-Kaluza model should form the basis of any future theory. The biggest problem with the five-dimensional model and the largest hurdle that it has to overcome is not exactly scientific, but human. We only think in three-dimensional terms since all of our conscious experience of the world is of that type, so it is extremely difficult to imagine or picture how the world would look and act from a five-dimensional perspective. We have no practical experience with higher-dimensional manifolds, so there is a bias of our ignorance against all such theories.

Imagining the un-imaginable

If our physical reality, the universe that we know and love, is actually five-dimensional then we cannot even hope to imagine what the fifth dimension would be like. All of our experience is based upon our material existence within the four-dimensional space-time continuum. This was Einstein's dilemma. Einstein abandoned his five-dimensional theory because he could neither understand nor justify the fact that we cannot perceive the fifth dimension in any way. But Einstein was wrong, or at least he never realized that we do perceive the fifth dimension, but not in any way that he would have accepted. Our normal five senses work within the four-dimensional continuum in which our material world exists, but we have an innate sixth sense whose existence is not recognized by most people. This sixth sense works completely within the fifth dimension. It can be shown that life, mind and consciousness are specific five-dimensional field structures, rendering our perceptions of the fifth dimension completely para-normal rather than non-existent as Einstein and everyone else has assumed. We do experience and interact with the single field in the fifth dimension of our universe in a manner not yet recognized, nor probably even suspected by Einstein, para-normally by our sixth sense.

Although we cannot even fathom a five-dimensional world, we can be taught or mentally conditioned to at least recognize some of the basic characteristics of a higher-dimensional space. Our imaginations can be primed through analogies with lower dimensional spaces. We can draw analogies to our three-dimensionality and how our three-dimensional space would appear to a two-dimensional creature and how the two-dimensional creature's life would appear to us as three-dimensional creatures. These methods are not new, but were developed between 1870 and 1900, when a few scientists first proposed that our common three-dimensional space might actually be four-dimensional.

Imagine a two-dimensional creature. Such a being could literally live in the plane of a sheet of paper, but it would not have any measurable or detectable thickness in the third dimension as does a sheet of paper. If this creature did have a thickness in the third dimension, it would be so small as to remain totally undetectable by the creature. As far as this creature is concerned, its whole life, its whole experience, and its whole knowledge of its universe and its being are two-dimensional within the sheet. The creature does not even suspect that his universe is just a sheet, since it is the creature's whole universe and he has nothing against which to compare it. It is only a sheet from the perspective of the higher dimension. It does not even suspect that a third-dimension is possible. In fact, the idea of a higher dimension has never even been raised in the brain of the creature and is so alien to the creature's existence that it could not even fathom such a possibility. His sheet is the whole of his reality. This creature, of course, can move around and do things in his sheet, as well as interact with objects and other beings and sense things within his sheet. The creature is alive and functions according to the physical laws that apply within its sheet. We, however, would see this creature as restrained by his two-dimensional existence, although he would not think he is so restrained since he knows nothing beyond the sheet.

The two-dimensional creature would have a skin that protects and separates him from the rest of the two-dimensional environment. He could also have a skeletal system and organs within his body. Yet we three-dimensional creatures could see each organ, the skeletal system, and each and every point within the two-dimensional creature's body from the perspective of our higher dimension because each mathematical point in his two-dimensional world would be visible or open to a three-dimensional being. Yet the two-dimensional creature could not see or otherwise detect us, no matter how close we would come to his sheet. Even if we pierced the sheet, say someone stuck a pin through his sheet; all the two-dimensional creature would see would be the point or extended point where the pin struck through the sheet.

If our two-dimensional creature was locked in a jail cell, with no exit through the walls or the bars, we would wonder why from our higher-dimensional perspective, because we would see that all the creature needed to do to escape would be to move into the three-dimensional world and around the two-dimensional walls in the sheet and then return into the two-dimensional sheet outside of the walls. The two-dimensional creature would have gone through the walls, from the two-dimensional perspective, without ever having passed through them, as if by magic instead of through the unsuspected third dimension. Nothing could be hidden in the two-dimensional world from a three-dimensional perspective. If something were locked in a two-dimensional vault, that thing would be open for viewing or taking through the third dimension. This is a strange world indeed, but the same would be true for the three-dimensional creatures that we are when considered from the perspective of a four-dimensional being.

On the other hand, if a three-dimensional being were to interact with the two-dimensional world of our new two-dimensional friend, the panic caused would surely disrupt his pleasant little life. A three-dimensional being could just poke a finger through the two-dimensional sheet with no trouble whatsoever. However, from the two-dimensional viewpoint, only a round or nearly round two-dimensional object, a slice of the finger, would appear when the finger penetrates the sheet. As the finger passes through the sheet, the portion in the two-dimensional sheet would change shape and size, with no apparent reason that the two-dimensional creature could distinguish. If that were not puzzling enough, the finger first popped into the two-dimensional world with no apparent cause, out of nowhere, like magic. The two-dimensional creature and all of his scientist and philosopher friends could not explain the strange events that occurred around the appearance of the finger in their world.

Nor could they move the finger by pushing or pulling the round slice of the finger that they witness moving through their world. If they tried to move the object, they would actually be trying to move the whole finger in the higher dimension, which they are not even aware of, and that would be impossible. So the round object that they perceive in their two-dimensional world would appear to be immovable, which would more than likely defy the tenets and physical laws of their two-dimensional science. The finger could even appear to have an infinite two-dimensional mass. They would be frightened and perplexed at this new development, and certainly unable to understand the physical characteristics of this event relative to their two-dimensional world.

If they tried to tie a rope around the round object they could not even form a knot in the rope. If they could form a two-dimensional equivalent to a knot, then that knot would still appear to be just a untied line of rope relative to the three-dimensional world, and the round object would just mysteriously slip out of their knot and away. The same would be true of a knot tied in a rope in our three-dimensional world. No matter how complex the knot in three dimensions, it would only be a simple line of untied rope to a four-dimensional creature or being looking into our three-dimensional world. In fact, everything closed in our three-dimensional world and hidden away inside boxes and safes would be completely open to the four-dimensional intelligence. Nothing in our three-dimensional world could be kept away from and safe from a four-dimensional being, just as nothing can be kept away from and safe in a two-dimensional world from a three-dimensional being. And just as every point in the body, every organ and cell of the two-dimensional creature is seen simultaneously and can be touched by a three-dimensional being, a four-dimensional being would have immediate access to literally every point inside our own three-dimensional bodies.

Now suppose that the two-dimensional creature wished to move around his two-dimensional world. His motions would follow the physical laws of his simple universe. His sheet would have a geometry where lines are parallel and the sum of the angels in a triangle was always 180 degrees. However, his two-dimensional sheet universe need not be flat. If it were not flat the geometry of his world would change and he could not explain how or why the geometry changed because he would not know or even understand the curvature of his sheet in a higher dimension, even if he knew about it. Suppose that the two dimensional sheet was actually the two-dimensional surface of a sphere. Then the sum of the angles of a triangle could actually be as much as 270 degrees, not just 180 degrees. Now if the two-dimensional world of the creature were very nearly flat, but had bumps and creases or varying curvature in areas of the sheet, the motion of the creature would be accelerated or decelerated as the creature moved over the varying curvature of those bumps and creases. We would see the creature moving through the curves at a constant speed, even though the creature would sense a force accelerating him at the exact same instant. The creature, not knowing about or perceiving bumps, creases and curves in a higher third dimension, could not attribute his changing speed to the curvature of the sheet, but would instead believe that unknown forces accelerated his motion.

And this is exactly how gravity is explained in general relativity. The presence of matter bends our four-dimensional space-time, so when other pieces of matter travel along and come to a regularly curved region of space-time they can experience acceleration in their motion due to the curvature caused by the presence of matter. The cause of this acceleration is interpreted as a force of gravity since we cannot perceive the changing curvature of our own space-time continuum. The question is whether our space-time is curved in a higher dimension, so the curvature would be an extrinsic property of our space-time continuum, or if the curvature is intrinsic to our continuum and thus no higher dimension is needed. Gravity does not care which case is true, although electromagnetism cannot be included in this model of the space-time continuum unless an extrinsic curvature in a higher fifth dimension is included.

Now suppose that the two-dimensional creature was somehow able to leave his sheet and travel into the third dimension before returning to the sheet by the other side of the sheet. In other words, suppose the creature could make a complete transit of three dimensions and come up on the other side of the sheet. In this case, the creature could return to the sheet exactly as he left it, or he could return with everything on the left of his body and the right side of his body exchanged. He could return as the mirror image of his earlier self, before leaving the sheet. If the path that he follows while in the third dimension is just along a circle and back again, then the two-dimensional creature would have reversed himself twice, once as he passed through each of the two poles of the circle. But if the path of the creature through the third dimension followed a twisting surface, like a Möbius strip, then the creature would have only reversed himself once through the single pole of the Möbius strip, and thereby returned to his own world reversed, left for right.

Again, the same is true for our three-dimensional space or our four-dimensional space-time if it is curved in a higher-dimensional space. According to general relativity, our four-dimensional space is Riemannian (if curvature is intrinsic) or the surface of a Riemannian sphere (if curvature is extrinsic). But there are two possibilities associated with the Riemannian sphere, it can be either single or double polar. We assume that the Riemannian sphere that is our four-dimensional space-time is double polar, but that need not be true. One of the great problems of relativity theory is accounting for a property of elementary particles called spin. Protons, neutrons, electrons and other particles have either a plus or negative half-spin. If science adopted the possibility of a real fifth dimension into which every point if our four-dimensional space-time were extended, and that fifth dimension were closed, meaning that the line from each point extended around a great Riemannian sphere and back again to the same point on the other side of the fourth-dimension, then that Riemannian sphere could be either single or double polar. If it were single polar, then the closed line around the Riemannian sphere would return to the same point twisted half way around, thus accounting for the half-spin of the protons, electrons and neutrons from which atoms in our normal universe are constructed.

The analogies that have been made between the two-dimensional sheet and our common three-dimensional space are not really new. This particular way of 'picturing' a higher-dimensional space is over one hundred years old. During the 1880s, Edwin Abbott wrote a book titled *Flatland*, in which a two-dimensional creature named A. Square learned about the third dimension. Abbott's book is still in print today and is used to teach students about higher-dimensional geometries by analogy to two-dimensional worlds. However, this method did not originate with Abbott. Others used the same method both before and after Abbott wrote his book, to help explain the theories proposed by W. K. Clifford in and after 1870.

Clifford proposed that our three-dimensional space is actually part of a larger four-dimensional manifold. He stated that what we perceive as matter is no more than hills and curves in our common three-dimensional space curved in the fourth dimension, and that what we know as matter in motion is nothing more than variations in the curvature. Others scientists and mathematicians then set about finding the basic geometrical properties of these higher dimensional spaces and manifolds. Yet today's scientists seem to know little or nothing about these common geometrical properties of higher-dimensional spaces and do not speak of them in their own higher-dimensional theories. Clifford's theories were well known, hotly debated and very popular among non-scientists. So popular, in fact, that Abbot wrote his book to help explain the basic principles of a higher-dimensional space to common educated and uneducated people who were interested in Clifford's theories.

Clifford was not trying to explain gravitation with his notion of space curvature, as Einstein did nearly fifty years later. Instead, Clifford was trying to better explain and expand the new electromagnetic theory that his friend J.C. Maxwell had just developed. Clifford hoped to eventually explain all forces, including gravity, within his four-dimensional model, but died very early at the age of thirty-four or consumption. So Clifford's theory was not a precursor of Einstein's theory of relativity, nor did he anticipate Einstein's notion of curved space-time. However, Clifford did try and develop a TOE (theory of everything) more than a century before modern scientists began working on that same type of theory. Of course it was not called a TOE in those days; Clifford was just "solving the universe". Clifford's work was actually a forerunner to today's five-dimensional theories as well as many of the Brane models of physical reality that are now being proposed by physicists. Under these circumstances, modern scientists would do well to take a more careful look at Clifford's work and the work of others who already explored these hyper-dimensional paths a century or more ago.

IF it seems difficult to imagine a three-dimensional space curved in a four-dimensional manifold (space), as Clifford proposed, or a four-dimensional space-time curved in a fifth dimension, as Kaluza proposed, you should not feel that discouraged. The task is difficult and it takes time for the notions to sink in. When Abbott wrote his book *Flatland* in the 1880s, he was trying to explain the concept of higher-dimensional space to the common people so they could understand what scientists and mathematicians were talking about. But he was not alone in this endeavor. Charles Hinton, a mathematician and geometer, as well as others also wrote on the subject in common terminology. In particular, Hinton tried to develop a method for actually 'realizing' higher-dimensional spaces. Hinton wake take a regular three-dimensional geometric shape, such as a cube, and count the number of edges, corners and surfaces. Then he would extend that figure into a fourth dimension and again count the number of edges, faces and corners. In this manner he developed a four-dimensional cube called a tesseract.

Hinton would then try to 'realize' or build up an image of that four-dimensional regular solid in his mind. Twenty or thirty years ago the same methods were again tried, but this time using computers to generate the four-dimensional images in a three-dimensional picture space. During the 1880s and 1890s, debates were conducted in the pages of popular scientific journals about whether such 'realization' was even possible, but no conclusions were ever reached. As far as anyone knows for sure, no one has ever successfully imagined a four-dimensional solid although some of us are still trying by using these same or similar methods. Even Hinton gave up the cause after years of trying without success, but he also tried to develop a scientific theory based on the concept in the first decade of the twentieth century.

One famous American Astronomer, Simon Newcomb, went so far as to call these higher-dimensional spaces the "fairyland of geometry" and modern historians have wrongfully concluded that Newcomb therefore thought that such attempts were no more than fairytales and fiction. However, if modern historians and scholars would just read Newcomb's essays on the subject, they would find that he was seriously considering the physical attributes, consequences and properties of higher-dimensional spaces which otherwise, if not considered scientifically, would seem like 'fairylands' and imaginary places because of their bizarre and wonderful physical properties.

One last word was said on the subject in 1919, when Robert Browne published a book titles *The Mystery of Space*. The book is a historical and philosophical treatise in which Browne argues that the realization of a higher-dimensional space is the next step in human evolution. So it would seem that human evolution is progressing toward an understanding of higher dimensions. Human fate might after all be to actually realize, learn of, know and understand the human existence in a higher-dimensional universe, which would constitute an evolution of MIND rather than body. We should not be saddened or frustrated if we do not yet 'realize' a higher-dimensional reality, learning of it is just the first step taken so that we will be prepared when we come face-to-face with that reality. And we will come to face that reality when we learn what LIFE, MIND and CONSCIOUSNESS really are.

Our five-dimensional Life

With these new conceptual tools for exploration in hand, we can finally look at the five-dimensional model of physical reality and the model of life that emerges logically from it. Since it is a natural consequence of our physical reality, life can be considered either a product of or a specific structural configuration within the five-dimensional single field, depending upon whether you are taking a four- or a five-dimensional perspective on life. To build a model of life, we start from the simple hypothesis of a real physical fifth dimension filled by the single field. Then we ask – what physical form or structure does our space-time take relative to this fifth dimension?

Our normal four-dimensional world has the structure of a 'sheet' suspended or embedded within the single field. But the 'sheet' is not separate from the rest of the fifth dimension as the various Branes would be according to the superstring and related theories. The 'sheet' is just a portion of the overall five-dimensional single field that is distinguished from the rest of the single field by its overwhelmingly greater field density. The 'sheet' is continuous with the rest of the single field in the fifth direction, just as space and time are continuous in the normal four dimensions of our world. The single field density decreases very rapidly the further the distance from the 'sheet' in the fifth direction. The difference between the densities in the fifth direction is so great that the 'sheet' has an 'effective width' of about the diameter of the proton times the fine structure constant (e^2/hc .1/137), rather than a 'true width'. A 'true width' would imply that the 'sheet' has boundaries that would render it separate from the single field.

The internal structure and properties of the 'sheet' relative to the fifth dimension are easy to determine since there are simple geometrical rules for governing the structure. The point-by-point extension from four dimensions into and along the fifth direction must be extremely small to guarantee that we do not 'normally' sense or otherwise detect the fifth dimension. This simple fact represents the paradox that tripped up Einstein and led to his failure to develop a successful five-dimensional theory. Quite simply, Einstein's wonderful physical insights into nature failed him at this point. We can rule now out both infinitesimally small (including the Planck length of about 10^{-40} meters) as well as infinite extensions or 'sheet' widths in the fifth direction. These cases lead to anomalies, physical impossibilities and/or logical paradoxes.

Kaluza's original mathematical model required a closed fifth dimension, allowing the duplication of the electromagnetic formulas from a single field. So if the extension in the fifth direction was infinite, and thus not closed, then the electromagnetic portion of the field would be lost. Instead, the fifth direction is macroscopically extended. So, while the overall extension in the fifth direction can be extremely large and still remain closed, there is no such restriction on the thickness of the 'sheet' that constitutes our four-dimensional space-time. We cannot perceive the fifth direction in any manner that resembles our perceptions of the other common dimensions because the 'sheet' is so thin in the fifth direction, not because our overall extension in the fifth direction is so small. Only the 'effective width' or 'thickness' of the 'sheet' itself need be microscopically small, not the complete extension in the fifth direction. This distinction was completely missed by Einstein and his colleagues, who were convinced that relative space and time could have no properties and therefore did not consider the possibility of density variations in the single field along the fifth direction.

Even with these density variations, the single field model fully complies with Kaluza's original restrictions on his five-dimensional mathematical model, such that (1) all points in four-dimensional space-time are extended to an equal length in the fifth dimension, and (2) the extensions form closed loops with four-dimensional space-time. Kaluza used a completely and purely mathematical model that was completely non-physical. Kaluza did not consider the possibility that a real physical fifth dimension existed, at least not in public in his publications on the theory. However, the present theory considers a real physical fifth dimension that is characterized by the mathematics of Kaluza's original model, except for the density variations along the fifth direction. In essence, the fifth dimension is physical although not material, since matter is a purely four-dimensional concept and has only a four-dimensional presence, as will be shown below. The words physical and material are not synonymous. Material signifies matter, which can only be measured by its mass, while fields are physical yet non-material, such that they have no mass associated with them. Fields are characterized by potential rather than mass.

The fifth dimension has the structure of a single polar Riemannian sphere, which explains many common physical phenomena and properties such as torsion and particle half-spin within the 'sheet', where the particles actually exist. It is difficult to imagine a four-dimensional surface forming a sphere in five dimensions because all of our visual and sensual experience is within our common three-dimensional space. So the best that we can do is imagine a two-dimensional surface forming a sphere or globe in three-dimensional space and extrapolating the geometrical properties of the five-dimensional curved surface by analogy. Unfortunately, a globe is not the best example because it is a double polar sphere. However, a Möbius strip presents a simple example of a two-dimensional surface that forms a single polar curve in three-dimensional space. So the concept of a single polar figure such as a sphere is not completely beyond normal reason. An object following a path around a Möbius strip will return to the same spot from which it started but reversed. In a similar manner, a closed loop originating from the top of the 'sheet' will twist half way around before returning to the same point at the underside of the 'sheet', thereby accounting for the half-spin of elementary particles such as protons and electrons.

Elementary particles are folds (protons), bumps (electrons) and very small bends (neutrinos) in the 'sheet'. These folds, bumps and bends are described completely by the concept of space-time curvature according to general relativity. It is just as Clifford described over a century ago: Matter is nothing but the curves of three-dimensional space in a fourth dimension and the motion of matter is nothing more than the variations in that curvature over time. The SOFT structure guarantees that individual particles do not form infinites and singularities as in classical general relativity because the overall 'sheet' is closed with respect to the fifth dimension in a Riemannian single polar sphere. Elementary particles are singularities in classical general relativity because the five-dimensional structure in classical general relativity is either open with respect to the fifth dimension or simply because classical general relativity does not assume a five-dimensional structure. This is a serious problem for scientists. The fact that there are no singularities in the SOFT model allows the first ever application of general relativity to determine the internal structure of elementary particles. In fact, the physical properties of elementary particles are derived from this closed structure. In this respect alone, if nothing else in the model were true, then SOFT would still represent an advance over classical general relativity.

Since material particles are made from the 'sheet' giving us what we perceive as matter, we can say that all matter is stuck within the confines of the 'sheet'. So, all living beings are stuck within the 'sheet'. The gravity, electric, and magnetic fields that we normally perceive and experience are just different structures of the five-dimensional single field as it manifests within the 'sheet'. Our five normal senses therefore operate totally within the 'sheet' and thus our direct knowledge of the fifth dimension is severely limited. As far as we know (at least within our normal existence), our total experience of the world and universe is inside the 'sheet' and thus independent of any higher dimensions. Yet it is from our four-dimensional experiences and observations, our four-dimensional sensory data, that all of our logic, reason and science, including physics, is constructed. So what we regard as normal science and normal physics is science within the 'sheet', whereas paranormal science would be the science of whatever is outside of the 'sheet', or just the science of the fifth and higher dimensions.

Normal science within the 'sheet' must therefore include modern quantum theory. In fact, the 'effective width' of the 'sheet' determines the quantum of action upon which all of the various quantum theories are based. Quantum theory only describes physics 'inside' the 'sheet' and is therefore 'incomplete' as Einstein argued in the 1930s. Otherwise, all particles follow the physical restrictions noted in the special theory of relativity with regard to the fifth dimension. When particles move at some speed relative to the 'sheet' they undergo Lorentz-Fitzgerald contraction along the direction that they are moving through three-dimensional space. This change squeezes the particle's five-dimensional 'effective volume' forcing the interior of the particle 'up' further along the fifth direction, increasing the particle's five-dimensional aspect (five-dimensional height) as a function of the increasing kinetic energy of the particle or object. As a particle slows down, its aspect or height in the fifth direction decreases until it attains its normal rest height position relative to the 'sheet' or simply relative to all other matter in the universe which influences the overall curvature of the 'sheet'. This mechanical like movement, 'up' and 'down' in the fifth direction, impresses or imprints fluctuations in the single field density just 'outside' of the 'sheet'. This imprint would be the trace or energy signature of the physical event in the single field.

In chemistry, reactions take place by swapping and sharing electrons between atoms or molecules. In other words, the shared electrons change energy relative to their changing positions in the reacting atoms and molecules. Any atom or molecule as a whole is characterized by its total energy, while the electron energies and their changes are a strong contributing factor to that total energy. While normal chemical reactions occur within the 'sheet', they still affect the single field density just 'outside of' or 'above' the 'sheet' whenever the energies of the particles involved in the reaction change. So chemical reactions can be characterized by the energy exchanges between the elementary particles from which atoms and molecules are composed. The changing aspects of the reacting particles moving up and down in the fifth direction more-or-less 'compress' and 'extend' the single field above the particle, thus imprinting a single field density pattern in the fifth dimension corresponding to the chemical reaction in four-dimensional space-time. All chemical reactions of the same type would leave the same energy traces imprinted in the single field.

These single field density changes are 'felt' or 'sensed' by other particles and objects due to the connections of all material particles afforded by the single continuous field. Variations and changes in the single field are essentially 'communicated' throughout the whole field. That 'communication' is not restricted by our normal physical laws, which apply only within the 'sheet'. This 'communication' process forms the basis of what scientists commonly call quantum entanglement, so we can say that 'entanglement' occurs completely 'outside' of the 'sheet', which is outside of our common four-dimensional space-time. We could also say that entanglement is a 'subtle energy' effect, depending on the type of phenomenon or event described, although this would not be a true 'energy' since no matter has been put in motion. There are only two types of energies, kinetic and potential. Kinetic energies occur when matter moves, while potential energies occur when an appropriate test particle is placed within an appropriate field. The test particle produces potential energy by interacting with the potential at that location in the field. So there is no true 'subtle' energy associated with entanglement since no test particle has been introduced into the single field in the fifth dimension. Entanglement would more accurately be described as a 'subtle potential' or 'subtle' variations of potential in the single field, since all physical fields are associated with potential and potential differences.

However, entanglement is the word commonly used in quantum physics and scientists have an idea what is meant by the concept even if they cannot truly understand the mechanism behind it, so the word entanglement will suffice. The single continuous five-dimensional field is the medium through which quantum entanglement acts as well as being the quantum entanglement itself. So, while the quantum occurs within the 'sheet', entanglement occurs between three-dimensional particles across the fifth dimension outside of the 'sheet'. Entanglement lies outside of quantum theory and could never be explained by quantum mechanics nor reduced to a quantum mechanical action, which necessarily occurs in the 'sheet'. Therefore, entanglement need not always follow the physical restrictions on material particles that lie inside the 'sheet'. Consciousness is quite like entanglement, so the same must be true of consciousness.

The chemical reaction imprint in the single field ends as the reaction runs to completion, making it a temporary non-repeating trace of the chemical reaction. However, equilibrium chemical reactions can go backward or forward, depending on the physical conditions of the reactants and the local conditions of the reaction. Under these circumstances, equilibrium reactions would imprint a repeating pattern of field density variations in the single field as time moves forward. It would be a recognizable reciprocating ordered pattern of aspect change along the time-line representing the reaction and would look like an irregular wave, wrinkle or crease along one direction in a tablecloth or bed sheet, except that it would be a four-dimensional form across the surface of a four-dimensional 'sheet'.

However, we can visualize our four-dimensional 'sheet' by making an analogy to a real two-dimensional tablecloth or bed sheet curved in our three-dimensional space. One axis of the bed sheet is time and the other represents the normal three directions within our common space of experience. An ordinary chemical reaction would look like a crease or wrinkle, as it is being made or developing along the surface of the bed sheet as time moves on. The wrinkle would stop forming when the particular chemical reaction reaches completion and ends. The height of the crease at any point along the line, corresponding to a specific time, would correspond to the intensity of the reaction at that moment. But an equilibrium reaction would not end quickly, so the wrinkle or crease would continue as the reaction went back and forth like a wave along one dimension of the surface.

We can then reason that it would be possible for several different equilibrium reactions to combine their 'efforts' for a common 'cause' and form a connected or interlinked (quantum entangled) interaction. The common 'cause' need be no more than the continued existence of their effort, their physical survival, rather than a conscious choice to exist. The survival would be a matter of physical convenience and no more. Normal equilibrium reaction patterns are subject to physical conditions outside of their three-dimensional physical environment, such as energy input and temperature variations, pressure changes, concentration changes, and so on, but those conditions could work in concert with other equilibrium reactions, separating the combined reaction from its environment and defining that reaction as a separate viable entity within the environment. The entanglement pattern of this chemical configuration is of greater consequence and far more interesting for this ensemble than the patterns of other simple chemical reaction. The special entanglement of these equilibrium reactions renders the pattern completely stable relative to the outside environment in a manner similar to a very complex standing wave in common physics, but in this case the standing wave is creating and sustaining itself within external input.

Under the special circumstances where different equilibrium reactions come together to form a common 'alliance' and form a structure characterized by shared energies between different reacting substances, a new special structure emerges. The overall reaction or structure does not need to rely on the three-dimensional conditions or the environment for its internal stability although it may need to take energy from its environment at times. Each internal reaction supplements and complements the other reactions, such that, each reaction feeds and/or contributes the energies needed for continued existence to the remaining common chemical mixture yielding a **mutual self-sustaining chemical complexity** and stable reaction 'state' over longer and longer periods of time. The resulting density variations are imprinted in the five-dimensional single field to form a single complex yet stable entangled pattern. This pattern is a fixed imprint in the single field and it could theoretically be modeled as a chaotic mathematical complexity by an attractor (equation). In this case, the cloth or bed sheet would have a complicated repeating pattern of changing wrinkles and ripples across its whole surface.

The wrinkling pattern in the bed sheet would look more like a very complex two-dimensional wave pattern over the surface of the water in a swimming pool or pond. If the two-dimensional surface of the water represented only the three-dimensions of normal space curved in a fourth direction, with a separate time, then the wavy pattern on the water would have an undulating appearance as time passes, but a distinct repeating set of sub-patterns would emerge if the undulating water surface were watched closely for a long period of time. Over a period of time, as the complex wave pattern undergoes shifts and changes, a definite pattern of repeating internal variations in the curved surface would emerge and seem to control the overall surface pattern of curvature variations. This emergent internal structure of curvature variations is the LIFE pattern as viewed from a five-dimensional perspective.

LIFE, the long sought 'something extra' associated with animate matter, occurs or comes into being when a specific mixture of different chemical reactions evolves into an organism by forming a special complex entanglement of different equilibrium reactions. This entanglement corresponds to a mathematical **complexity** that represents a **self-sustaining** single field density pattern in the five-dimensional extension of the reacting substances that form the material living organism.

Any single chemical reaction would appear chaotic from an atomic/molecular point of view. Thousands upon thousands of atoms and molecules would be reacting in what would seem a random fashion from the microscopic perspective. The same would be true for a complex of interacting equilibrium reactions as described above. However, the process would seem very orderly, proceeding from an initiating event to an intense reaction and then tapering off from the macroscopic perspective of viewing the whole chemical reaction at one time. Chemical reactions do occur in a timely order; therefore any chemical reaction could theoretically be represented as a mathematical complexity of chaotically reacting elements. The more complex the combination of reactions that contribute to a single whole organism is, the stronger the complexity formed by that group of reactions. So the more stable and complicated the interacting chemical reactions, as in the SOFT model of LIFE, the more assured the possibility that a mathematical complexity has been reached. Such a mathematical complexity can be represented by a specific equation of formula called an attractor. So theoretically, the LIFE or Biofield of a simple living organism could be represented by a single attractor equation. Such an equation must necessarily exist since the living organism exists and the chemical reactions upon which its existence depends are chaotic on the microscopic scale, although it would be difficult to find the proper attractor equation for representing any given LIFE form, except possibly the simplest LIFE form.

In other words, LIFE evolves or emerges from the four-dimensional chaos of interacting atoms and molecules in a chemical reaction as a five-dimensional complexity. LIFE then acts as an organizational principle for the entangled chemical reactions from which it emerged. LIFE further organizes and controls the organism's internal chemical reactions conform to the internal consistencies of the pattern in such a way that it improves the normal functioning of the body and enhances further evolution, forming a two way dependence of LIFE on the material body and vice versa. So LIFE is a physical (five-dimensional) but non-material (non-four-dimensional) field structure with a real physical existence. The LIFE pattern can also be called a Biofield, a term the many scientists are now adopting to describe life as an organizational principle, although the term Biofield would better describe only the special entanglement that exists in the fifth dimension 'between' the internal chemical reactions of a living organism.

Over time, the living organism comes into contact with a greater portion of its material surroundings and undergoes a greater variety of material interactions with different components of its four-dimensional environment. The Biofield regulates the living organism to compensate for and conform to an ever-growing array of variations in the organism's physical environment and conditions of existence. New internal chemical reactions may even become necessary for ordinary survival within the context of the changing external environment, but only those chemical reactions that are compatible with, help sustain or enhance the function of the complexity (those that are compatible with and do not disrupt the attractor) survive. As the number and variety of organism/environmental interactions increases, the internal structure of the organism becomes increasingly more complicated in both form and function. The organism can even absorb new chemical reactions from the environment if they are beneficial to the organism's existence, compatible with the emergent LIFE complexity and do not disrupt the Biofield. However, such absorptions become less common the more complex the organism becomes.

After an appropriate period of evolution, the organism begins to differentiate its internal functions through the development of specialized functional organs. The internal structure of the organism eventually becomes so complicated with the emergence of functional organs that it must develop a new specialized organ solely for controlling and regulating the other developing organs. This specialized control organ is the Brain. The Brain has a strictly four-dimensional existence. However, the development of the Brain coincides with the emergence of an even more refined complex pattern of field density variations. This new pattern is imprinted on the LIFE pattern. Since it is a refined field imprint pattern on the LIFE pattern, the emergence of this new five-dimensional extension of the Brain could be described as a secondary complexity yielding a secondary attractor. This new secondary complexity is MIND.

In spite of the logic of the evolution of MIND and Brain in some species, they have not developed in all living organisms. So, one major group of living organisms (the animals) differs from another major group (the plants) by the fact that plants have not evolved Brains and MINDS. At some earlier point in the evolution of LIFE, a second evolutionary path was initiated that did not lead to the development of a Brain. This alternative evolutionary would have had some physical trigger. Perhaps it was triggered by a difference of sources of energy from the environment or by the internal development or absorption of some new beneficial chemical process. Whatever the case may be, evolution split into two major tracks based upon the requirements of the form and function that living organisms can take. An emphasis of evolution on function as opposed to form resulted in the development of complex internal organs and the Brain and the development of the animal kingdom However, in come cases the emphasis of evolution was on form instead of function, so a different route for the further evolution of LIFE was taken leading to plants and the vegetable kingdom.

In the mathematical system called chaos theory and the corresponding physical system called non-linear dynamics, such a split would be called a bifurcation, literally a splitting of the complexities. In the complexity of LIFE, this bifurcation occurred due to the alternate needs of LIFE on form and function. It is no coincidence that the outward forms and shapes of plants can be modeled using simple algorithms in chaos theory. If you apply the correct algorithm you can easily draw a fir tree or a fern, a maple leaf or an elm leaf. This coincidence can be explained by the bifurcation of LIFE at some point in time, with the emphasis on form over function being followed in the development of plants. Therefore, plants do not have Brains, but still conform to the SOFT model of LIFE.

In animals, the other half of the bifurcation, MIND corresponds to the whole body, but acts primarily through the Brain since they emerged together. MIND is purely five-dimensional even though it is in contact with the four-dimensional material body through the LIFE pattern. In our cloth analogy, LIFE is represented by the pattern of ripples and wrinkles in the cloth, while the weave of the cloth represents the MIND. The weave would not be uniform across the surface of the cloth. It would be irregular with stretched areas and gaps as the cloth conforms to changing pattern of wrinkles and ripples that correspond to LIFE (the Biofield).

The role of MIND is twofold. After emerging, MIND acts as the overall organizing principle for both LIFE and the material body just as its acts directly through the material four-dimensional Brain for the more mundane and common purposes of interpreting sensory data and making decisions. It is in the MIND's role of interpreting sensory data and making decisions that the MIND acts as a storage device for memories. The Brain continues to evolve under the organizational principles and guidance of the MIND as new contacts with the physical environment are formed and experienced. The Brain also continues to collect new sensations and data as electrical impulses that the MIND stores magnetically in the single field as memories (specific patterns) each and every time that the material body encounters and interacts with an ever-wider variety of physical phenomena in its local four-dimensional environment. The five normal senses work wholly within the four-dimensional space-time environment to input data through the body to the Brain, thus affecting the further development of the MIND, and consequently the further development of the Brain itself, LIFE and the body over which the MIND has organizational and functional control.

The growing collection of memories eventually allows the MIND, through its function as an interpreter of data and memories, to learn that physical objects outside of the **local** range of its five senses continue to exist even when they are not directly sensed. The MIND begins to cognize or become aware of the **non-local** nature of its external spatial environment.

In the next phase of mental evolution, the MIND begins to comprehend, or rather becomes aware of, the flow of time itself. The MIND develops an awareness of the past, present and future. In other words, MIND becomes aware of its own **local** position in time as opposed to the temporal **non-locality** of the past and future. The memories stored in MIND at this juncture, which led to an awareness of both spatial and temporal **non-locality**, precipitate the formation of a new and subtler five-dimensional single field density pattern within the five-dimensional MIND pattern. This new field density pattern emerges as an imprint over the MIND pattern. In other words, when MIND cognizes or realizes the non-locality of space and time, a third complexity emerges which we call CONSCIOUSNESS.

The emergence of CONSCIOUSNESS corresponds to a living organism's development of an awareness of its unique place within a greater universe, which has a history as well as a future. In this manner, the organism distinguishes between its 'self' and the rest of the universe at large. The 'self' is the greatest abstraction possible of the physical concept of local position in space-time as opposed to the non-locality of position in space-time, which constitutes the rest of the universe outside of the 'self'. The recognition of 'self' has long been the hallmark of consciousness for scientists who regard any living being that can recognize itself (say in a mirror) as having consciousness. CONSCIOUSNESS evolves or emerges within MIND, so it is a refinement of MIND, but CONSCIOUSNESS then acts to organize MIND from its superior position of external interactions in the five-dimensional single field continuum. The

MIND, LIFE and all that is the conscious being is then 'rewritten' in a sense to conform to the rest of the universe with respect to its local environment. In the cloth analogy, LIFE is represented by the complex structure of wrinkles and ripples in the cloth and MIND is represented by the weave and variations in the weave of the cloth, so CONSCIOUSNESS would be analogous to the strands and fibers of pure cotton that comprise the threads that constitute the weave.

Once it has emerged, CONSCIOUSNESS organizes the MIND and memories in the MIND according to its new awareness of space and time with respect to the rest of the universe. Brains become hardwired for CONSCIOUSNESS in the newborn. In other words, each step in the evolutionary development of a species leads to more complex single field patterns which are reflected in the genetic materials that are passed on to the next generation of that species. The primary function of CONSCIOUSNESS is the orientation or organization of memories in MIND, relative to the universe or the order of the universe. Yet, organizing memories and facts within the context of a knowledge of the universe is the basis of all logic, reason and science itself. CONSCIOUSNESS also organizes and controls the MIND and LIFE directly as well as the material body through the intervention of the Brain. CONSCIOUSNESS is a completely five-dimensional entity, whose connection to four-dimensional space-time is mediated by MIND, so it has the ability to directly sense its five-dimensional environment and the totality of the universe, without the mediation of the four-dimensional Brain. This new sense amounts to what we commonly call intuition and PSI, the basis of our sixth sense.

CONSCIOUSNESS is a physical entity, but it is not material since it is not four-dimensional. Since CONSCIOUSNESS is not a four-dimensional construct, it cannot be explained by the modern theories that constitute quantum physics. The scope of these theories only incorporates our four-dimensional world. In this manner, CONSCIOUSNESS is similar to the concept of entanglement, whose explanation also lies outside of the scope of quantum physics. Both CONSCIOUSNESS and entanglement are both invoked in quantum theory to account for different physical results and phenomena that lie beyond the ability of quantum theory to explain, but they must always be introduced from outside of quantum theory when they are called upon to solve riddles and paradoxes in the physical world. While MIND and CONSCIOUSNESS act on their four-dimensional environment through the Brain and thus body, they correspond to and envelop the whole organism, not just the Brain as is commonly believed. Since CONSCIOUSNESS is not localized within the Brain, paranormal phenomena such as ESP will never be explained by theories of the material Brain and its inner workings, and since MIND is not localized in the Brain either, scientists will never find a location in the Brain where all of our memories are stored.

In this model of the evolution of a sentient life form, such as a human being, there are four tiers or levels corresponding to stages of evolutionary development.

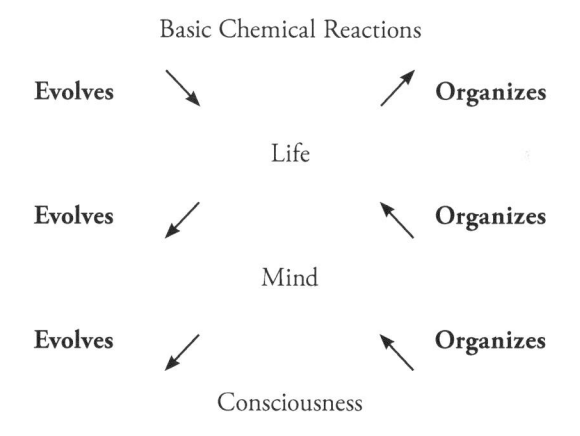

Basic Chemical Reactions

Evolves ↘   ↗ **Organizes**

Life

Evolves ↙   ↖ **Organizes**

Mind

Evolves ↙   ↖ **Organizes**

Consciousness

The emergence of CONSCIOUSNESS is different from the other levels since the transition from MIND to CONSCIOUSNESS is electromagnetic and thus only indirectly associated with the material being of the living organism. Also, once each complexity has evolved, it organizes the next lower level in such a way that the next and all later generations are chemically (i.e., genetically) 'hardwired' for that complexity in subsequent births. In other words, future generations are then born with a propensity for consciousness once consciousness or any higher level of complexity has evolved in a single being. Otherwise, each level reinforces and stabilizes its neighbors above and below in the hierarchy, or, in words that have more meaning in physics, each level is entangled in a very special way with the other levels adding extra structural stability and a very high level of coherence to the living being.

LIFE: The body inside out

All of our normal experiences in life are moderated by the physics of the 'sheet' because the phenomena and events upon which we gain those experiences occur entirely within the 'sheet'. Our everyday common material world exists within the four-dimensional 'sheet' as depicted below.

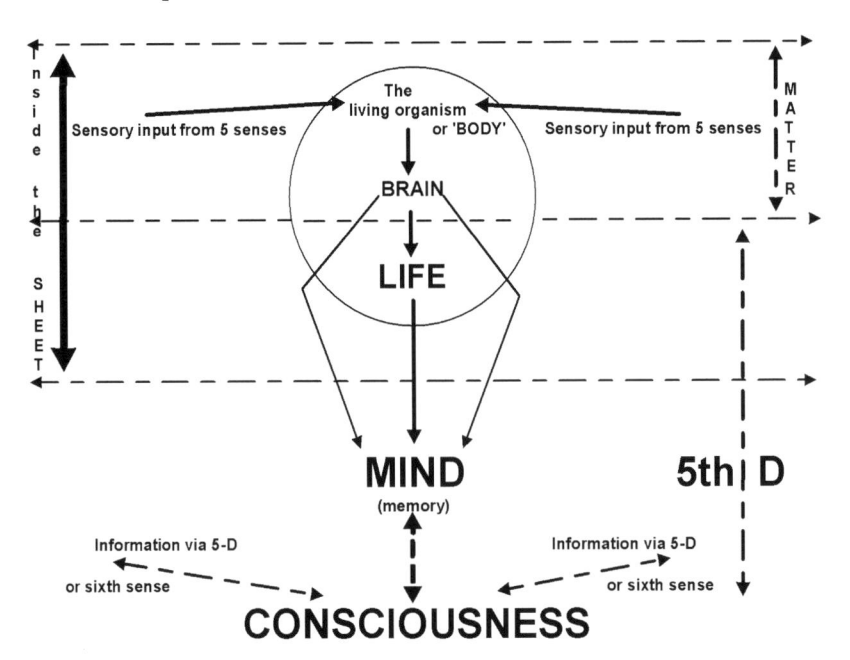

However, the four-dimensional 'sheet' is still a portion of or a cross-section slice of the five-dimensional single field. So, while everything occurring in our four-dimensional universe has a five-dimensional component, the opposite is not necessarily true. Things can occur in the five-dimensional single field that have no counterpart in the 'sheet' and thus do not directly affect the four-dimensional universe. So according to the circumstances of our physical reality, human beings are extended into the fifth dimension where they have an unsuspected physical but non-material existence, beyond their normal everyday four-dimensional world. While LIFE extends across both four and five dimensions, MIND and CONSCIOUSNESS are fully five-dimensional field structures. This explains why LIFE, MIND, CONSCIOUSNESS are beyond any possible perception by our normal five senses. However, this model implies the existence of a very special sixth sense that is completely five-dimensional.

In other words, PSI and paranormal phenomena can be easily explained within this model as an entanglement between different CONSCIOUSNESSes (ESP) or between a CONSCIOUSNESS and a material object (Psychokinesis). We experience PSI when we become 'consciously' aware of our greater connections or 'entanglements' with other material bodies via the single filed in the fifth dimension and utilize those connections to our benefit or the benefit of others. Our CONSCIOUSNESS is in constant contact with the single field at all times, but we are not normally aware of this contact, which is subliminal and clouded by our five normal senses, just as we are unaware of a clock ticking in a room until somebody mentions the ticking clock. Under normal circumstances, we can only perceive or detect the fifth dimension indirectly through intuition, thought and other normally intangible qualities commonly associated with the MIND, in other words, via our five-dimensional CONSCIOUSNESS since CONSCIOUSNESS is the most refined or detailed, and thus sensitive, of the field density patterns. Only the CONSCIOUSNESS has a subtle enough construction to 'feel' or 'sense' other variations in the overall single field that emanate from material bodies in the normally and physically inhabited fourth dimension.

The material body as a whole is an antenna and the Brain only works as a tuning/amplifying circuit for PSI, so PSI cannot be localized in the Brain. For that reason, all attempts to localize PSI in the Brain are predestined to fail. PSI phenomena are essentially five-dimensional sensations, acting through CONSCIOUSNESS and back to MIND, as opposed to our normal five senses which act through the material world to the body and the Brain, imprinting thoughts and memories into MIND. However, these paths can be reversed. Just as MIND can affect (locally) the body via the Brain, MIND can affect (non-locally) other objects in the single field via CONSCIOUSNESS under the proper conditions. So PSI can also be interpreted in the language of normal four-dimensional quantum theory as "consciousness acting non-locally", even though PSI is a five-dimensional effect and therefore quite beyond any greater or more detailed explanation by quantum physics, which is strictly limited to four-dimensional space-time by its very nature.

Our five normal senses evolved through the living organism's contact with its four-dimensional environment in the 'sheet'. But our total universe is five-dimensional, so the same evolutionary process by which LIFE developed, and continues to develop, 'requires' the evolution and emergence of a sixth sense through the living organism's contact with its five-dimensional environment. This sixth sense, which results from the direct interaction of CONSCIOUSNESS with the rest of the universe via the single field constitutes what we normally call PSI and explains paranormal phenomena.

Our sixth sense is an evolving property of living organisms and always acts according to the physics or nature of the five-dimensional environment. The sixth sense of human beings will continue to evolve or emerge into our conscious awareness as we gather more information about our world and our 'total' physical environment, which includes the five-dimensional elements of our total reality. In this last respect, the single field model conforms to the Buddhist and mystical concepts of enlightenment. Enlightenment occurs when a person develops a waking or conscious awareness of his or her five-dimensional connection with the rest of the universe, when that person's consciousness has evolved beyond the normal predilection for consciousness that all humans are born with. Although enlightenment of this kind can happen spontaneously, it can also be trained to some extent through practices such as Zen and Chan Buddhism. Practitioners are taught to meditate, a form of deep concentration that minimizes their immediate reliance on their five senses and their mental contact with the four-dimensional space-time environment. This enhances their ability to 'listen' to their sixth sense and become consciously aware of constant contact with the single field. They are taught to let go of the 'self' which is considered an illusion from the point of view of a higher consciousness and reality, which would be an apt description of a five-dimensional viewpoint of our four-dimensional material 'self'. But this model also opens up a possibility for explaining other phenomena such as NDEs (near death experiences) as well as the very concept of death and what happens when conscious beings die. Spontaneous enlightenment could be no more than the result of having a spontaneous NDLE, a near death-like experience.

So what we call LIFE, the 'something extra' that goes beyond the chemical reactions in an organism or body is a complex pattern of interlinked density variations in the single field, 'outside' of the 'sheet'. Many scientists and scholars also call this pattern the Biofield. Unlike other structures within the single field, such as the 'sheet' itself, magnetic and electric fields as well as material particles, no 'forces' of the Newtonian type are associated with the Biofield or LIFE. The Biofield is more of a control mechanism or an organizational principle that influences other common forces, such as magnetic and electrical, and thereby influences the internal workings of the organism. There is no 'life force'. There is instead a LIFE 'influence' over the mechanisms within the body. Nor is LIFE a form of 'energy' in the normal sense of the word. It cannot be a kinetic energy since nothing is moving, it is a pattern in space-time, and it is not a potential energy although it is associated with a single field potential. LIFE would therefore defy any attempt to explain it within the Newtonian physical paradigm.

The paradox of this model of LIFE is that LIFE is simultaneously 'inside' and 'not-inside' (or 'outside' of) the body. LIFE acts 'inside' the four-dimensional 'sheet', but LIFE 'exists' 'outside' of the 'sheet'. That is why scientists have never been able to locate or identify LIFE by all of their past reductions of the living organism. Material reduction, as normally practiced in science, occurs only within the 'sheet'. For the same reason, quantum theory will never be able to explain LIFE. LIFE is 'inside' the material body because each and every point within the volume or confines of the three-dimensional body is extended in the fifth direction, while each and every point in the body contributes to the pattern that is LIFE or the Biofield 'outside' of the four-dimensional material body in the fifth dimension extension of the body.

It is just like the three-dimensional being's finger passing through the sheet world of our hypothetical two-dimensional creature. The finger is analogous to LIFE in its three-dimensional existence both inside and outside of the two-dimensional sheet. But it is also the circular two-dimensional object that the two-dimensional creature perceives within his two-dimensional space as the finger passes through the sheet. The circular object is real and material to the two-dimensional creature, but the greater reality that the two dimensional creature knows nothing of and could never even comprehend, is the whole of the three-dimensional finger. The whole finger is the reality, not just the two-dimensional slice of the finger that is only the intersection of the finger with the two-dimensional sheet world that the two-dimensional creature perceives and interacts with. So LIFE is in every point 'inside' our four-dimensional bodies, but it is actually a five-dimensional extension and therefore 'outside' of our four-dimensional bodies, just as the finger is outside of the two-dimensional world that it intersects. LIFE controls the body, just as the whole fine controls the part of the finger that intersects the two-dimensional world of the sheet. In order to get to, find or discover the essence of LIFE in a scientific manner, we would have to go 'inside' our 'selves', through each and every point in our bodies, because LIFE is 'outside' of our four-dimensional material bodies. That is the paradox of LIFE that has puzzled and perplexed science and human thought for centuries.

The next complexity higher, MIND, is another matter altogether. MIND did not evolve from the mechanical actions of chemically active material particles on the single field, as did LIFE. Instead, MIND evolved from the coupling of the internal actions and functions of the specialized organs within the body. So MIND, which evolved in conjunction with the Brain, is a strictly internal pattern or refinement of the LIFE pattern. The MIND pattern evolved from LIFE, the five-dimensional extension of the body that is LIFE, not a four-dimensional quantity or event. It emerged from a five-dimensional field structure to begin with, so MIND is not directly connected to our four-dimensional material body, as is LIFE. While the MIND pattern exists in the five-dimensional extension of our material being, just as the four-dimensional sheet is part of the five-dimensional single field, MIND is not directly attached to the four-dimensional material world. MIND is a strictly five-dimensional entity or thing. This is an important feature of MIND and some of the characteristics of MIND within the context of the whole single field, such as its role as the storehouse for memories, are consequences of this feature of MIND.

CONSCIOUSNESS, an even more refined complexity within the single field structure of LIFE and MIND, is yet another type of thing altogether. CONSCIOUSNESS is not derived from the mechanical actions of elementary particles that have affected the density of the single field and thereby imprinted a pattern in the field, as is LIFE. Nor is CONSCIOUSNESS completely like MIND, although, like MIND, it is strictly a five-dimensional thing that suffers no direct connection to the four-dimensional sheet and body even as it passes through them. MIND evolved as an interaction between the functioning single field sub-patterns associated with body organs, but the CONSCIOUSNESS pattern was imprinted or impressed on MIND, the pattern of MIND, by electromagnetic means. CONSCIOUSNESS evolved as a complexity of magnetically stored memories in MIND and those memories correspond to the electrical transmissions of information within the Brain, and to a lesser extent within the whole body.

Our three-dimensional space with matter is electrically structured, or rather it is three-dimensional point by three-dimensional point electrically connected, while the fifth direction is magnetically constituted point-by-point. That is why a magnetic field is generated by any moving electrical charge in direct proportion to the speed of the moving electrical charge, a commonly known fact in normal electromagnetic theory. However, from the five-dimensional point of view, a charged particle's aspect in the fifth direction increases giving as it increases speed generating a proportional magnetic field around the particle. The changing aspect of the particle activates the magnetic point-to-point connection in the fifth direction that causes the magnetic field. The strength of the magnetic field associated with the moving electrically charged particle is proportional to the aspect change in the fifth direction. So, in essence, the memory portion of the MIND is a magnetic imprint, very finely structured, along the fifth direction and overlapping the MIND pattern. Yet it is electrically connected three-dimensionally to contiguous points across the rest of the single field, so the electromagnetic structure of CONSCIOUSNESS in five-dimensions is not exactly the same as normal electromagnetic structures in our four-dimensional world. The imprinted memories are magnetically induced in the single field by complex magnetic field variations within the Brain. The probable source of the imprint within the four-dimensional Brain is not that hard to find.

Memories are made of this

Specific criteria must be met when looking for the Brain mechanism by which memory is stored in the MIND and retrieved from the MIND. First of all, there must be enough basic elements in the

Brain for imprinting extremely complex memories. A simple digital camera can take and save pictures of more than ten megapixels. Such a picture would depict a common scene in our world using ten million little bits of memory. Yet the resolution of standard 35 mm film is over twenty megapixels. That much information is necessary for just one simple static high-resolution picture. Yet biological optical systems, based on the eye, are far more complex and would need far more memory for a simple picture. So whatever part of our Brain acts in the storage and retrieval of memories would need to have billions, upon billions and billions of individual memory elements just to be able to store individual static pictures of the world that we see. Moreover, these elements would need to have a dynamic aspect to their storage/retrieval capabilities to account for the constant streaming of data and information to and from the MIND. Our view of the world, in fact all of our sensations of the world around us, is not static, but an extremely complicated interplay of movement and change. The MIND is not a camera that stores multi-megapixel static photos of our world. It is a dynamic interactive device that stores phenomenally large amounts of constantly changing data.

Given the necessity of such a large number of individual memory elements, the next requirement must be that there is a magnetic or electromagnetic component in the structure of the individual material elements in the Brain. Since memory is magnetically stored in the single field, each of these elements must be a magnet or have the ability to generate individual magnetic fields. Using magnets to store bits of information is actually an old and well-proven technology. Every time you record your voice on a tape recorder, you are utilizing a magnetic storage medium. The wave pattern of your voice is codified on small magnetic bits within the plastic tape. When you play back the recording, the recorder/player just reads the patterns of magnetic bits, analyzes them and sends the signal to the amplifier and the speakers to produce your replicated voice. The same is true for video tape recorders, although the picture and sound patterns are far more complex and thus the stored information requires a much larger and more concentrated packing of the magnetic bits on the plastic tape. Video storage tape uses a denser array of magnetic bits. Hard drives in computers as well as computer floppy disks use similar technology.

On the other hand, DVD and CDs encode the patterns digitally and optically using laser beams, which produce coherent electromagnetic waves, to read micro-dots of information that have been encoded into the plastic surface in the discs. They use the same general principle, but the storage retrieval technology is based on the optical properties of electromagnetic waves rather than the magnetic properties of matter. However, the optical discs can store far more data in a smaller area by using light waves that measure in the hundreds of nanometers (a billionth of a meter). This human made storage device only begins to approach the storage capacity necessary in the Brain, so the magnetic storage elements in the Brain must also be extremely small, at least in the nanometer range if not smaller, to accommodate the huge amounts of data and information that the Brain processes.

So, whatever biological elements in the Brain are used for the storage and retrieval of memories they must be extremely small, measuring at least in the nanometer range, and they must occur in extremely large numbers as well as have magnetic properties. Now, believe it or not, there are components in the Brain that fulfill all of these requirements. And, as an added feature, these biological bits are already associated with the sensory system and sensory data transmission in the body and Brain. These bits are called microtubules. They form the cytoskeleton of the neurons, cells that are specialized in the electrical transmission of sensory data to the Brain and signals within the Brain as well as control commands from the Brain to the organs and muscles.

It is an accepted and well-known fact that neurons are associated with thought, memory and the transmission of information within the Brain. Information travels from the synapses, along the axon to the neuronal cell body, and vice versa, in the form of electrical potential changes. It is generally assumed in science that the information that becomes memory is somehow codified in the neurotransmitters that carry signals across the synapses between neurons during the transmission process.

The Neuron

(From Giancoli, 273)

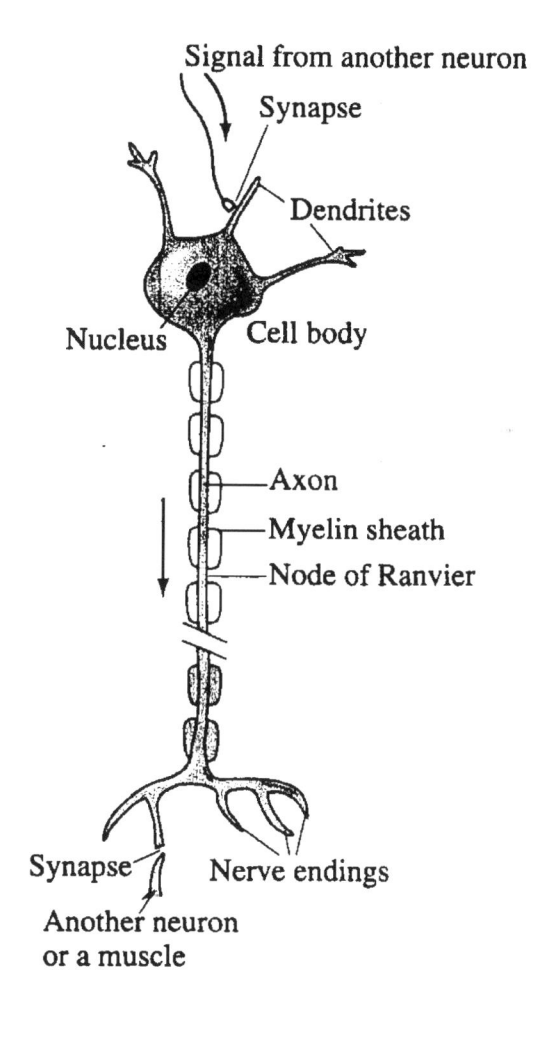

Therefore, it is assumed that memory and thought are linked to this particular chemical process.

The axon has its own particular structure that figures into the transmission of information as electrical potential differences along thee length of the axon. The outer wall of the axon is already known to act as a capacitor as electrical potential changes proceed up and down the length of the outer surface of the axon/neuron.

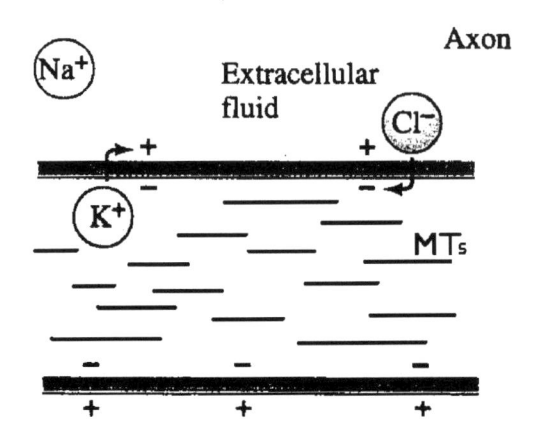

The Axon Wall as a Capacitor with a Microtubular cytoskeleton (Adapted from Giancoli, 274)

However, it is also known that the interior of the axon has a cytoskeletal structure constructed from microtubules (MTs), which are protein cylinders, connected by microtubulin-associated proteins (MAPs). The MTs also function as an internal transport system to move ions within the axon, thereby proving evidence of their electromagnetic characteristics.

The Microtubule

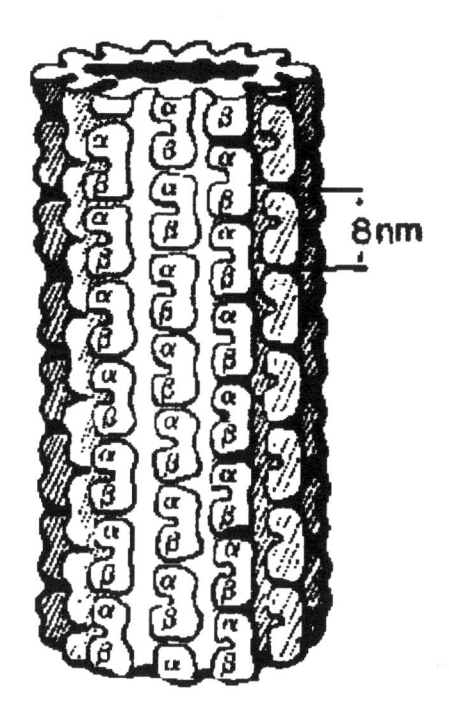

(From Amos and Klug, 1974)

In turn, the MTs are made up of individual tubulin proteins. Each tubulin protein is about eight nanometers in length. The proteins join together to form a sheet or surface that then curls around to form a cylindrical surface of the proteins, which is the microtubule. The tubulin proteins follow a helically shaped path around the MT cylinders. However, in normal consideration of information transmission the MTs do not seem the play any role. The real 'action' potential occurs on the surface of the axons.

The Brain itself is a massive structure of neurons, neuron bundles and glial cells. Within this mass there are different areas and sub-structures that control different functions, such as body mechanics and the functions of internal organs, while different areas of the Brain are associated with specific portions of memory, such as sound, sight, smell, taste, and touch. Different types of thought and emotional responses have also been localized in different areas of the Brian. But the basic mechanism of memory and the location of the memory have not yet been discovered.

Many scientists believe that the material Brain is not physically capable of storing the massive amount of memories that each individual has, at least not on a one-to-one ratio for elements or memory to bits of data as in computers, so they are looking for other mechanisms of storage. In other words, memories do not seem reducible to individual neurons or chemical reactions in the Brain like memories that are encoded in the magnetic bits on videotapes. Instead, memories have a certain collective or holistic character to them and seem to be holographic in nature. This fact has led to speculation that memories and thoughts are the product of a neural net, a vast complex of interconnected neurons that somehow store images and memories as holograms by utilizing larger portions of the Brain instead of microscopic storage elements. Such a mechanism, if real, is far from being confirmed let alone understood.

A major problem with all theories of mind and or consciousness is that coherence is needed among the different neurons to maintain a single thought, let alone a stream of thoughts in the Brain. In other words, how do so many different neurons 'fire' in specified sequences in great enough numbers or strength to maintain a thought within the Brain? A certain threshold of neurons 'firing' together is necessary for a thought to be brought into our waking conscious awareness. All physical models of mind and consciousness must address this problem. In fact, it could well be the most important problem faced by any theory. For example, quantum mechanical models of consciousness have been developed based on the MT systems in neuronal axons. In such theories, the 'firing' and the coherent sequence of 'firing', choosing between the " or $ electronic state in the individual tubulin proteins, is decided at the quantum level.

In ordinary quantum action, the 'firings' would be totally random. However, in a coherent state the 'firings' would come in a highly structured sequential order. This forms the first level of coherence. In the second level, all of the MTs in an axon would need to fire sequentially. And in the third level of coherence, all of the MTs in neighboring axons would need to 'fire' in the same order to develop enough intensity to reach the threshold and thereby establish an awareness of a single thought. This level of coherence is extremely difficult for quantum theory to explain. This would mean billions of tubulin proteins, in billions of MTs, in billions of axons need to 'fire' in a specified order, all explained by a quantum theory that ultimately claims that the 'firings' should be completely random and thus completely independent of one another. In fact, critics of the quantum models claim that the Brain cannot sustain the quantum coherent state necessary for the quantum models to work because it is 'warm and moist'. The quantum theory cannot supply the necessary coherence and thus cannot explain the thought processes occurring in the Brain, let alone completely model mind and consciousness.

If a classical electromagnetic model is used instead, then the coherence is easy to explain within the SOFT model of physical reality and LIFE. The SOFT model suggests a different physical mechanism for memory and thought. In the SOFT model, individual memories are extremely intricate five-dimensional single field density patterns within the MIND pattern. These memory patterns are imprinted within the overall MIND pattern by electric and magnetic interactions in the Brain, and to a lesser degree the whole body. So a particular process, which is extremely complex and thus involves an extremely large number of components, is necessary to form this imprint, and that process very likely occurs within the axons of the neurons.

The individual tubulin proteins in the MT can exist in either one of two electronic states (α or β).

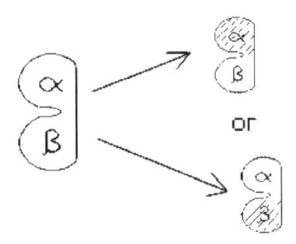

As the tubulin proteins in a MT go toward a single state, say they all 'fire' by attaining an " state in the proper sequence, the MT would become a small electromagnet or inductor. Although this function of the MTs is a recent discovery, the basic principles and concepts involved in electromagnets have been known for about two centuries.

The first electromagnets were designed and built in the 1820s and then used by Michael Faraday to develop the first theory of electromagnetism. Today, electromagnets can be found at the core of electrical transformers, coils, door buzzers, fire alarms, electric motors, automobile starters, electric generators and alternators, as well as a myriad of other devices. The principles by which they all work are quite simple and based on attraction and repulsion within a magnetic field. A magnetic field around a permanent bar magnet can be depicted by a series of imaginary field lines.

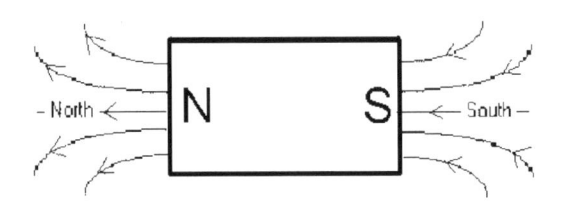

Inside the bar magnet the field lines are parallel to each other and very dense. The density represents the field strength, so the field lines should be very dense inside the magnet where the field is the strongest. The parallel field lines indicate uniform magnetic field where the magnetic potential is equal at all points and in then same direction. If a current flows through a coiled wire, wrapped helically, the space inside the coil will mimic the uniform field inside a bar magnet.

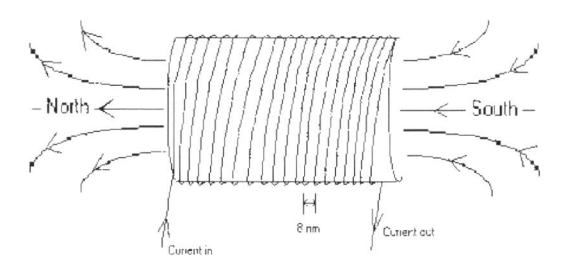

The magnetic field in the interior of a coil that is properly wound will be uniform just like the field inside of a normal bar magnet. Children all over the world have built and played with similar electromagnets for generations. Magnetism seems to fascinate children and adults alike. All you need to do is wrap an insulated wire around a roll of paper or cardboard, or better yet around a nail or screw, and attach the two ends of the wire to a small battery. Then you can play and experiment with your own electromagnet.

Now picture the electrical current in the wire, like water flowing through a pipe. But that analogy is technically wrong. Actually, the electrons flowing through the wire do not go from one end of the wire in the same manner as water goes through a pipe. The electrons actually bump their way through the wire, each traveling only a short distance before colliding with or interacting with the atoms of the conducting material, knocking a different electron loose to continue the journey and supply the current. The electrical current ion a wire is thus due to the collective motion of many electrons moving for short distances in the same general direction. Now if we look at the contiguous proteins in a MT cylinder 'firing' in sequence or just entering into a single electronic state, say an " state, in the same helical sequence that a wire wrapped around in a coil follows, then a microscopic bio-magnetic field is established inside the MT in the same manner as the individual electrons moving in a wire around a coil create a magnetic field inside the coil.

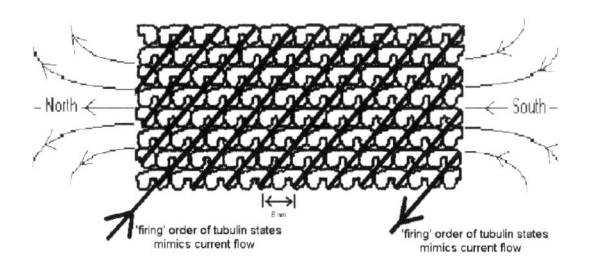

The magnetic field is induced inside the coil by the collective magnetic fields of all the moving electrons just as a magnetic field is induced in the MT by the collective action of all the changing electronic states of the individual tubulin proteins.

The existence of small electromagnetic inductors inside the axon, when linked to the surface of the axon that acts as a capacitor, implies that the axon itself can act as the microscopic equivalent of and LRC (inductor-resistor-capacitor) electronic circuit. This suggests that the axon is a bio-transmitter/receiver of electromagnetic waves commonly called a radio tuner. In a simple LRC electronic circuit, electrical potential energies are swapped between the inductor and capacitor so they draw no energy from the circuit. They charge and discharge in opposition to each other. So as the capacitor charges, the energy to charge it comes form a discharging inductor, and vice versa. Together they charge and discharge as a specific frequency called the circuit's resonant frequency. An LRC circuit, which is set to its resonant frequency, will either emit or absorb electromagnetic waves at this frequency, so the LRC circuit is the basic tuning circuit for all electronic devices, such as radios, TVs and radar units.

This means that the axons are small radio transmitter/receivers.

As an electrical potential, for example an incoming sensation of visual perception, travels along a single neuron, a group of neurons or a neuron bundle in the brain, the movement of the potential along the axon portion of the neuron causes the sequential firing of the tubulin proteins in the MT along their helical pathways around the MT cylinder. The movement of the potential along the outer wall of the axon actually induces the tubulin proteins in the MT to 'fire' in the sequential order in the same direction as the potential difference is moving.

The Neuron as an LRC Circuit

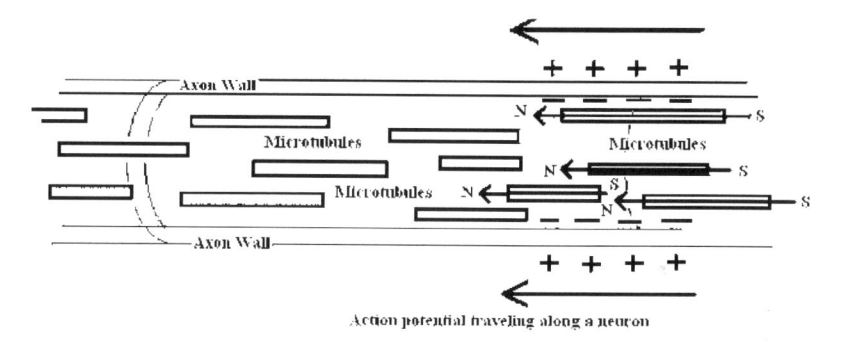

Action potential traveling along a neuron

The MTs act just like simple wire coil conductors transmitting a current along the helical surface when the tubulin proteins in the MT 'fire' in a sequential manner. And, just as in an electrical inductor coil, a uniform magnetic field is established in the core of the MT. The sequence of a vast number of MTs being magnetically charged and discharged as potential differences race up and down the outer surface of a single axon, creates an extremely complex magnetic signature or pattern which represents a single component of a sensation or thought. But the axon wall/capacitor acting in conjunction with the individual MT/inductor coils also produces and transmits an electromagnetic signal within the Brain that couples with other axon/MT systems to build a far more complex magnetic signature or pattern that is no less than a coherent thought in the Brain.

The extremely large number of MTs firing in a specific sequence within the axon, corresponding to a specific sensation that began the potential difference racing along the axon wall, couples or electromagnetically entangles with other MTs in other axons/neurons to form a complex thought. The mechanism of entanglement or coherence within the Brain is thus identified as the axon wall/MT complex within the neurons. Together, the axon walls and the MTs form an LRC (inductor-resistor-capacitor) circuit, which is the basic tuning circuit for all electronic devices. So, each and every axon is thus a small electromagnetic bio-transmitter and/or receiver. Electrical potentials moving along the axon wall trigger the MTs inside the axon to emit electromagnetic waves. Different neurons couple together electromagnetically to provide coherence once a signal is transmitted, forming a total image or sound or thought in the Brain. So this classical electromagnetic model of the neuron supplies the coherence in the Brain that the quantum models of mind and consciousness are missing. But accounting for coherence in the Brain is not enough; it only explains the process and awareness of a 'thought' and says nothing about 'memory', which is essential to consciousness. This is not a problem however since the magnetic properties of the MTs can also account for both the storage and recall of memories within the SOFT model.

The five-dimensional single field where the memories are stored in MIND is magnetically induced. Both electricity and magnetism depend on the ability of a substance to allow or permit the fields to pass through it. This ability is represented by constants or fixed quantities that are called the permittivity (for electricity) and permeability (for magnetism). Even empty space is characterized by its permittivity and permeability, which together control how electromagnetic fields and waves move through empty space or a vacuum.

In the SOFT model of the single field, we can further characterize electrical permittivity as a point-to-point connectivity constant in the normal three directions of space, while the magnetic permeability can be considered a point-to-point connectivity constant in the direction of the fifth dimension. We could alternatively say that the fifth direction constitutes a magnetic space component while our normal three directions constitute the electric space components.

Working together, the link between the electric and magnetic components in five-dimensional space-time thus constitutes the electromagnetic field and variations in the electromagnetic field through time, the fourth dimension. Since the fifth direction of the space-time continuum is magnetically induced, it is closely associated with magnetic fields established in our four-dimensional material reality by moving electrical charges. So the complex pattern of individual MT/inductors in an axon and the far more complex pattern of the larger group of MT/inductors in different neurons that are electromagnetically entangled to form a single thought or, over time, a stream of conscious thought, also imprints a unique and extremely complex magnetic field density pattern within the five-dimensional single field. This magnetic pattern is what we call memory. The magnetic fields of the MTs are the encoding device for storing and retrieving our memories. When all of the individual MTs charge and discharge, they leave a magnetic signature or trace pattern within the overall MIND pattern in the fifth dimension. Each and every thought that we have and sensation that we feel is thus imprinted as a special magnetic pattern in the five-dimensional single field.

This process is analogous to simple magnetic storage techniques for memory in computer hard drives, computer core memory, and audio and videotapes. So memory exists outside of the Brain, in a sense, and is holographically stored as magnetically induced complex density patterns in the single field extension of the four-dimensional organism. When we remember something, we merely search out that stored pattern in our MIND and replay it back to our Brain.

There are many advantages to this model of memory. For example, this model can be used to explain the simple concept of recognition. What is called recognition in the Brain is no more than 'pattern matching' from the MIND as transmitted to the Brain. When a person sees an object for the first time, the electromagnetic pattern that the act of seeing generates in the brain is stored or imprinted in the person's MIND.

Microtubules in an Axon

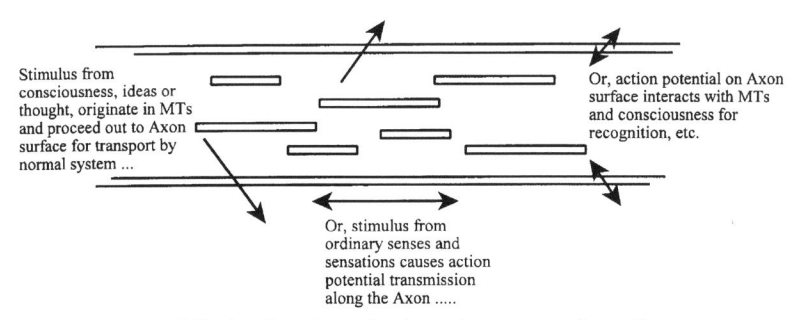

Stimulus from consciousness, ideas or thought, originate in MTs and proceed out to Axon surface for transport by normal system ...

Or, action potential on Axon surface interacts with MTs and consciousness for recognition, etc.

Or, stimulus from ordinary senses and sensations causes action potential transmission along the Axon .....

While the signal travels along the outer surface of the Axon and thus to other Neurons, the MTs are still interlinked with MTs in other Neurons via the Transmission/Reception accorded by the LRC circuit thus forming a cohesive network of thought, or conscious thought

When the person sees the object again at a later time, the electrical input signal from the previously seen object enters the MIND, only at a different position due to the difference in time. But the new pattern is 'pattern matched' to the stored pattern in MIND and the object is recognized in the brain where it generates a new electromagnetic pattern that is again imprinted in the MIND. You could say that the two complex patterns resonate together, but that analogy is only partly correct and can be misleading. A resonance occurs over time, or it has an element of duration, but the pattern is already imprinted in time and thus the duration of the original sensation has already been taken into account in the pattern itself, so a true resonance is technically not possible. That is why the phrase 'pattern matching' more accurately describes the process than the word resonance could, even though people talk about 'resonances' when referring to psychic examples of recognition and communication. They are just sensing the patterns ads they develop over time rather than sensing the complete patterns in space-time. During the 'pattern matching' process, the MIND plays back the stored magnetic memory pattern by reactivating or remagnetizing the same sequential pattern of MTs in the same or equivalent axons that originally imprinted the memory in MIND during the original thought or sensation. Since the patterns match, recognition of the scene or object is brought into a person's conscious thought or awareness.

Some circumstantial evidence for this model of recognition already exists. Suppose a person sees a telephone from some angle and the memory of the telephone is stored in the person's MIND. When that person again sees the telephone at some later time from a different angle, the person still recognizes the telephone even though it was not originally seen from that angle. For years, scientists have been trying to develop programs for computers to recognize three-dimensional objects, but they have failed because of this single feature of the Brain. The computer programs cannot extrapolate a three-dimensional 'image' or memory of the object from just a two-dimensional viewing, although the human Brain/MIND can. For humans, the image or memory of the telephone is stored in the five-dimensional MIND and just as knots in a rope in three-dimensional space are straight lines in four-dimensional space, the telephone is rotated in the five-dimensional memory pattern, remembered as seen from all angles, so that we can recognize the telephone from different angles even though the original sighting of the telephone was only from a single angle. A mental extrapolation of this kind requires a higher-dimension than the three dimensions of common space, which implies that memory occurs in a higher-dimensional space than the object for which the memory was created.

Further circumstantial evidence of this model comes from the paranormal phenomena called 'remote viewing'. In remote viewing, the observer is expected to 'view' an unknown object that is not present, but exists at some distant location. The observer has no previous knowledge of what the object is and probably only knows where it is, or the observer may know what the object is but not where it is located. In either case, the observer or remote viewer must 'see' or 'view' something unknown at a great distance. The remote viewer is placed in a simple sensory deprivation environment so that he or she can tune into the signals picked up five-dimensionally by their CONSCIOUSNESS. When the remote viewer 'senses' the object or its location, he or she must somehow 'recognize' the object or location without having any previous memory of them.

So the remote viewer begins with the very simplest memory patterns in MIND, such as squares, triangle, circles, lines, and other simple geometric figures that form strong memories in the MIND of the remote viewer and slowly builds up an image of the object or place to be observed. The remote viewer is merely 'pattern matching' with simple patterns that are common to all of the physical objects that we normally see since he or she has no previous memory of the complex pattern that they are 'viewing'. These simple patterns are then put together and thus progress to a more complicated image that can be more easily recognized. The remote viewer slowly builds up an image of the target in his or her MIND since there is no precedent of 'seeing' the target and recognizing it from past experience as recorded in the MIND. So, the remote viewer need only 'imagine' how the target would look, as if he or she had seen it before. In essence, this explains one method of how imagination works in the SOFT model of MIND and CONSCIOUSNESS. Remote viewing clearly fits the five-dimensional SOFT model of physical reality

The electromagnetic storage of memories in the MIND, as a more subtle density variation pattern in the single field, is different from the method by which the LIFE and MIND patterns originally evolved. The LIFE and MIND patterns were formed by the mechanical actions of five-dimensional aspect changes resulting from the energy exchanges between material particles during chemical reactions. Since the methods of field density pattern development are different in these cases, the CONSCIOUSNESS is qualitatively as well as quantitatively different from the MIND and LIFE complexities. This difference gives CONSCIOUSNESS its unique characteristics with respect to the single field as a whole. In other words, only the magnetically induced single field density variation pattern that is CONSCIOUSNESS is 'subtle' or precise enough to 'sense' or otherwise detect the extremely 'subtle' variations in the whole of the single field caused by other physical events in the four-dimensional environment, such as other people's thoughts or physical events.

These 'subtle' variations in the single field constitute what we call PSI and their detection and manipulation by CONSCIOUSNESS are known as PSI phenomena, the source of the paranormal. The SOFT explanation of remote viewing is only one case of how the paranormal works with regard to our physical reality. The same would be true of other paranormal phenomena, such as ghosts, apparitions and communication with the dead. But the sensing of ghosts, apparitions and communication with the dead directly implies that we survive death, or at least some part of us survives the death of the material body, which is altogether a different question that needs to be addressed now that we know what LIFE is.

CHAPTER 6
THE NATURE OF DEATH

What is matter?—Never mind.
What is mind?—No matter.

Thomas Hewitt Key—1855

IRIMI—Entering Death

In spite of several decades and even centuries of anecdotal evidence and personal stories that attest to the survival of some part of us when we die, including the observations of many scientists and other respected observers, science will not, and in fact cannot, ever accept phenomena associated with survival after death as real because science has no theoretical basis for understanding or explaining them. Quite simply, science cannot explain these reported observations or even begin to understand the survival of consciousness, as a few have now come to call the phenomenon, within the present worldviews and paradigms by which it analyzes and explains other physical phenomena. On the other hand, science cannot deny their validity because our present theories and paradigms do not specifically say that survival of consciousness is impossible. Some scientists even believe that if something, some unknown type of phenomenon, is not denied by our natural laws and theories, then it should be possible. So a few scientists have slipped between the cracks in the great edifice of science and have actually attempted to verify the survival of consciousness by investigating the reported observational phenomena as best they can under the prevailing circumstances. To some their defiance of scientific orthodoxy is a brave and noble act, a fight to find and support truth, but to others their attempts are wasteful tomfoolery if not outright heresy.

Some scientists just frown on this research as 'unscientific' and turn their backs, but their opinions are nothing more than personal opinions that are not based on sound scientific evidence and fact. Nothing in science rules out the possibility of the survival of consciousness, in fact nothing in science even addresses the issue, and we all know that science is not yet complete in its total explanation of nature so that different types of phenomena that science might not like to consider real at present could be considered real at some future date. So there is no logical reason to deny the possibility. If the possibility of survival is denied, then it must be for other than logical reasons. Perhaps the complete rejection of the possibility of the survival of consciousness is based instead on personal animosity, prejudice or bias. Or it could be denied because it directly challenges prevalent religious views, which science tries not to offend. Whatever the reason, denial will not benefit science in the long run. Science as a whole will never accept the paranormal or the survival of consciousness, which implies the possibility of ghosts, apparitions and communication, until there is a theory to explain them and relate them to the rest of the scientific worldview and the paradigms that constitute that worldview. The only other possibility would be that science would be forced to accept the concept through some profound and spectacular event or phenomena that could not be denied, and it is doubtful that Einstein could rise from the grave, whole again and alive, and castigate science for its own shortsightedness.

The question is not 'Is there life after death?' Of course there is no life after death. Death is the end of life, plain and simple, so the question is silly. If people would stop asking that particular question, in exactly those same words, perhaps they would be taken more seriously by scientists who must carefully define everything as a first step toward understanding everything. However, that does not mean that something other than life, something extra that was associated with life while a person lived could not have survived. In religion the soul survives while some people talk about the survival of a spirit, which is a more of a religiously neutral term. And then we have apparitions and ghosts. But such terms are scientifically inaccurate and quite misleading. These terms are often accompanied by unscientific connotations if not outright bias and prejudice, all excess baggage when it comes to talking about a science of death or a theory to explain what happens and what survives when we die. Again, that is not to say that nothing survives. It only means that science needs to determine and define, within its own terms and parameters, what could possibly survive. And it seems at present that the most popular scientific idea on the matter is the possibility that consciousness might survive the death of the body and the ending of life.

After the late 1960s, Eastern philosophies and religions made large inroads into western culture. Religions such as Buddhism and Taoism became popular, as did the practice of Yoga, meditation and other aspects of Hinduism. The concepts upon which these practices are based slowly began to influence all of western thought. The surge in western interest in consciousness and consciousness studies in the 1970s, whether popular or scientific, was directly related to the new interest in Eastern philosophies and religions. Simultaneously, Eastern forms of the martial arts became a growth industry within the USA and Europe.

One of those martial arts, hailing from Japan, is called Aikido. Aikido is the fairly recent development of the Japanese Master Ueshiba. He developed Aikido by borrowing the essences of other forms of Japanese martial arts. So Aikido is rather unique among the martial arts. Aikido is completely defensive, although all of the movements are based on aggressive methods, and Aikido uses the forces and energies of the attacker to foil the attacks. In simpler terms, Aikido turns linear attacks into circular/rotational motions to confuse, defuse, disrupt and control the attack. Done properly, Aikido utilizes no energy or force of its own, it just redirects the energy and force of the attack, throwing the attacker's 'ki' or 'chi' harmlessly away. This philosophy has earned Aikido a reputation as the most esoteric of the martial arts.

One of the basic movements in Aikido is called IRIMI (ear-ree-mee). Literally translated, IRIMI means 'entering death'. Instead of moving away from the attack or blocking the attack, the defender moves directly into the attack and thus 'enters death' to defeat 'death'. When an Aikido student makes an IRIMI move, he or she moves into the attack and then blends with the 'ki' of the attacker to redirect the flow of the attacker's 'ki' and the energy of the attack. The metaphor of 'entering death' or IRIMI is significant since science must also enter death, at least through investigations and intuition, to completely understand death and our new state of being after death. No one is proposing that someone need die to study death, but there are people who have died and been revived in the normal course of human events. These people have eerily similar stories of what happened to them in the brief moments while they were dead. You would think that scientists and investigators would listen carefully to those stories. But the investigators find excuses for the 'visions' and try to say that those experiences are not related to a continued existence after death, when in fact science should be shouting the merits of this evidence from the rooftops. Science must accept the possibility of survival and confront these reported events with an open and unprejudiced collective mind if it truly wishes to understand physical reality and our relationship to physical reality completely. Science must 'enter death' by accepting the validity of the evidence at hand and then redirect the energy of this evidence into a scientific theory of death.

Another relevant tale has also emerged from the same Zen Buddhist philosophies: The story of the Japanese Tea Master. The idea of a Zen Master is a person who has reached the highest possible level of his art and even enlightenment through the mastery of his particular art form, in this case the Japanese Tea Ceremony. In a very broad sense, a Master in one art form need not study any other art form because he has already reached a state of enlightenment and thus knows all there is to know. Better still; he knows what is necessary to know of life and that is all that he needs to know. So there is no real purpose in studying other art forms. The purpose of a Master is then to live out that enlightenment in everything that he does, in all actions that he takes for the rest of his life, and thereby be 'one' with all things. This way of conducting oneself throughout life is the essence of enlightenment.

Now this particular Tea Master had a terrible run in with a famous Samurai warrior and through a terrible misunderstanding quite mistakenly insulted the warrior. The warrior, to uphold his own honor, challenged the Tea Master to a sword fight to the death several days hence. The Tea Master was appalled. He would surely die, but he had no desire to die. He could not run away from the dual either. Running away would be a slight to both the warrior's honor and his own honor. So the Tea Master went to his friend the Sword Master for help and advice. There was just no helping him. The Tea Master would have to fight the Samurai Master and would surely die. Besides, the Tea Master was already enlightened so there was nothing that the Sword Master could teach him so that he could save himself. So the Sword Master did the only thing possible to give the Tea Master some peace of mind. The Sword Master taught the Tea Master one simple sword movement, but not really to defend himself since there was no direct defense from an attack by such an accomplished Samurai warrior without years and years of special training. This special technique would allow the Tea Master to kill the Samurai warrior as the Samurai warrior killed the Tea Master.

The day of the dual finally came and each opponent took his position on the field of honor. They eyed each other carefully and set themselves ready for the battle. The Tea Master took the position that the Sword Master had taught him and the Samurai warrior stopped dead in his tracks, recognizing the stance that the Tea

Master took. The warrior thought for a moment and then realized that the Tea Master was truly enlightened and would surely kill the Samurai warrior as he himself died by the Samurai warrior's sword. But the Samurai warrior was not willing to die at the hands of the Tea Master, even though the Tea Master was willing to enter death, an act for which he had no fear since he was enlightened. The Samurai warrior, master of the sword in battle that he was, was not an enlightened Sword Master. To be a Sword Master was to learn the use of the sword simply for the art of movement and the sake of the sword, not for the sake of winning battles and glory. So the Samurai warrior retired from the battlefield and the Tea Master won the battle by default, saving his life and his honor.

This story is also a metaphor for the course that science must take to understand death and dying. Science must face the issue of dying directly to understand the concepts and issues involved in the process. Science cannot run from or dodge its responsibilities, just as the Tea Master was obligated to face his own death without retreating or running, or even making excuses. In order to understand death, life must be put on the line. This suggestion is not an invitation for anyone to actually die for the sake of science. Science needs to become 'enlightened' after its own manner and be willing to accept death, or at least to face the concept and the real possibility of survival, to even begin to reach an understanding of death. Life must be defined and understood as a scientific concept before death can be confronted as a scientific issue. So scientists must stop making excuses and start working on a theory of life. Scientists should also take a serious look at the circumstantial evidence for survival in the hope of finding new clues to the nature of life. Science can no longer turn the other cheek, and allow religion, personal prejudice, superstition, mythology and outright greed and avarice, to control what society and culture think about death. Scientists need to develop working hypotheses at the very least, if not theories of life. Since no other theories or scientific descriptions of life have come forward, the SOFT model should be taken seriously and put to the test. Only the SOFT model of LIFE is up to the task of explaining death, while spelling out exactly what survives death and how it does so. Other theories would be welcomed, but none are forthcoming.

We enter death SOFTly

For all practical purposes, no theory of LIFE and the Biofield could be considered complete without at least a consideration of the question of what happens when the LIFE of an organism ceases. More to the point, does any part of the living being survive when the material body dies? We know and understand that the basic chemical reactions that constitute life cease all action and the LIFE complexity or Biofield is irreparably disrupted when a living being dies. I the terms of SOFT, the field density pattern of LIFE in the fifth dimension loses coherence, which means that the Biofield structure collapses and LIFE ends in every meaning of the word. So there is no LIFE after death. Metaphorically speaking, LIFE has been scattered like ashes to the wind when the body dies. However, MIND and CONSCIOUSNESS are separate from LIFE. They are unique to themselves. Even though MIND evolved from LIFE, it is not LIFE itself and different from LIFE, but in a sense also a part of LIFE that all LIFE has a propensity to develop. So there is no logical reason to assume that MIND would also be disrupted and, in essence, dissipate away to nothing upon death of the material body.

In reality, both MIND and CONSCIOUSNESS are protected from the disruptive influence of the dying material body by the LIFE pattern. They are shielded by LIFE from the disruptive influence of the destruction of cooperative chemical reactions in the body, so they survive as a unit, linked together by the very nature and physics of their mutual evolution. MIND cannot just dissipate away because it is organized by CONSCIOUSNESS, while CONSCIOUSNESS lends to MIND the coherence that it originally received through LIFE and the chemical reactions in the material body. CONSCIOUSNESS automatically takes over and supplies coherence to the MIND when the material chemical coherence, otherwise supplied by the body before death, ceases. Quite literally, MIND and CONSCIOUSNESS reinforce each other and thereby maintain an internal pattern of stability within the five-dimensional single field. However, this new existence of the 'person' is substantially altered from his or her previous experience in the four-dimensional world. MIND and CONSCIOUSNESS were only connected to the four-dimensional material world by LIFE, so they need not remain connected to the four-dimensional material world after death. MIND is freed from the shackles of its previous four-dimensional material prison. How MIND now interprets its new physical conditions of existence determines the future course of that existence.

The more evolved the CONSCIOUSNESS of the individual before death, the greater the complexity of the CONSCIOUSNESS pattern and the more stable it would be upon death. But the greater the complexity, the more CONSCIOUSNESS is able to 'tune-in' to the rest of the universe through the single field and density variations in the single field. The greater the complexity of CONSCIOUSNESS is, the greater the sensitivity of CONSCIOUSNESS to detecting the whims and fancy of the rest of the universe. This level of complexity first developed during the life of the living person, so how a person lived his or life, the quality of his or her life, determines the state of being of what survives at death, a situation that needs for further clarification. When a person dies the chemical reactions that sustain LIFE are irreparably disrupted and normal energy interactions within the body cease over a period of time. Chemical equilibrium in organs as well as individual cells is disrupted. The loss of this underlying chemical basis of LIFE initiates a ripple effect of decoherence that propagates throughout the LIFE pattern in the fifth dimension. With no organizational direction or control from the LIFE pattern and MIND, individual organs no longer coordinate their activities for the survival and health of the body. Those organs and processes that dependent most heavily on the interaction process stop functioning immediately, while those semi-autonomous organs and processes that previously worked independent of the other organs and processes continue a while longer. But smaller and less complicated sub-structures continue to function for some time, as exemplified by hair growth and fingernail growth, both of which continue after the body dies.

In other words, LIFE ends when the primary complexity pattern of five-dimensional field density variations corresponding to chemical energy exchanges in four-dimensional space-time is disrupted. But LIFE itself had been quite stable and durable under normal conditions and could withstand small disruptions, so the patterns of field density variations that are LIFE could not have been so easily derailed. Minor chemical disruptions due to disease and some forms of physical impairment within the body are not enough to disrupt the primary pattern of LIFE in the fifth dimension. All diseases disrupt the chemical balances within the body to some degree, but only the most serious lead to death and among those only a few cause a rapid and irrevocable march toward death or disruption of the primary LIFE pattern.

In fact, it should be theoretically possible for MIND and CONSCIOUSNESS to help cure diseases from their superior vantage point since they are organizational principles and diseases are disruptions to the organizational pattern of LIFE. Recent research has indeed shown that the MIND and CONSCIOUSNESS could have some power to alter the course of diseases, but no definitive or conclusive evidence of their direct effect on diseases has yet surfaced. The diseases and physical conditions that lead to immediate death are those that harm the internal organs or the brain directly, because the brain and organs such as the heart are absolutely necessary for the continuity of LIFE. The Brain and heart have are represented by major sub-patterns in the LIFE pattern that would disrupt the whole LIFE pattern if they were to be disrupted themselves. Such major organs are either chemical processing plants absolutely necessary to LIFE or special regulatory agents that provide a vital function for the body as a whole and thereby contribute significantly to the LIFE pattern.

No matter how the body dies, the cessation of brain activity and function represents the most significant symptom of death because the Brain is the single most complex organ in the body and therefore has the strongest influence on the LIFE pattern itself, independent of its relationship to the MIND pattern. When a person is Brain dead, the person is dead just as those in the medical field have concluded. So the medical community and legal systems have correctly accepted this relationship between Brain and Life as a measure of death, although for more practical rather than theoretical reasons. Although this conclusion is correct, it is however incomplete since the medical community does not completely understand the special role that the Brain plays in LIFE. So the medical and legal systems now find themselves having to modify this 'marker' or 'measure' of death to take into account those people who are 'miraculously' brought back to LIFE after all Brain function has ceased. MIND survives without the Brain and is independent of Brain while MIND also organizes LIFE independent of material body and the chemical reactions of material life, so people can be revived in some cases and brought back to LIFE. If the organizational ties of MIND to LIFE have not been completely severed by the loss of Brain function, then LIFE can be restored to the body in spite of Brain death. The manner of dying is important in this case. Bringing a person back to LIFE after Brain death is not a miracle, just good science demonstrating a more complete knowledge of LIFE and science of death.

The cessation of Brain activity not only disrupts LIFE directly, but also cuts off the most direct and primary connection between the body and MIND, as would be the case in the event of a major stroke. However, when the body dies by freezing or drowning in a freezing ice-covered lake, the chemical reactions sustaining and feeding into the LIFE pattern are only slowed evenly and not completely disrupted. So the contact between LIFE and MIND are not severed even though Brain activity slows to a virtual standstill. In such a case, a person can be revived without Brain damage, even if Brain activity has ceased for a short time, since the MIND is still interwoven with the LIFE pattern. When the body actually dies, normal sensory input to the MIND ceases because the electro-chemical interactions in the Brain stop processing incoming sensory data from the immediate physical environment of the body. If this connection can be reestablished, then a person can be revived by normal physical means. Medical science is now supporting this fact by its admission that a person is not dead until they are both 'warm and dead', because the extreme cold conditions when some people die slows the chemical reactions to the point where they do not completely disrupt the LIFE pattern and the MIND does not break away completely from LIFE and body as is the case with true death.

In any case, when a person dies only the four-dimensional body actually dies, or rather disrupts the pattern of LIFE in the fifth dimension due to the discontinuance of chemical processes at the four-dimensional level. MIND and CONSCIOUSNESS, both of which are completely five-dimensional entities, are shielded or insulated from the body by LIFE itself, which is also five-dimensional and directly connected to the four-dimensional body. When LIFE is disrupted to the extent that death occurs, MIND and CONSCIOUSNESS undergo a process of detachment from the chemical interactions from which LIFE originally evolved. Beyond this point, and only beyond this point, a person cannot be brought back to LIFE by normal material or medical means.

Some of the most basic five-dimensional sub-structures or internal patterns that contributed to the LIFE pattern before the body died also remain intact, but disconnected from the other patterns as the MIND/CONSCIOUSNESS complex is freed to exist independently within the fifth dimension. The MIND/CONSCIOUSNESS complex retains its identity (the person still remains) after a manner in the fifth dimension, in so far as self-identity is not a material but still a physical quantity or quality. However, the extent to which the complex is 'conscious' or mindful of its own existence, its being, would depend upon the extent to which it was 'conscious' or aware of its five-dimensional connections before the death of the four-dimensional body and what is perceived by the MIND as 'self' while the body still lived and functioned.

The MIND/CONSCIOUSNESS structure or complex (a neutral title) survives, rather than a soul, spirit or ghost. These terms carry too many preconceptions and false impressions to be of any objective scientific value. Upon bodily death and the subsequent disruption of the LIFE pattern and Biofield, the MIND and CONSCIOUSNESS survive together as a mutually cohering complexity. Perhaps this concept can be better understood in terms of cloth analogy that was presented earlier.

The vast pattern of ripples and wrinkles that represents LIFE in the tablecloth, which represents the four-dimensional space-time continuum 'sheet' itself, becomes irreversibly disorganized when the body dies. The pattern of undulations and waves begins to break up so that no recognizable sub-patterns can be found, or in extreme cases the cloth just goes flat altogether, showing only the curves due to individual non-reacting atoms and molecules. This flat state would represent a total and complete dissolution or breakdown of the LIFE complexity. But the woven threads (MIND) and the fibers (CONSCIOUSNESS) from which the threads are made still exist independent of the four-dimensional space-time. They were just imprints over the LIFE pattern and the 'sheet', never actually part of the 'sheet'. So they are freed within their five-dimensional realm and the single field. This freedom and detachment of the surviving MIND/CONSCIOUSNESS complex from the 'sheet' determines a very special existence for the 'person' after death.

In a slightly different analogy, picture LIFE as a blank sheet or page of paper, perhaps the paper upon which your morning news is printed. The blank sheet before printing represents the living body. The paper only seems flat if you look at it from a distance. If you look at it from a much closer position, or perhaps through a magnifying glass, you would see a structure of small bumps and curves spreading over the surface of the page. These represent the LIFE pattern. Then print is applied to the paper. The writing itself represents MIND and has its own patterns, such as words, sentences, syntax, paragraphs and even different sizes of fonts and print. It also has pictures representing perhaps visual memories. Again, if you look closer at the writing and the pictures, you will see that the ink is made of very small and even minuscule particles of black colored material. These particles would represent the CONSCIOUSNESS. But the ink and writing is not actually part of the page or the paper. The ink was only laid over the page. It would be possible to lift the ink off of the page if it was perhaps latex ink. Suppose this were true, that the ink could be peeled off of the page intact. This is what happens when death occurs. The ink is peeled off of the page intact, while the page itself goes (eventually) flat as all internal motion and chemical reactions in the body cease. But the print and ink are separate now, peeled away, just as the MIND and CONSCIOUSNESS are 'peeled away' from the four-dimensional space-time 'sheet' when the material body dies.

While the organism was alive, all sensory input to the Brain and through the Brain to the MIND resulted from four-dimensional contacts between the body and its material environment. The four-dimensional 'sheet' is the realm explored by the five senses of sight, sound, touch, taste and smell. Information and data originally entered the MIND as new electrically induced magnetic patterns from the Brain. After the death of the body, the pathways for normal sensory input to MIND via the Brain no longer exist. The five senses become inactive, as the electrical impulses to the Brain are short-circuited. However, the MIND still expects mental input and will accept input from any source available. Yet an extremely subtle sensory input had always existed below the threshold established by the five normal senses even as it acted independently of the normal five-senses: The CONSCIOUSNESS has always been aware of the single field, sensing variations throughout the single field as the sixth sense. In this sense, it would be safe to say that the MIND searches for new sources of memory input and these can only come from the CONSCIOUSNESS after death via its five-dimensional connectivity in the single field. How the MIND conducts itself at this point in time depends upon how high a level of CONSCIOUSNESS the person had achieved during his or her lifetime, or rather how aware the person had become of his or her five-dimensional connectivity while living. The sixth sense may not have been recognized as five-dimensional, but the connectivity was still sensed by the conscious MIND and interpreted as special forms of 'intuition' and other abstract sensations that were considered 'qualities' associated with MIND.

If the person had achieved a higher level of CONSCIOUSNESS, such as enlightenment, then the MIND would already have memories of the five-dimensional experience and would then merge with less difficulty into its new state of being. This would be the case of the Tea Master who was not afraid to die since he was enlightened and suspected what awaited him on death. However, if the MIND had no memories or even the slightest idea of its five-dimensional existence during life, such that the person had only attained the lowest minimal level of CONSCIOUSNESS before death, the surviving MIND might not accept its new reality and continue expecting input from the Brain and the four-dimensional world. Under these circumstances, the MIND might be 'stuck' in its four-dimensional reality even though it is materially cut-off from that reality and not realize that the body is dead. Or the MIND might not accept the death of its host body and experience a total blackness or 'nothingness'. At a slightly higher level of CONSCIOUSNESS, the MIND, not yet realizing its new state of reality, could look to its own internal memories and go through a past-life review, like a body cannibalizing its body fat when no food or nourishment is available. In some cases, the MIND could get stuck in its own memories thinking that they are material reality and suffer through having to endlessly repeat all of its past memories. Whatever the case may be, at some point the MIND will hopefully become cognizant of its new five-dimensional single field existence and then cope with its new situation. The MIND will 'see the light', which is why such a higher awareness is called 'en-light-en-ment'.

In this case, the MIND /CONSCIOUSNESS complex, liberated from its material body by death, would realize or become aware (mindful) of its new situation as a strictly five-dimensional entity. By doing so, the MIND would begin to accept input from the CONSCIOUSNESS and form new memories, a reversal of its learning methods while living. The sixth sense would become permanently active as the primary and only real source of input for the MIND. The MIND would have a sense of entering the 'light', which is just a four-dimensional interpretation of realizing or becoming aware of contact in and with the single field, which is completely filled with electromagnetic waves of every imaginable frequency and wavelength. All previous experience of the MIND had come from it five senses and the material world alone, so the MIND had learned to interpret all incoming information in the context of its four-dimensional environment. Therefore, the MIND would interpret its conscious contact with all of the different frequencies of electromagnetic waves, both visible and invisible when the body lived, as a brilliant 'white' light, beyond anything seen or known during life. In our normal four-dimensional world, inside the 'sheet', white light is a combination of all frequencies of visible light. The Biofield, the very essence of LIFE, would end at death in a new and wonderful beginning of a higher physical yet non-material existence.

Since the MIND is intact after death and still functioning, it can still make choices. The MIND might choose to remain in contact with the four-dimensional material world after death for one reason or another. But that contact would be minimal at most because the MIND is no longer attached to specific material particles, such as it was through LIFE to the material body. Such contact would constitute a haunting. The degree of effectiveness of the haunting would depend on any residual connection that the MIND might have retained to the four-dimensional world upon death. But the probability of a ghost having any real material affect on the four-dimensional world is unlikely.

Alternately, a ghost or apparition could retain contact without having made a conscious choice to do so. At the lowest possible level of pre-death CONSCIOUSNESS,' awareness of five-dimensional connectivity is minimal if it exists at all. The surviving pattern complex, an echo of the field density variations of LIFE that existed in the past, could unknowingly maintain contact by not recognizing or not accepting the death of the body. A slightly higher degree of CONSCIOUSNESS could, in principle and theory, utilize those connections to its own benefit, as in a haunting, but in the other extremes the ghost would just be a mindless zombie stuck to a place from its own past and reliving its own memories without knowing it is dead. Whether one cares to call this level of MIND/CONSCIOUSNESS complex a ghost, spirit or apparition is immaterial, it will always be just a complexity consisting of the MIND, CONSCIOUSNESS and a few disconnected fragments of LIFE, virtual echoes of the original living being, remaining after death in the four-dimensional material body. It would perhaps be accurate to say that this complex is not a ghost or apparition unless and until it interacts in a detectable manner with the four-dimensional material universe, until it haunts some location or person. Although ghosts and apparitions fit the SOFT model, the haunting phenomenon does not offer much information about the physical properties of the afterlife. On the other hand, NDEs are the best indicators to what happens to the MIND/CONSCIOUSNESS complex after material death. NDEs offer a brief look at our direct connection to the single field via CONSCIOUSNESS, as interpreted afterward by our four-dimensionally handicapped MIND. MIND's total experience up to the moment of the NDE was completely dominated by a four-dimensional material bias so MIND will try to interpret the NDE within that limited context when the body is revived and returned from death.

What happens to the surviving MIND/CONSCIOUSNESS complex upon death in any given case is open to question since its own awareness of its being within the fifth dimension is subject to the state or level of CONSCIOUSNESS at the time of material death and that is an immeasurable quantity. So each and every person's NDE would be slightly different but still conform to the limits of a single framework, just as each and every person's

LIFE is different before death. A living person who has developed a higher level of CONSCIOUSNESS during his or her life stands a far better chance of becoming a free agent, so to speak, in his or her five-dimensional existence after material death. So their NDE would be more vivid and realistic as well as more complete. Truly enlightened people float to the surface and move about freely over the water, metaphorically speaking, when they die. Those known phenomena usually associated with death, such as NDEs, actually conform to and thus enhance the SOFT model of death and the afterlife. So the SOFT model of physical reality and LIFE would seem to be accurate, given this evidence.

SOFT NDEs

In the most extreme NDEs, the patient actually dies for a short period of time and is revived. The duration of death does not seem to matter since many NDErs experience a lifetime of memories in just a few moments while dead. This fact implies that they are not within the normal space-time continuum and not experiencing time as it flows within that continuum. NDEs associated with these extreme events are the most reliable for death studies, but do not supply all of the needed data on death. With a momentary death, the lack of sensory input from our normal world of four-dimensional sources induces a never before experienced state of quiet and peace in which MIND and CONSCIOUSNESS can function as a single structural unit independent of LIFE and body. Some people try to mimic this state of mental quiet and peace of mind through meditation so that they can become enlightened, aware of the true nature of their CONSCIOUSNESS and its place in the five-dimensional single field.

During the NDE, the MIND/CONSCIOUSNESS structure is momentarily freed from the four-dimensional 'sheet' in what is normally called an Out-of-Body Experience (an OBE), but it is still physical so the 'entity' or complex can still 'sense' portions of the material world from which it has just emerged. These are not true sensations, since they do not come through the normal five senses. These are just remnants of the five senses. The CONSCIOUSNESS can play tricks on the MIND, which may not yet realize what has happened. The sixth sense is just mimicking the normal five senses, at least until the MIND begins to understand or become aware of what is happening, at which time the MIND 'moves' on to tunnels, the 'light', and other phenomena that characterize the NDE. The OBE is the simplest of the NDE characteristics to explain, since the MIND and CONSCIOUSNESS, which exist in five-dimensional space, are not within the body inside the four-dimensional 'sheet' of material reality.

The whole experience gives the MIND a 'sense' of being dead as well as the feeling of being 'out-of-body'. These last aspects of the NDE also follow from the SOFT model and seem to be related. Since the MIND/CONSCIOUSNESS complex separates from the physical body at death, as a physical entity of its own devices, it should realize the death of the body as a result of that separation. The failure to receive sensory input from the brain would be a sure sign that the physical body had died, hence the 'sense' of being dead. The separation of the MIND/CONSCIOUSNESS complex would be interpreted as an OBE when the body/Brain system is later revived because, quite simply, it had very nearly separated from the material body in four dimensions within the fifth dimension for a moment, or rather its connection to the material body had become so tenuous that it could be considered separate for all practical purposes although not in fact. The complex had all but separated.

Traveling to different places in three-dimensional space and time while 'out-of-body' would merely refer to a conscious awareness or logical interpretation of how the CONSCIOUSNESS realized its five-dimensional connectivity. Once again, the MIND/CONSCIOUSNESS complex would first seek to interpret its new surroundings in the five-dimensional realm in terms with which it is familiar, i.e., in terms of its immediate four-dimensional pre-death existence and post NDE situation. So CONSCIOUSNESS could follow a specific but random linear and familiar path in three-dimensional space before it realized its connections with the same known and recognizable four-dimensional objects, but before it came to terms with the totality of its five-dimensional connectivity. When the complex realizes the totality of its connectivity, that realization would be interpreted as experiencing or 'seeing' the 'light' that normally comes after the OBE, if the NDE even proceeds that far.

Freed from the background chatter of stimuli from the four-dimensional material world, the MIND can more easily 'hear,' 'see' or otherwise 'sense' the input of data from its five-dimensional connections with the rest of the world. The words 'hear,' 'see' and 'sense' are actually inadequate to describe the way of sensing and interpreting what the MIND and CONSCIOUSNESS now experience, but no other words are available to describe this state of being. These words represent a four-dimensional description of a perception that is strictly five-dimensional, based on a lifelong collection of four-dimensional experiences and memories. Input of data from connections of the CONSCIOUSNESS to the rest of the universe was constant even during 'life,' but extremely subtle so it would not have registered in the person's conscious awareness under normal conditions. So, the single field could not have been normally detected and brought into the conscious thought (mindfulness) of the living person without specialized training or through some unique and usually random situation (spontaneous enlightenment). The effect is similar to our inability to see stars in the sky during the bright sunlight of the daytime even though the stars are still there during the daylight hours, but the difference is even more extreme than this simple analogy might imply. Sunlight would represent the normal senses and starlight the awareness of five-dimensional contact or the sixth sense. The momentary death during which the NDE occurs would be similar to the sun setting followed by a clear night when all of the stars are seen in all of their bright and beautiful glory. Then the person is revived, brought back to LIFE, and again the bright sunlight drowns out the beautiful starlight just as our waking five senses drown out our awareness of our sixth sense.

Immediately after death, the surviving MIND/CONSCIOUSNESS complex is isolated from contact with the four-dimensional material world by the disruption of LIFE. With the loss of the normal sensory input that has dominated all of the previous existence of the individual's MIND and CONSCIOUSNESS, the MIND searches for new connections to orient and stabilize its new 'self' in its new environment. The MIND will, of course, first search for input that it is the most familiar with, such as the normal sensory inputs from the Brain that are the most recognizable within its own patterns of memories and knowledge. Even an unrecognizable input would be interpreted in light of the already existing knowledge patterns which correspond to the dominant four-dimensional reality of the person's previous living experience, his or her worldview, philosophy of life or religious belief system. This search and the MIND/CONSCIOUSNESS complex's need to reestablish its contact with the surrounding physical environment explains the common characteristics and central aspects of the NDEs, which are wholly and completely experiences that transcend the cultural, religious and educational limits of those who experience them.

For some individuals, the MIND would first seize upon its own inherent memories—its past—to stabilize its new 'self' after the field variation disruptions caused by death. From the simple physics of the five-dimensional perspective, there is no qualitative difference between three-dimensional space and single-dimensional time as explained by special relativity. Both are linear measurements within four-dimensional space-time. All of four-dimensional space-time is laid out 'below' the MIND. No spatial or temporal position is favored relativistically over another. However, what we regard as normal time and space are irrelevant to the MIND once it is separated from its inferior four-dimensional material existence because MIND resides in the fifth dimension outside of linear time and space. MIND and CONSCIOUSNESS now transcend four-dimensional space-time. So, the MIND grasps for something to stabilize its 'self' by searching for reference points in its new physical reality and reverts to what it knows the best, its own past time-line of memories relative to the overall four-dimensional portion of the universe. The MIND acts in a way similar to a body that receives no food or nourishment for a long period of time and begins to cannibalize its own stored body fat. This cannibalization of past memories is the past life review that many NDErs have reported. However, it is more than just the momentary realization of the memory patterns from which the MIND itself was formed because the MIND actually has the ability to 're-view' or once again experience the events of the past within normal space and time. In a sense, the MIND actually becomes its own past experiences, lending a unique quality to the experience. It is far more than just a normal remembering of past events that were experienced during one's lifetime.

Those who experience the past-life review report that it happens in an instant, which is in strict keeping with a physics and geometry of a 'superior' position in five-dimensional space outside of the normal four-dimensional space-time 'sheet.' However, not all experiencers undergo the past-life review. This review is neither an absolute nor even a necessary event since some people have a greater experience with the five-dimensional extension of physical reality before they die and thus their MINDs do not need the orientation provided upon death through a past-life review. Others may not be advanced enough in their own personal paths of conscious evolution to warrant the past-life review, and still others may not mentally accept what has happened (they deny their death) and thus 'sense' nothing at all. In other words, people's MINDs seize upon the most familiar surroundings when they enter the new environment of the five-dimensional universe, but can still reject the experience completely depending upon their mind set and mental priorities at the time of death.

Those people who witness the past-life review have died, if only for a moment or two, without ever having completely developed their CONSCIOUSNESS very far past the minimum degree which does not allow for more than a minimal knowledge of the greater natural connection with the rest of the universe afforded by their new five-dimensional existence. Yet their CONSCIOUSNESS has evolved enough to accept their death to some degree. People with a more extensive knowledge of their five-dimensional connections, and this supposition does not require that they have any 'conscious' awareness or specific knowledge of the existence of the fifth dimension, only that they have some experience of (or belief in) the connections themselves, i.e., a more highly developed CONSCIOUSNESS, could conceivably skip the past-life review. The surviving MIND would rely on previous knowledge of the five-dimensional connectivity to establish a new framework, a non-material non-four-dimensional reality, upon which to focus its new existence or being. These people would have reached some stage of what is called enlightenment in the Eastern mystical sense of the word. In this case, CONSCIOUSNESS begins to absorb input directly from its greater five-dimensional extensions of the universe without a great period of disruption upon the loss of four-dimensional sensory and data input to the MIND. CONSCIOUSNESS itself reacts via its five-dimensional connectivity immediately after death rather than MIND through its inherent memories and knowledge patterns. People with a more developed CONSCIOUSNESS immediately utilize the connectivity of their CONSCIOUSNESS to other portions of the five-dimensional single field to orient and prioritize the MIND rather than using the MIND to expand CONSCIOUSNESS within their new five-dimensional habitat.

Other characteristics and properties of NDEs are just as easily explained within the SOFT model of physical reality. One of the more common features of the NDE is the apparent movement through a blackness or tunnel at the end of which one has the experience of entering a region of bright white light. In physics, black is a total lack of color that results from the absence of any waves of visible light. The surviving MIND might not recognize or understand its connections to the rest of the universe through its own 'conscious' effort when a person dies. The darkness is experienced since the MIND is expecting input from the normal five senses, but receives nothing. The MIND must interpret this lack of input in some manner and an interpretation of darkness fits the MIND's previous experience in the material world, an eternal nighttime. The person would be totally cut off from the 'light'. Being cut-off from the 'light' would be interpreted by the MIND as a vast and consuming darkness or the blackness due to a lack of 'light'. CONSCIOUSNESS could literally 'encapsulate' and isolate the complex as separate from anything and everything else. The MIND may not even be aware that the person has died or it could be rejecting that possibility as unacceptable.

In any case, there would be a specific sphericity to what the MIND senses of its new habitat through CONSCIOUSNESS and it would seem as though the MIND were surrounded, at equal distances in all directions, by some indefinable or un-sensed thing. That thing is the rest of the single field, but the MIND has rejected that reality. Since the MIND thinks it is still mentally attached to the four-dimensional space-time by its past experiences, but not by any real material connection as during LIFE, the MIND would sense 'time' moving normally 'below' it. The experience of MIND and CONSCIOUSNESS in the pre-death material world was one of motion or change from position to position in space and time. The MIND could only cognize change within their previous four-dimensional existence. In the material world, that 'change' would constitute change in spatial position. That is why normal physics is based upon 'matter in motion' as the lowest common denominator for all physical phenomena. So, for the complex to realize its five-dimensional connections as opposed to the 'encapsulated' isolated self would be interpreted by the MIND (which is still primed for four-dimensional input) as a change in spatial position, movement through the black tunnel or a similar dark void into the 'light' representing the new existence and environment of the complex.

This sensation would nearly be the same as sitting at a stoplight, motionless, next to another car or a bus. But when the vehicle next to you moves, the relative motion tells you that you are moving and the other vehicle is stationary, even when you are not moving. The relative progress of time in the four-dimensional space-time continuum tricks the MIND into a sensation that it is moving in time even though it is not. The combination of a sensed sphericity and progress in time builds the sensation of moving in a dark tunnel in the MIND. The bright light at the end of the tunnel marks the MIND's first awareness or 'glimpse' of the five-dimensional single field. The bright 'light' is not actually 'seen' by the MIND since the MIND has no eyes, but 'seeing' a bright 'light' is how the MIND interprets or mentally copes with the sensation of an awareness of its connectivity to and continuity with the single field. This awareness would occur when the MIND comes to realize that it is not actually moving, either in time, space or the single field, just experiencing an eternal moment 'outside' of four-dimensional space-time. It would be the same as the person who thinks he is moving as the bus pulls away from the stoplight, and then jerks into action when he realizes that the bus is moving and he is still stationary. The light at the end of the tunnel is that 'jerk' back into a realization of the true reality of the situation at hand.

The various forms of experiencing the blackness are not required for all experiencers. Indeed, each person's NDE is unique and consists of different representations of the common characteristics colored by individual variations to the theme. In other words, some people might experience the darkness without any sensation of motion and thus interpret their NDE as a bad experience upon their revival, as if they were completely lost and disoriented during the NDE.

A lesser although substantial number of people who have had NDEs report that theirs was an unpleasant experience and perhaps even 'hell'. Of course their NDEs are colored by their mind set and mental biases, but the blackness could also be interpreted as unpleasant. However, a person's MIND could also fixate on particular bad memories during or in place of the past-life review. Since 'bad' events are usually highly emotional, they would have created strong or intense memories and would thus still exist as very intensely imprinted memory patterns. If the MIND were to fixate on such 'bad' memories, then the person would perceive their NDE as an experience in 'hell' or at least a 'hellish' experience upon revival and return to LIFE. Indeed, a number of people have had bad NDEs and these have defied explanation by scientists even though they need to be explained to give those NDErs some peace of mind during the rest of their lives. These NDErs would not have seen the 'light' and would thus not have the same positive after effects of the experience as those who saw the 'light'.

There are different variations upon this theme, but in general those variations are inconsequential relative to the geometric structure and physics of the fifth dimension. In fact, the white light is quite easy to explain. At some point, CONSCIOUSNESS should become the dominant and primary element of the complex in the survival of death, if not immediately then soon after death. When this occurs, CONSCIOUSNESS would seem to reach out and embrace or touch the surrounding single field. Again, this is a four-dimensional interpretation of events in the five-dimensional continuum. CONSCIOUSNESS would begin to interact with its new environment – the single field and the rest of the universe that it fills – as the MIND becomes open and accepting to the idea of its five-dimensional environment. After death, CONSCIOUSNESS should become aware of those connections to some extent, whether or not MIND had any awareness of those connections before death. Whether or not MIND accepts this physical reality decides the future course of events. So CONSCIOUSNESS begins to absorb information input directly from the rest of the universe in the absence of sensory input from the Brain and filters that input through to the MIND.

In one sense, CONSCIOUSNESS spreads its feelers or awareness out over the universe via its five-dimensional connectivity, even while remaining an intact entity within a stable structure with MIND. Awareness would spread out equally in all directions, like an expanding balloon, since CONSCIOUSNESS has no real 'material' appendages and thus we have a sphericity of awareness emanating from the MIND/CONSCIOUSNESS complex. Every point in the four-dimensional continuum is filled with a field of some type. Most empty three-dimensional spaces are filled with electromagnetic waves of some frequency and this is extended to five-dimensional space. It is well known in physics that white light is merely the sum total of all colors or wavelengths of visible light. So CONSCIOUSNESS quite literally experiences all wavelengths of both visible and non-visible light simultaneously. CONSCIOUSNESS 'senses' the brilliant light as it realizes its connectivity to all points and particles in the universe that are filled with real light waves. The brilliant light which is experienced is diffuse and complete beyond anything ever experienced while the a person is alive because it represents the sensation of all wavelengths in the electromagnetic spectrum, something with which a person is not normally familiar. Therefore, the 'light' experienced is brilliant beyond comparison.

The event of sensing the 'light' is many times associated with being greeted by loved ones who have already passed away or even greeted by religious figures. It would be logical to assume that loved ones who had already died would have undergone a similar process of realizing their own connectivity of thought and CONSCIOUSNESS with the fifth dimension and their own MIND/CONSCIOUSNESS structures had survived within the single field, so they would appear to come out of the light when contact is made by the newly 'arrived' MIND/CONSCIOUSNESS complex. The surviving MIND/CONSCIOUSNESS complex of the NDEr would literally pick out memories within its own MIND pattern with which it was already familiar, such as family members, loved ones or friends and this would in a sense 'call up' the external MIND/CONSCIOUSNESS complexes of those loved ones. This process would be analogous to the normal physical process of resonance between different objects with the same resonant frequency except that this process occurs outside of time so there can be no true material 'resonance'. So the phrase 'pattern matching' better and more appropriately describes the phenomenon. The person would later interpret this selection process as people coming to greet or direct the person into the 'light' – the single field where they still exist. When an NDEr is revived, he or she must 'fit' the experience into his or her own mental pattern of knowledge: It must be made to match the worldview that represents their living MIND. This fact explains why NDErs describe the person that they met as they remembered that person, not as that person appeared at the time of his or her death.

Having sensed or 'seen' the 'light' and thus having begun to experience the whole of the universe in an instant, an experience which has no living four-dimensional counterpart (although mystical enlightenment comes very close), the MIND might interpret this event as having experienced God or some particular representative of God depending upon the person's worldviews and religious biases. This explanation does not mean nor does it imply that God has been incorporated into a scientific model of the universe. Everyone is free to determine what God is or is not from his or her own perspective. Nor does this explanation mean that one view of God is better than any other. Whatever God is, God is 'not' a concept that is amenable to scientific analysis of this or any type. The NDErs may have actually encountered the religious figure that they report, by the same manner in which they met relatives or loved ones, but that fact neither supports nor denies the reality of their religious beliefs.

If the NDEr thought of any particular religious figure, which was once a real person that died long before the present, the NDEr would have 'called up' that person from the single field. There is no logical reason why an NDEr would not experience meeting a particular religious figure as long as that religious figure was once a real live person (rather than a mythical religious being who had never existed in the four-dimensional material world). The religious figure from the past that greets a person during the NDE would simply be the religious person who the NDEr thought of during the experience. When the NDEr thinks of a person, any person, during the NDE, the MIND of the NDEr stresses that memory pattern on that person which causes that person to appear. In fact, many NDErs have later explained that they do not consider their NDE as a confirmation of their own religious beliefs, even though they have 'met' a religious figure that they believed in during the NDE, but rather a confirmation of their own spirituality.

But the NDEs are not symptoms of a final and irrevocable death. The experiencer is brought back to LIFE by the efforts of a medical team, by chance or by a mental choice to return. The reanimation of the body and LIFE occurs in spite of the momentary Brain death of the person. When a patient is revived, he or she is usually yanked back into the normal material four-dimensional linear reality by a medical team or by some active four-dimensional stimulus. In some cases the NDEr could also be 'told' that his or her time had not yet come by those that are met in the 'light' and is thus forced back to the normal reality of four-dimensional space-time (within the 'sheet') by the MIND itself. Just as people can 'view' the past portion of their time-line as a whole during the NDE, they could theoretically become cognizant of the future portion of their time-line from their new five-dimensional vantage point, although this particular occurrence is far less likely. During the initial exploration of one's connectivity within fifth dimension, the NDEr could possibly become 'conscious' or 'mindful' of the continuity of their time-line into the future as well as the past. Their MIND, upon learning that the time-line had not ended in the death that initiated the NDE, but actually continued into the future, would be 'literally' forced by that knowledge to return to the material body and the continuance of their LIFE. The MIND realizes that the patterns in the fifth dimension, which are LIFE, were not completely disrupted and it is able to reorganize or reconstitute, from the top down, the fundamental chemical and mechanical reactions within the body, which jump-starts the living process.

Once again, all who have reported NDEs do not experience each and every one of these aspects. Whether or not a person experiences any one particular aspect of the many different aspects that constitute NDEs would depend upon the quality and degree of CONSCIOUSNESS that the person reached during life before the NDE. The individual aspects of any given NDE are a matter of a person's psyche or psychology rather than space-time geometry and physics. Geometry and physics only offer an explanation of the physical background or arena in which the NDE takes place. In spite of the individual details of any reported NDE, the experiencer must remember the details when revived to communicate that information to the investigator. This begs the question "What would the person remember of the experience?" The experiencers' memories of the NDE would have to fit the organizational pattern of memories and knowledge existent before death in the MIND, the person's living worldview, which is controlled by specific memories except for the common characteristics of all NDEs. So the remembered details of the NDE would be biased by the person's preexistent belief system and the difficulty of translating five-dimensional experiences into words and concepts developed to describe four-dimensional phenomena. But the MIND/CONSCIOUSNESS patterns would also be remapped or rewritten to some degree by the experience. After all, the experience itself is a new memory and a very intense memory at that. Scientists have been careful to isolate those properties that are constant across cultural, educational and societal boundaries, the most general characteristics, which are not colored by social or cultural biases, so that they might gain an understanding of the NDE. But these scientists must still be careful in their attempts to explain the events common to NDEs so that their explanations do not include their own personal and physical worldview biases. The only problem remaining is how to describe something which is inherently indescribable in terminology that can be understood.

Post NDE SOFT landings

NDEs are transformative events in the LIFE of those who have experienced them, and the changes that they cause are far from superficial and thus are quite evident. The personality changes that a person undergoes after the NDE are literally a transformation of the 'self', in some cases a remaking of the MIND and CONSCIOUSNESS patterns. The nature and intensity of the changes are extremely subjective and differ from person to person, but they do share some common characteristics, as do the NDEs themselves. These changes can include a loss of the fear of death, increased compassion, an increased ecological awareness, and an increase in paranormal abilities, but are not limited to these characteristics alone. Since the changes are so subjective, the greatest personality changes can only be summarized on a case-by-case basis. Kenneth Ring, one of the foremost experts in NDE research, describes the situation far more eloquently and completely.

> To give some indication of these changes, it is necessary to recall here that the NDE is not only a revelation of the most profound and soul-shattering beauty, but, as I have said, it is also something that has the power to drastically alter and improve the lives of those who are visited by one. For example, we now know that the NDE tends to bring about lasting changes in personal values and beliefs -- NDErs appreciate life more fully, experience increased feelings of self-worth, have a more compassionate regard for others and indeed for all life, develop a heightened ecological sensitivity, and report a decrease in purely materialistic and self-seeking values. Their religious orientation tends to change, too, and becomes more universalistic, inclusive and spiritual in its expression. In most instances, moreover, the fear of death is completely extinguished and a deep-rooted conviction, based on their direct experience, that some form of life after death awaits us becomes unshakable and a source of enormous comfort. In addition, many NDErs say they come to develop powers of higher sense perception, increased psychic ability and intuitive awareness and even the gift of healing. In short, the NDE seems to unleash normally dormant aspects of the human potential for higher consciousness and to increase ones capacity to relate more sensitively to other persons and the world at large.

The NDE, then, appears to promote the emergence of a type of functioning suggestive of the full human potential that is presumably the birthright of all of us. In a phrase, whenever the blessings of the NDE are fused properly into one's life, the individual comes to exemplify what a highly developed person would be and act like. (Ring and Valarino, Introduction)

Dr. Ring both proposes and defends the view that these personality changes are the result of having attained a higher level of consciousness during the NDE. He has also determined that these qualities are inherent abilities in humans that normally lie "dormant", and the NDE only brings them out or raises them to the conscious level of the experiencer. Both of these conclusions fit the SOFT model. Ring's research further indicates that whatever portion of the NDE causes the changes it is only acquainting the experiencer to an ability that already exists. The NDE awakens sleeping abilities of the CONSCIOUSNESS, or in terms of SOFT, the NDE merely renders the person aware of his or her inherent single field connectivity with the rest of the universe. Although the NDE is interpreted in four-dimensional terms and judged against four-dimensional experiences after the experiencer is brought back to LIFE, the NDEr may never realize the true five-dimensional nature of the experience.

Each and every living being is and has always been connected to the rest of the universe by way of its five-dimensional extensions, but people do not normally become aware of these connections in their everyday lives prior to death or the NDE. Those people who do become aware of them usually do so through the use of their sixth sense in various paranormal practices, or by various contemplative practices such as meditation that led to enlightenment. It is claimed that those who are enlightened in this fashion have enhanced 'psychic' or paranormal abilities, so the two are related. However, these psychics do not really understand how their paranormal abilities work or the connection with higher-dimensional reality that they imply. So the value of paranormal experiences may or may not increase the probability or intensity of having an NDE if the opportunity ever arises. Otherwise, people who are enlightened through meditation in the Eastern mystical sense of the concept are far more aware of the connectivity accorded with the single field even if they do not know anything of the five-dimensional geometry or space. Therefore, these inherent abilities (stemming from the awareness of the five-dimensional connectivity) exist and function at all times rather than lie dormant in the MIND and CONSCIOUSNESS of each and every human and 'sentient' being. They are in continuous operation throughout the LIFE of the individual, but they are subliminal and act well below the threshold set by the normal five senses, which literally blinds us to the existence of these connections and our sixth sense. So, it is the precise nature of the NDE and the connection made with the 'light', the single field, that causes an awakening of an individual's CONSCIOUSNESS to its connectivity within the five-dimensional single field, initiating or turning these abilities on. Quite simply, the NDE alters the memory and knowledge patterns in the MIND just through the realization of the connectivity that results from seeing the 'light'. The deeper and more intense the NDE, the more profound the alterations to MIND and CONSCIOUSNESS are and the more lasting the changes to personality after the NDE and the return to LIFE.

Melvin Morse another well-known researcher of NDEs has confirmed this prediction of the SOFT model of death and the subsequent explanation of NDEs. Morse has concluded through his own researches that it is the experience of the 'light' alone, among all of the aspects of the NDEs, that induces these changes in the individual consciousness of the experiencer and thus contributes to the transformation of the experiencer's psyche.

> The transformative part of the experience is seeing the light. If a person has a paranormal experience such as leaving their body but it is not accompanied by the light, then the experience is not usually transformative. If the light is experienced then there is a transformation. The transformative powers are

in the light. That is what our research tells us. (Morse, 175)

And,

> I have found the experience of the light to be the keynote event of the near-death experience, the element that always leads to transformation. I do believe that the light seen by NDErs comes from a source outside the body (Morse, 212)

Personality transformations induced by an initial contact with the 'light' are indicative that a person had experienced an acute awareness of a far greater and more comprehensive physical universe than is currently suspected by scientists, theologians or scholars. Both science and our common perceptions generally limit our physical reality to the four-dimensional space-time continuum, but only the existence of a fifth additional extended dimension of our common space-time, as specified by SOFT, can account for these transformations.

Sensing the 'light' would be the key event, since it marks recognition and an acceptance by the MIND of the connectivity characterizing the relationship between CONSCIOUSNESS and the single field. The final step from which a return to LIFE would pose a significant problem would be entering the 'light', which would block any attempt to the body being revived. Therefore, seeing the 'light' and thus surrendering to the connectivity between the CONSCIOUSNESS and the single field would be enough to change the MIND pattern should the person be revived. Seeing the 'light' is such an overwhelmingly intense experience that the memory that is imprinted in the MIND pattern easily dominates the intense memories from normal life in the four-dimensional material reality. Seeing the 'light' would be like opening a window and seeing the sunlight for the first time. Only the sunlight in this case is the whole of the universe. Dr. Morse has further stated that "the transformative part of the near-death experience, the portion that leads to positive changes in personality, is somehow contained in the light" and cited one small girl's statement that "all of the good things are in the light" (Morse, 70-71) to support his statement. All of the "good things" are in the 'light' because therein lies the whole truth of our physical universe, including every moment of all of our lives.

If it is assumed that the 'light' represents a direct conscious awareness of the connection between the MIND/CONSCIOUSNESS complex with the rest of the universe, as illustrated or explained by SOFT, then all of these transformative changes are easy to understand. After death, whether it is permanent death or the momentary death experienced during an ND event, the surviving MIND/CONSCIOUSNESS complex experiences its own structure as well as its connectivity relative to the whole of the universe as a portion of the single field. This process is interpreted by MIND as entering the 'light.' The new form of knowledge gained by the experience is internalized, as are normal memories and information, by a corresponding pattern alteration within the MIND and CONSCIOUSNESS themselves. After the experiencer is revived, returned to a living state, he or she will retain what was learned of the experience in their MIND because the MIND has been physically altered by the new intense memories of the NDE. The MIND has not been materially changed since it is physical and non-material, nor has the Brain been altered to any measurable extent in any normal concept of measurability. The change is purely physical and far too subtle to measure in the grossly structured material Brain.

Since the MIND is a complex pattern combining the thoughts, memories and knowledge of the individual, the experience of the 'light' will have transformed the very thought processes in the MIND and increased the overall knowledge of the universe possessed by the person. Quite literally, the person would have had a strictly material worldview against which all events are judged before the NDE, but after the NDE the experiencer would judge and make choices against a more complete and truer view of the whole five dimensions of the universe. The person would have developed at least a very strong intuitive sense of the truth of his or her own being, completely relative to a whole and comprehensive universe. All choices would then be made against this background by taking into account the truly perceived universe beyond normal space-time that most other people cannot comprehend or even imagine. The person will have a new "zest for life," as Dr. Morse would say, (Morse, 8) because he or she would have a better innate understanding of the relationship between LIFE, MIND, CONSCIOUSNESS and the material body due to the internal density pattern alterations.

The alterations in the complexity pattern of MIND would over time filter downward to the individual level of LIFE as well as the physical body since each higher level of complexity is an organizational principle for the next lower level of being. The amount and overall effect of such a process of downward reorganization would depend, of course, on the quality and duration of the NDE, although even the simplest NDE would affect such reorganization as long as the experiencer sensed the 'light'. This top down process could easily affect the physical health of the experiencer as well as give the experiencer a new "zest for life." The new knowledge of what awaits one in death, connecting with and becoming one with a complex universe that could also be described, although a small bit inaccurately, as 'resonating in tune' with all of the varying single field patterns in the universe, would tend to ease anyone's fears and anxieties of dying. Not only would death loose its mystique, but also it would become a friendly and pleasant state of being, to say the least.

A subsequent increase in compassion is also a direct result of the connection made with the 'light' as well as other living and sentient beings connected with each other through the 'light.' Whether or not there is a conscious awareness and mental acknowledgment of the mechanics of these connections, and such an awareness is not necessary (we do not need to understand gravity to fall down, it is the experience of falling from which we learn to stand and walk), the NDE would still alter the MIND and CONSCIOUSNESS patterns sufficiently to cause the increase in compassion and other lasting effects on the personality of the experiencer. The experience itself alters the MIND and CONSCIOUSNESS patterns by subtly realigning the patterns to be in 'tune' with other beings and bodies in the universe regardless of a person's awareness of how the change occurred or its mechanics. The pattern alteration could be directed toward connections with all humans, all sentient beings, all animals, all plant life, all living beings who lived during any period of time, or any combination of these depending on the person's attitudes and beliefs before the NDE. A person would simply become more compassionate and respectful of other people and living beings due to the structural changes in MIND and CONSCIOUSNESS, not by any choice in the matter, while an innate intuitive feeling of the intimate connection with other people would permanently bind the NDEr to them. This particular effect would change completely the NDEr's relationships with other people, sometimes in unpredictable ways, depending on the NDEr's other attitudes. However, the NDEr would always change in a moral fashion, answering to a higher morality that he or she would intuitively sense since the NDEr is more 'in tune' with the universe as a whole as well as any concept of 'universal truth' which exists, although 'universal actuality' would be a better description of the reality of the concept than 'truth'.

In a sense our individual lives and LIFEs would be shared with others since we are all part of the same single field. Doing any type of harm to another living being would alter that being's internal patterns of MIND that would be communicated back to the individual causing harm by the five-dimensional connection between the two. So harming others would not sit well with anyone who had had an intense NDE. The NDEr would have a new respect and tolerance for other people and their views. In a sense, as MIND/CONSCIOUSNESS patterns alter on seeing the 'light,' a little of the other person becomes a part of each and every one of us as we become a little part of them. And this connection and sharing of being would also be true to the extent that all of our interactions with others during our lives is also a part of each and every one of us via our subtle collective five-dimensional connectivity while we are still alive.

Many experiencers have associated the 'light' with love as well as compassion. Within this five-dimensional model, love can now be defined as an intimate and purposeful or directed five-dimensional connection with another being or beings. Love is essentially tuning or aligning one's own personal field patterns to the corresponding patterns of the object of one's love. True love would amount to a complete knowledge and sharing of connection between two people. To a psychiatrist, neuro-biologist, neuro-physiologist, biochemist, endocrinologist, or other reductionist scientist, love may be no more than a specialized chemical interaction in the Brain or body, but that chemical interaction has a five-dimensional extension and thus forms an alteration in the complexity patterns of one's MIND and CONSCIOUSNESS in the fifth dimension. But this specific type of love is directed 'down to' the chemical level of Brain and body through CONSCIOUSNESS', MIND and LIFE, so it differs in context from normal material based love. A love initiated by direct five-dimensional contact would be considered a case of 'love at first sight'. Two people can fall in love at first sight because the appropriate patterns in their MINDs naturally 'resonate' together and immediately establish a common link between the two.

A full and natural pattern matching between two people would constitute what many call 'soul mates'. The concept can be explained as a product of the natural connectivity the CONSCIOUSNESS enjoys with the single field, out of which this particular special connection emerges. 'Soul mates' are rare, but not unheard of. However, whether the fact of the relationship is recognized by the individuals who experience this particular pairing of patterns or not is a matter of the level of CONSCIOUSNESS that each has attained. It would be possible for either one or both of the 'soul mates' to completely miss recognizing the connection or in recognizing the connection miss its full extent and meaning. Also, this pattern matching is not to be confused with pure lust, which is as much a choice made by MIND, rather than coming from CONSCIOUSNESS as it is a chemical/hormonal reaction to another person. Too many people mistake chemical reactions for real love, which is far beyond just body chemistry, and have problems coping when they ultimately realize that it was not real love to begin with.

We do not normally know why we love any particular person, although we could think up hundreds upon hundreds of reasons why, but we still love that particular person completely, unconditionally and passionately if it is true love. This example would represent an extreme case of love, at the level of a pure connectivity issue. On the other hand, we may already know a person and over a shorter or a longer period of time come to love that person whether we were originally attracted to that person or not. In other words, there are cases where we can learn to love a person through a process of constant contact and familiarization. In this case, the patterns are more slowly altered through material/physical contact between the lovers that initiate chemical responses in the body and Brain and thus form new imprints in MIND and CONSCIOUSNESS.

When any two people touch, when they look at each other or make other contact through their common senses, specialized chemical reactions occur in their bodies, depending on the type of relationship they have with the other person. Normally, the reactions mean nothing as when we just see strangers on the street. However, with some special people who are attracted to one another for some reason, these chemical reactions can either alter our field density patterns or enhance those patterns from below, from the four-dimensional material world 'up to' the fifth dimension where the patterns exist. Continued for a long period of time, this process can produce love. This process of love imprinting could work as well for two brothers as for two lovers, each type of touch and material contact (including 'seeing') generating its own specific chemical reactions in the body, even though familial love and sexual love are two different things. So the SOFT model fits all kinds of love; emotional, sexual, spiritual, and familial. And then the pattern imprinted into the MIND and CONSCIOUSNESS of each person forms the special intimate pattern matching connection that is love. This process marks the other extreme, so love comes in two flavors with an infinite number of in-between mixes.

Compassion, on the other hand, comes from a similar physical process although it is more of a blanket effect that is not directed toward a specific individual, but generally toward all other living beings. There are specific similarities between the patterns of every one in a specific species and general similarities between all living beings. Compassion would represent an alteration of one's MIND/CONSCIOUSNESS complexity patterns in such a manner as to 'tune-in' or 'align' with either the specific or general patterns of other beings. This process can be initiated during an NDE by just seeing the 'light', but it can also be initiated by attaining a mystical type of enlightenment. When seeing or experiencing the 'light' as 'love', we become more compassionate as a result of the intimate five-dimensional connection which is established between all beings and ourselves, an action which permanently alters the MIND/CONSCIOUSNESS patterns in the individual.

Many of the great religious Masters of human history have taught either love or compassion and in some case both, as a way to reach a higher level of CONSCIOUSNESS. This higher level of CONSCIOUSNESS would be a form of preparation for entering the 'light' when an individual dies. Teaching and practicing love and compassion as a precept during LIFE would help a person recognize and accept the 'light' upon death, or according to the SOFT model, accept one's connectivity to the single field and subsequent role in the universe as a whole. Entering the 'light' upon death would then be interpreted as 'going to heaven'. The religious Masters have taught this principle for millennia as interpreted through their own social and cultural contexts as well as mediated by their own four-dimensional material perceptions and worldviews.

By the same token, 'hate' would result from the purposeful 'encapsulation' or isolation of one's MIND/CONSCIOUSNESS complex from others or from a specific person, effectively insulating a person's CONSCIOUSNESS from the direct five-dimensional connections that are interpreted as 'love'. 'Hate', carried far enough to an extreme, could also entail a loss of compassion by blocking the general five-dimensional connectivity with other beings, a possibility that is part and parcel to our existence within the single field, if it were strong or intense enough. It would be difficult if not impossible to block the person's complete connectivity with the single field. Since the single field is the precursor to all physical fields such as gravity, electricity and magnetism, and to the best of our knowledge it is impossible to block gravity, then we could also deduce that it would be physically impossible to block any connections associated with the single field. This fact implies that 'hate' could not be the result of throwing up any kind of barrier around the MIND/CONSCIOUSNESS structure. Instead, it would seem that 'hate' must therefore be internal to the field density patterns that form the MIND/CONSCIOUSNESS structure. 'Hate' must consist of an intentional scrambling of one's own MIND and CONSCIOUSNESS patterns so they would no longer 'align' with the patterns of the person or persons toward whom the 'hate' is directed. Therefore, 'hate' would do far more harm to the person who 'hates', if the emotion is allowed to fester and get out of control. A person's 'hate' alone could not harm another person via the connectivity afforded by the five-dimensional single field, but the person with the 'hatred' could harm another person materially within the four-dimensional world if that 'hate' distorted the person's MIND and CONSCIOUSNESS patterns so intensely that common morality was diminished or altogether scrambled in the MIND of the person.

By 'scrambling' one's own internal MIND patterns in the cause of hatred, the MIND as a whole might be affected and suffer because of the continuity between all of the patterns and memories from which the MIND is constructed. An intense enough hatred could conceivably scramble or disrupt other patterns in the MIND affecting everything from a person's health to various moral precepts and decisions. In an extreme case, such an altered or scrambled pattern could harm other links with other portions of the single field and the universe, leading to very severe mental problems. The person initiating the hatred could conceivably throw himself out of alignment with the rest of the universe. If a person were to die during this state of affairs, the ensuing problems could be catastrophic for the person in death. The person could theoretically create his or her own hell based on their hatred, their scrambled interpretation of the universe and their discontinuity with the single field. They could not possibly 'see' the 'light' with their connectivity scrambled to such a destructive degree.

The same connectivity with the single field that explains love also explains the increased ecological sense that NDErs have reported. While experiencing the 'light,' connections are made between all material objects in the universe, inanimate as well as animate, but the connections with animate matter would be the strongest or most intense. We gain compassion for other humans, even for all sentient beings and animals, but the connections also extend to the plant life in the ecosystem. Humans evolved on this Earth by a spontaneous utilization of a special chemical and physical balance within the Earth's ecosystem. That balance is enhanced by our five-dimensional connectivity within the single field. So humans have a special connection with all LIFE on the Earth, which also evolved within the same ecosystem. When the NDEr realizes this special connection, he or she becomes more ecologically sensitive.

Just as it can be argued that a part of each person exists in every other person, it can also be argued that we are 'in tune' or intimately connected with the entire ecological system in which we evolved. Cherie Sutherland reports Michael's interpretation of his NDE during a surfing accident that alludes directly to this aspect of the phenomenon. "I was looking around and I sort of got this feeling, it's hard to explain, as though I was part of it all. I felt as though I was part of everything around me. I just felt as if everything was in me and I was in everything." Without knowing it, Michael was describing the mutual connectivity that he shares with other material objects in the universe as sympathetic patterns in the single field. The ecological system in which we live and thrive is an integral part of our CONSCIOUSNESS, just as we are a material part of the ecological system.

If a person develops a special sensitivity to the connections with other people, or one person in particular, after their NDE, and their connectivity became so sensitive that their MINDs can become momentarily linked, the transfer of thought can occur. The same is true for clairvoyance, precognition, telekinesis and other paranormal phenomena, the only difference being the object with which the momentary contact is established. The NDEr has had a unique and wonderful experience that alters the MIND and CONSCIOUSNESS patterns rendering them accessible to the rest of the universe via the fifth dimension. So it becomes possible for the experiencer to later utilize the newly acknowledged connectivity in a manner that increases the experiencer's 'paranormal' abilities.

In cases where information is passively communicated between an NDEr and another person or material object, we have extrasensory perception and clairvoyance. In other cases, a five-dimensional 'signal' from a person that propagates as a single field density variation could initiate an active change in the normal 'state of matter' or 'motion of an object' in some way which could be interpreted as psychokinesis. It is also possible that a five-dimensional 'signal' resulting from some material action occurring elsewhere in the single field could be interpreted as precognition or some form of remote sensing depending on where in space-time the 'signal' originated. In any case, the person who has become aware of their five-dimensional connections with the rest of the world should, theoretically, undergo an increase in psychic abilities, as has been the case reported by Drs. Morse and Ring as well as other researchers.

It is quite evident, from this discussion that experiencing the 'light' is essential to the transformations that occur after the NDE, as Dr. Morse concluded. Without experiencing the 'light', a person otherwise undergoing an NDE may not have become aware of his or her inherent (and seemingly but not really dormant) extra-dimensional connectivity and would thus not undergo the corresponding to intense pattern alterations in MIND and CONSCIOUSNESS that cause such radical personality changes. 'Seeing' or experiencing the 'light' induces a single field pattern alteration in the NDEr's MIND and CONSCIOUSNESS that opens the NDEr to further 'conscious' and 'subconscious' sensations within the single field after the NDE is completed. Having an NDE that does not include the 'light' can still cause personality changes, but not of such profound intensity as those in which the 'light' was experienced. These are permanent alterations that affect a person's character and personality. However, this explanation should not be interpreted to imply that an NDE is absolutely necessary to expand human CONSCIOUSNESS resulting in these transformations. There are other methods, all of which are closely related to NDEs and a knowledge of death or the single field, by which human CONSCIOUSNESS can expand to the point of developing these same characteristics.

NDLEs

OBEs and other characteristics are not unique to the NDEs alone. Mystics, psychics and others claim to have experienced OBEs without the corresponding NDEs and without dying for even a few brief moments. Furthermore, some people have had Fear-Death Experiences (FDEs). According to Dr. Morse, the FDE is "a dissociative reaction to a life-threatening situation" (Morse, 214) and shares many characteristics with its close cousin, the NDE, but the FDE seems to lack the experience of sensing or 'seeing' the 'light'. In all of these cases, there seems to be a subsequent expansion of CONSCIOUSNESS, but not to the same extent as when the full NDE occurs. Yet some of the incidents and events reported by people who have experienced these related phenomena can be even more profound and worldview altering, especially in the case of mystics attaining enlightenment, than the incidents and results claimed by NDErs.

An OBE could occur under specific conditions without the occurrence of an NDE according to the SOFT model. Perhaps the most common such example of such an OBE would proceed by intentionally minimizing the four-dimensional chatter of sensory stimuli through meditation or similar self-initiated techniques, literally going into some type of trance or trance-like state, or even though meditation and other self-induced altered states of consciousness. However, an OBE is extremely difficult to obtain using these various methods of simulating the conditions of an NDE. During death all five senses are shut down totally and completely. An equivalent state of sensory deprivation is very hard to duplicate without outside of death, while the body still lives. Even when a person is unconscious from sleep, a blow to the head or drugs, the senses are still functioning to keep the body alive and the senses are still sending data on environmental conditions to the Brain. The MIND is just not conscious or aware of the newly input data from the senses during these un-conscious states or sleep. During meditation, the MIND remains conscious, but generally ignores ordinary sensory input to the Brain. The MIND runs on automatic during these times and does not consciously sense things in the environment.

The same is true for attempts to attain 'enlightenment' as sought by certain mystical, philosophical and religious practices such as Chan or Zen Buddhism. In such instances, a state mimicking some of the physical and biological characteristics of 'dying' are reached by blocking sensory input to the MIND, literally a stilling of the MIND that simulates the NDE. This induced state of CONSCIOUSNESS might lead to an awareness of the specialized connectivity between one's CONSCIOUSNESS and the rest of the universe without the final step of actually entering (or becoming one with) the 'light' as occurs in real 'death' and the ability to have OBEs on command. However, many practitioners of this type of meditation never truly reach the state of CONSCIOUSNESS that they seek although they still profit from other benefits of the meditation and the subsequent lower level changes in their CONSCIOUSNESS and MIND patterns. Some people even claim to have 'seen' the 'light' during these states of deep meditation, but could not 'enter' the 'light' because that would be death. These techniques and practices do not utilize any special knowledge of five-dimensional mechanics or physical models of death. Instead, they utilize an individual's innate intuition as a method of exploring physical reality rather than reason or logic as reflected in scientific inquiry and its models. So, these OBEs do not have the same quality as the OBEs that occur during NDEs because LIFE is left wholly intact and still connected to MIND during this process.

FDEs though are a different matter. These can be induced by no more than a situation where death is threatened or thought imminent, rather than actually dying as in the full NDE. So the FDE is normally an experience of a much lower quality than a true NDE. Some scientists do not even like the term FDE, preferring to call this experience something entirely different. However, a more generic name would be preferable, perhaps a Near Death-Like Experience or NDLE. The title NDLE would then incorporate a variety of related phenomena that have vastly similar characteristics such as the OBE. The FDE would then rank as a situation-induced NDLE.

When the MIND is completely convinced that death is imminent, it can elicit a defense mechanism by which it spontaneously enters a low level NDLE. The mind starts to shut LIFE down in preparation for what it believes will be an inevitable death. The situation-induced NDLE (or FDE) is a purely protective mechanism of the MIND and is probably part of the 'flight or fight' mechanism which animals of all types display under conditions of extreme danger. Under threat of death, certain chemical reactions in the brain 'prepare' animals for their impending death. In beings with CONSCIOUSNESS such as humans, this chemical process can spontaneously initiate a chemical cascade in the Brain resulting in an abridged form of an NDE. However, this NDLE differs from a full NDE in that LIFE is not disrupted or terminated and the connection between body, LIFE and MIND is not broken. In a very real sense, a true NDE occurs from the 'bottom up', from chemical disruption in the material body to pattern disruption of LIFE and the subsequent freeing of MIND and CONSCIOUSNESS from the Brain, body and the five senses. The FDE is an anticipatory 'top-down' reaction in which the MIND is fooled into believing that the LIFE pattern is about to be disrupted. So MIND prepares itself for the expected separation of the MIND/CONSCIOUSNESS complex from the physical body. This process may involve some of the primary aspects of the NDE, even to the point of a low level awareness of the five-dimensional connectivity, but it does not usually allow CONSCIOUSNESS to sense or 'see' the light.

On the other hand, the attainment of a state of 'enlightenment,' as attempted in some religions, is very similar in many respects to the NDE. The 'enlightenment' experienced in these cases would therefore amount to a self-induced NDLE. These lofty goals are usually associated with the Buddhist, Taoist or Hindu religions, although mystics in Christianity, Judaism and Islam also attempt to reach similar states of consciousness. Shamans and others have also reported similar accomplishments and many people practice meditation or similar methods of concentration to achieve altered states of consciousness in the hope of inducing an NDLE and attaining enlightenment. In most cases, the preferred method of attaining states of MIND that alter CONSCIOUSNESS and produce enlightenment requires lengthy and repeated periods of intense meditation and then the results are still extremely hard to attain and never guaranteed.

During meditation, the MIND is stilled as much as possible as sensory input from the four-dimensional world is minimized to the lowest extent possible. The objective of the meditation exercise is not an OBE as described above, but rather a state which is often referred to as 'sartori' or 'nirvana,' depending upon the practice being used. The goal is to 'realize' or experience the illusory nature of the material 'self' during deep meditation when the MIND and CONSCIOUSNESS are either deprived of or ignoring the sensory input from the material four-dimensional world. In a sense, this would amount to the CONSCIOUSNESS realizing its 'self' independent of the body, experiencing CONSCIOUSNESS as separate from but still connected to the body and the material world. In terms of the SOFT model, all sensory input is minimized so that the MIND can become aware of the sixth sense acting through CONSCIOUSNESS. Concentrating on CONSCIOUSNESS its 'self' rather than the 'self' as a material four-dimensional being would invoke a differentiation between the MIND/CONSCIOUSNESS field patterns and the single field itself, causing a conscious realization of the connectivity between the CONSCIOUSNESS structure and the whole single field in the fifth dimension of space.

Once this point of realization has been reached, there comes an understanding of the intimate connection of a person to all things, not a logical understanding as reached through normal contemplation and reason as in science, but an intuitive experience of the fact of that connection or rather an experience of the connectivity itself. There is a subsequent loss of ego and self with increased compassion and love for other beings. In other words, the results are quite similar to the transformation of the MIND and CONSCIOUSNESS resulting from an NDE. It is from just such religious practices that the *Tibetan Book of the Dead* came into existence. This book describes the state of death from the Tibetan Buddhist point-of-view and is meant to serve as a spiritual guide for those who are about to die. The major difference between the *Tibetan Book of the Dead* and the SOFT model is the social and cultural context through which the experiences are interpreted. SOFT is a scientific theory based upon the present state of scientific inquiry. Science and human thought have come a long way since *Tibetan Book of the Dead* and similar treatises were written and used, but they were the appropriate venues for the presentation of a knowledge of death within their own cultures, given the low level or lack of science at the time.

Even scientists have tried to duplicate the conditions of isolating the MIND from its natural sensory pathways through such techniques as sensory deprivation. These attempts represent a class of artificially induced NDLEs. There have been some limited but intriguing results from these studies, but nothing even approaching the quality of an NDE has been achieved. More recently, 'remote viewing' and other paranormal events have been induced using the

'Ganzfeld effect.' A patient will lie quietly in a sound proof room with ping-pong ball halves placed over his or her eyes to disperse the light. Sometimes special monotonic sounds are used to induce altered states of consciousness, as is diffuse red light. These various methods blur or neutralize the senses of sound and sight. Rather than stilling or cutting off the constant 'chatter' of normal sensory data from the four-dimensional world, constant unchanging sensory input is used to anesthetize the MIND to its four-dimensional senses and environment, thereby mimicking the isolation of MIND and CONSCIOUSNESS from the sensory background 'chatter' of the material world.

In a very broad sense, these are just methods of confusing the sensory input so that they are ignored by the MIND. The MIND normally depends upon its awareness of change in isolating and interpreting sensory data, so when sensory input is constant and unvarying the MIND tends to ignore it or block it out. In this manner, CONSCIOUSNESS can be opened to the far more subtle five-dimensional signals that are received by the normally 'dormant' channels of CONSCIOUSNESS. In remote viewing and perception experiments using the Ganzfeld method, the viewers have attained some startling results. The practice is similar to an OBE, but the viewer would normally never leave the body. Only a mental vision of distant locations and objects is attained, so remote viewing depends more directly on the sixth sense without a realization of the connectivity between CONSCIOUSNESS and the single field, or its equivalent experience.

There is still another class of NDLE that seems to have escaped scientific inquiry even though this class of phenomena may well prove the most interesting because it sometimes leads to psychic experiences of other kinds. These are the spontaneous NDLEs, which have no apparent cause. These particular NDLEs are hardly ever reported in the literature on the subject and are little understood. In fact, they may be misreported and thought to be symptoms of different psychological trauma and conditions. Some people just become psychics after they have had some type of traumatic psychological experience. Such experiences could be very low-level NDLEs. On the other hand, some people undergo spontaneous enlightenment. The causes of this phenomenon are unknown and will probably remain unknown since the details crucial to explaining this turn of events are lost to history. Hui Neng, the seventh patriarch of Chan Buddhism, reached enlightenment spontaneously without any religious or other training. His enlightenment was confirmed by the sixth patriarch who forced to keep Hui Neng's enlightenment secret from the other monks of his order for political reasons. The more advanced monks would not have understood how a poor peasant boy could be enlightened without trying and they could not after studying and practicing for so long. So when it came time for the sixth patriarch to choose a successor, he chose Hui Neng who was the only other person in the order that had reached true enlightenment. The sixth patriarch gave Hui Neng the symbols of the office and told Hui Neng to run for his life for the other monks would surely kill him when they found out. This act caused a severe split or schism in Chinese Buddhism, which thereafter developed along two different lines.

The last major category of NDLEs will come as a surprise to nearly everyone. This NDLE is actually externally induced in the MIND and CONSCIOUSNESS of the experiencer or as they are more commonly known, the abductee. Strangely enough, this class of extremely controversial phenomena that seem related to NDEs and the state of CONSCIOUSNESS induced by the experience are usually associated with UFO activity. Whatever the root cause of 'abduction' events, regardless of whether they represent the actual physical kidnapping of humans by aliens or not, some 'abductees' have clearly experienced the same transformative changes that are experienced by people who have had NDEs. This fact has not gone completely unnoticed. Dr. Kenneth Ring and Dr. John Mack have both written on the correspondence of the two types of phenomena. Dr. Ring has described the effect from the point-of-view of the NDE researcher, and Dr. Mack has noted the correspondence from the point of view of the Abduction researcher.

In one case, investigated by Dr. John Mack, an abductee called Carlos described his own abduction experience as if it were an NDE.

> At their core Carlos's encounters have brought about a profound spiritual opening, bringing him in contact with a divine light or energy, which he calls "Home," which is the source of his healing and transformational powers. In our sessions, when he comes close to this light he becomes overwhelmed with emotions of awe and a longing to merge with the energy/light being. Space and time dissolve, and he experiences himself as pure energy and light or consciousness in an endlessness of eternity, "a pure soul experience ... I go back to the source because I am not *just* human. I need to go back to the source to continue."
>
> Carlos, like so many abductees, has developed an acute ecological consciousness. He is deeply concerned with the earth and its fate. (Mack, 362)

The "profound spiritual opening" and the new ecological awareness experienced by Carlos after the fact have both been reported by NDErs. In fact, if the word "abductees" had been left out of the final sentence, the whole episode could have been mistaken as a description of an NDE. Curiously, Carlos has experienced both NDEs and abduction and compares them favorably, regarding the abduction phenomena as "having far more transformative power." He also noted that "through the alien encounter there is 'access to the bliss, and the near death experience is ... a momentary place in between. It is a soul place, to gather up.'" (Mack, 363-364)

Dr. Mack has reported many others encounters which sound a great deal like NDEs. Aliens entering people's bedrooms and floating them out of their beds and through walls to the alien ship sounds almost like a form of OBEs, except for the fact that the abductees claim that they were physically and materially transported to the ships. An OBE is a physical but completely non-material transport of conscious awareness from one location to others, so it is not known if the abductee's experiences represent true OBEs or not. In spite of the similarities, the abductees still claim to have been materially transported. However, this fact does not mean that the alien encounters are real material abductions. Nor is it an endorsement of either the validity of an alien presence on earth or the physical reality of extra-terrestrial spacecraft. The fact that some abductees make claims that they were floated through solid walls or traveled over great distances instantaneously certainly implies that the experiences were extra-dimensional. The primary similarity between the two seemingly different types of phenomena actually resides in their power to expand CONSCIOUSNESS. The abductions certainly seem more material and physical than NDEs, just as the NDEs seem more mental and spiritual by their characteristics, but both seem to expand the CONSCIOUSNESS of experiencers in the same manner.

Given the SOFT explanation of chemistry and the naturally occurring evolution of 'life', it would seem likely that sentient beings would already have evolved elsewhere in the universe. Our Solar system and the planet Earth are relative newcomers to the universe as compared to the many older star systems in our own galaxy. It would also seem likely that such sentient beings would have attained a higher degree of consciousness than humans and should thus have realized at some time in their own past that 'consciousness' supports extra-dimensional connections as described by SOFT. It would thus seem logical that under these circumstances they would have evolved the ability to communicate with humans and other 'conscious' beings by utilizing their knowledge of the field density variations and 'pattern matching' in the fifth dimension in a manner that is similar to the descriptions given by Dr. Mack. This scenario is highly conjectural, but it does have a certain logic given the characteristics of a five-dimensional universe described by SOFT. So, SOFT implies that sentient beings elsewhere in the universe could presently be capable of communicating with humans extra-dimensionally.

The alien abduction and UFO phenomena are well beyond the scope of this study, but still pertinent to the science of LIFE, MIND, CONSCIOUSNESS and death. Most scientists readily accept the very high probability that life exists elsewhere in our galaxy and the universe. The study of life elsewhere even has a name within the scientific community, exobiology. Science, in light of evolution theory, accepts the fact and existence of life as a common development in nature. Given a source of water and energy, it is inevitable that life will evolve, so life should naturally evolve in many different locations in the cosmos, not just on Earth, or so science believes. Then, give the statistically large number of planets that science estimates are Earthlike, the probability that life exists elsewhere in our own galaxy is very high. Science believes that life should have evolved on millions of planets in our galaxy alone, and the evolution of life, since it should follow the same developmental pattern everywhere, seems to insure that MIND and CONSCIOUSNESS should also have evolved on other planets.

Science discounts UFOS and visitations by alien life forms, not because the laws of physics says interplanetary and interstellar trips are impossible because they 'are' possible even given our present knowledge of physics. Such trips are just impractical. The real possibility that aliens could be visiting us at the present time is discounted by most scientists because the probability of planets that are close enough to us to send emissaries to our little green planet as well as having just our own level of learning and knowledge would be extremely slim at best. Since the universe and the galaxy are so old, and our Sun and the Earth so young, it would seem that other alien civilizations with the capabilities to send spaceships to Earth would have evolved past that stage in their development or just died out eons ago. More advanced races and species of sentient being would be so far beyond us in evolutionary terms that they would not be interested in us if they even realized that we exist here a the edge of the Milky Way Galaxy. Most scientists discount the possibility of alien visitors because they would probably be too advanced for us to be of interest to them, if they exist at all.

Now that very fact implies an interesting possibility. This possibility, the very crux of the scientific argument against visitors from other worlds, actually opens the possibility and even the probability that there are alien races of sentient beings that are more evolutionally advanced than humans. Now if the path of human evolution were the rule in the universe, rather than the exception, then those other sentient beings would probably have reached a higher level of consciousness than humans quite some time ago. They would have had their own debates and studies of NDEs and NDLEs and would have developed their own SOFT model of LIFE and death within the context of their own scientific inquires. By now they would be masters of using and manipulating the single field density patterns that are LIFE, MIND and CONSCIOUSNESS. They would surely be enlightened beings and would not need to travel to our little and inconsequential planet because they could communicate with us by their equivalent of our sixth sense. Perhaps that is now occurring and perhaps that extra-dimensional communication is the reason why NDEs and abduction phenomena are so similar. As farfetched as this idea may sound, there is at least circumstantial evidence of this possibility. Both types of phenomena seem to affect humans at the cellular level of our being, a fact that can be easily explained within the SOFT model of LIFE.

The SOFT view of death is further supported by reports that experiencers sense or feel the material effects of the NDE at the cellular level, many times as vibrations. These are not true vibrations in the material sense, but are instead identified as physical field resonances as described by SOFT. From the four-dimensional material point-of-view, the lowest common unit of living beings would be the cell, but the cell is also a product of the most fundamental pattern of LIFE in five dimensions according to the SOFT model. When energy changes occur in the cell they affect the five-dimensional extensions of the individual elementary particles that constitute the cell by increasing or decreasing their extension into the fifth dimension. These changes entangle within the five-dimensional single field to determine the overall field density variation pattern which is LIFE itself. However, these changes are actually variations in the field density of the single field as a complete continuous whole. As such, the cells act like antennas to pick up changes of field density from other sources in the five-dimensional single field environment of LIFE, MIND and CONSCIOUSNESS.

The individual cells could theoretically be 'in tune' with other elements and objects in the universe at large, at least in so far as those sympathetic 'resonances' do not disrupt the pattern of LIFE. So, the 'light' portion of the NDE would conceivably be felt at the cellular level of LIFE. Some NDErs have described this very action at the cellular level.

> Everything was suddenly light and vibrating. I was vibrating. It was like a rocket, like I was ready for takeoff. I was in this incredible light, and then I found myself in this place, just of love. It's like every cell either was love, or is love, or is loved. It was a vibrancy, you know, like when the New Age people say 'I am light. I am love.' They don't have a clue what that means. There were these beings in front of me. One predominant one told me that it was not my time to cross over. Then the being said: 'Your spirit wishes to give you this unfolding of information for the mission which you are here to do.' That was telepathic, cellularly. It was like, everything became a part of me. (Rommer, 69)

This description by Tara came from the investigations of Dr. Barbara Rommer. The references to "light," "incredible light," "every cell," "vibrancy" and "cellularly" certainly imply that the experience of 'light' in the NDE was both overwhelming and affected every cell of Tara's body as well as the LIFE pattern directly. Janet, another experiencer stated this obvious fact even more clearly, as reported by Dr. Morse: "I became aware of every cell in my body. I could see every cell in my body." (Morse, 150)

Yet this description of feelings at the cellular level has also been reported to Dr. Mack by abductees in a number of different cases. In these reports, the cellular disruption caused by the alien sentient beings was quite unpleasant if not outright painful. However, the purpose of cellular manipulation by aliens is self-evident. If alien sentient beings are far more advanced than humans, perhaps they have found methods to artificially induce alternations in human CONSCIOUSNESS and are causing such alterations so they can communicate with us. The situation would be similar to our finding a new race of beings that had the ability to talk but had not yet developed language or language skills, so we could not communicate with them. We would still want to learn from them so we would teach them how to speak our language in order to communicate and learn from them. By the same token, consciously advanced beings would still want to communicate with us and learn from us.

If someone wished to physically engineer an alteration in CONSCIOUSNESS, then the cellular level would certainly be a good place to start, at least according to the SOFT model. Since the individual cells are the basic and most fundamental living unit, they would each have their own individual field density LIFE patterns. The overall sum of all the cellular LIFE patterns make up the total LIFE pattern of a human or other being. We are beginning to understand that changing CONSCIOUSNESS from the 'top down' necessitates an NDE or an NDLE, but they are not completely efficient. So why not engineer the same alterations in CONSCIOUSNESS in the same way that CONSCIOUSNESS first evolved, from the cellular level of LIFE 'up to' MIND and CONSCIOUSNESS. The cellular engineering of members of our species may be exactly what alien sentient beings are trying to accomplish so that they can alter our CONSCIOUSNESS and communicate with us in their own way. This scenario is purely conjectural, but the possibility is not completely outlandish and it does explain the reported phenomena.

Now why would other sentient beings in the universe need to alter human CONSCIOUSNESS to communicate? We just cannot expect other sentient beings to understand English, French, German, Japanese, Russian or any other Earthly language. They might not even understand the concept of language. In our own egocentric way, we humans believe that all LIFE should be the same as our own and should have evolved in a material environment similar to our own. Our view of the universe is far too egocentric and sometimes delays our scientific and intellectual progress. Perhaps the CONSCIOUSNESS of alien sentient beings evolved in an atmosphere or ecosystem so completely different from our own that they are totally different from humans in a material sense. So if sentient alien beings wish to communicate with us there would be no framework of ideas and concepts for a mutual understanding of our languages, terms and concepts, if indeed they even use sounds to form languages as we do. So other sentient beings might conceivably alter our MIND and CONSCIOUSNESS patterns to develop a proper framework or background for us to understand their concepts of communication. This idea is very speculative, but does seem to fit the stories told by abductees and experiencers alike. Our experiences with NDEs and NDLEs might just open the human race up to our first communication with aliens and other sentient beings with which we probably have no basis for communication. Perhaps we could even learn to 'speak' to dolphins and other animals on our own planet.

'Where' is death?

The concept of place or location is so ingrained in our way of thinking that we assume it carries over the threshold of LIFE to death. Indeed, the concept of place is the very basis of the concept of 'self' itself. The 'self' occupies its own place and is aware of its 'self' as opposed to everything else because the 'self' realizes if is located here (locality in physics) and everything else is located there or elsewhere (non-locality in physics). The same is true for the state of death. If we survive the material death of the body, or some part of us survives death, whatever survives must be somewhere and it is certainly not here. It is difficult if not impossible for us, at our present level of CONSCIOUSNESS, to conceive of our 'spirits' not having their own domain or realm of existence after death. The view that death must be a special place has been ensconced in our collective human psyche by religions and mythologies, which existed long before science became a factor in our search for knowledge since before we even began to write down such tales.

Nearly all, if not all religions describe a place where we go when we die. The western religious traditions of Judeo-Christian-Islamic heritage posit a heaven and a hell. People who lead a good life go to heaven when they die and people who sin go to hell. According to Greek mythology, those who are good go to the Elysian Fields after their journey across the River Styx and those who were bad during life go to the realm of the dead where judgment is passed. It is nearly the same in Viking and Norse mythology. Warriors who die in battle go to a special hall where they will await the end of the world to help the Gods fight the Frost Giants.

During the middle ages of European civilization, the Roman Catholic Church equated the astronomical heavens to the spiritual heavens and thus imagined the concept of a material place for spirits after dead. Everything beyond the orbit of the Moon amounted to the heavens where God, the angels and the spirits existed, while Hell was at the center of the universe at the center of the Earth. People lived on the surface of the Earth so they could be good and strive upward to the heavens or people could be bad and go beneath the surface of the world, where grubs, worms and the lower animals live, eventually to Hell. So the Catholic Church actually reinforced the idea that people go to an actual place when they die.

On the other hand, the Eastern religions followed a somewhat different path. In both Hinduism and Buddhism, people live according to their Karma. When people die, they are reborn or reincarnated back to earth to live a better or worse life according to the Karma that they accrued during previous lives. Siddartha Gautama, a Hindu reformist, was enlightened in about 650 BCE and came to believe that people could break this cycle of birth, death and rebirth by becoming enlightened. If a person were to die after enlightenment, they would not be reborn again since enlightenment breaks the reincarnation cycle. But that would raise the question of what happens to the CONSCIOUSNESS that has broken the cycle of life, death and rebirth, after the body dies. Some Buddhists believe that the Buddha, Siddartha Gautama, went to some kind of heaven when he died, but others believe that he just ended. So these Eastern religions, no matter how esoteric or mystical, have not completely gotten rid of the notion that the dead must reside somewhere in some place. Whatever the case may be, it is quite certain that these religions have also found a place or location for people to go when they die.

Now the SOFT model presents an actual place where the MIND/CONSCIOUSNESS complex exists when the LIFE of a person ends. However, this 'place' is the 'same place' that it has always occupied, which is both 'here' and 'there'. The surviving MIND/CONSCIOUSNESS complex is essentially freed from its material bondage in the four-dimensional 'sheet' that is our material world and continues to exist as a free entity in the five-dimensional single field. However, this place where the MIND/CONSCIOUSNESS complex now 'resides' is not a completely separate place from the four-dimensional material world. From a three- or four-dimensional point-of-view, the fifth dimension is a point-by-point extension of our four-dimensional 'sheet', but they are continuous and therefore not separate. Yet from a five-dimensional point-of-view, the four-dimensional 'sheet' and that part of the 'sheet' that makes up the material body and the pattern of LIFE is just a slice across the five-dimensional continuum. IN either case, MIND and CONSCIOUSNESS are five-dimensional field density structures both before and after the living being dies. It is only in this sense that the 'spirit', if you wish to call it that, 'dwells' in the single field. But the single field fills the entire universe, so the 'spirit' exists in the same complete universe that the living being occupied, only it is a far more complete universe that the 'spirit' occupies.

Even while religions and mythologies have found their own places for the dead to travel to, those NDErs who have actually experienced death have described a real physical place as the setting for the NDE. The NDEr's own descriptions of their NDEs lend support to an extended awareness of our common space-time continuum, as described by the SOFT model of physical reality. Mr. Spencer has described his experience as a "passage ... a transition into another realm." (Morse, 74) His statement would seem to indicate that he came to be located, in one form or another, in a real physical place other than our common four-dimensional space-time. Olaf described this place in even more detail.

> Olaf felt as though he were 'floating in a universe with no boundaries." He saw the universe as a system of shrinking soap bubbles, one in which the bubbles appeared in spherical, concentric trains that moved in intricate patterns that he completely comprehended. ...
>
> On the verge of death, this fourteen-year-old boy with a mediocre school record felt as though he had been handed the keys to the universe. "I felt I had a total comprehension which made everything understandable," he wrote. In his near-death experience, Olaf stood at a "bright orange light." He called this light "the point of annihilation," a frightening place to be but one that gave him universal understanding. (Morse, 11)

Olaf described the 'location' of his NDE as a "universe without boundaries," a phrase which certainly would not seem to fit our four-dimensional reality. According to general relativity, our four-dimensional continuum is a Riemannian sphere and according to SOFT our four-dimensional continuum is curved in the fifth dimension and forms a single polar Riemannian sphere. In either case, these geometric structures are both unbounded and finite in extent, so they would have no boundaries as Olaf perceived. The "floating" reference would just as certainly imply some physical reality to this place and certainly he was either beyond the gravity of our space-time or his new 'body' had no mass to interact with gravity. Since gravity is a product of the four-dimensional 'sheets' overall curvature, once you 'leave' the 'sheet' you would not experience gravity and would thus seem to 'float' in the single field, or rather that is how a person's four-dimensional Brain and four-dimensionally oriented and trained MIND would perceive or describe the sensation. Dr. Raymond Moody also came to the conclusion that the new 'body' was 'weightless' because of the descriptions that experiencers "find themselves floating right up to the ceiling of the room, or into the air." (Moody, 45)

The fact that Olaf saw the universe as a system of "shrinking soap bubbles" could easily refer to the fact that physical fields spread out spherically from their sources implying that the basis of physical reality is the field rather than the discrete particles of quantum theory. This assumption is further confirmed by Olaf's insight that "the bubbles appeared in spherical, concentric trains," a description which fits a picture of successive wave fronts of light passing toward or from a light source. A "tunnel of concentric circles" was also described by one of the patients in Dr. Moody's original study of NDEs. (Moody, 33) And finally, Olaf's description of "total comprehension" could refer to the intuitive sense of CONSCIOUSNESS upon the realization of the MINDs continuity and connectivity with all things within the single field. In his experience, Olaf seems to have progressed toward and witnessed an even deeper level of reality than living cells, all the way down to the relativity of the waveforms of material particles themselves. Olaf did not see the universe as hard little particle of matter at the most fundamental level of reality, or perhaps he only approached perceiving that level, but he saw the level as patterns and waves, as continuous field structures such as those predicted by SOFT.

We, as humans, explore our world and understand its innermost workings through reason and logical analysis. In so doing, we come to understand how our world works, in so far as science is concerned, within the four-dimensional 'sheet' where gravity, electromagnetic waves and material particles are real to us and have substance. To understand or explore the universe outside of the 'sheet,' the five-dimensional portion of the universe, we have only our natural intuition. So the extra-dimensional realm of physical reality is normally beyond our perception and worldview as is the infinitesimally small fundamental regions of physical reality. Both our four-dimensional existence in the 'sheet' and the rest of five-dimensional space-time are physical, but only that part within the 'sheet' could be termed material and thus subject to logical analysis and reason with respect to our common experiences of our material lives. Yet reason and intuition are both ways of knowing and coping with our world, universe and environment. Perhaps Olaf found a way to explore fundamental reality at a deeper level than reason can hope to achieve by utilizing an enhanced intuition resulting from his NDE.

MIND interprets our sensed world and environment using reason, the cumulative result of real experiences of the material four-dimensional world placed within a specific mental framework or worldview, while CONSCIOUSNESS deals more with intuition, our innate feelings and subconscious understanding of the larger five-dimensional framework of physical reality. CONSCIOUSNESS knows and understands the universe within the context of its own environment through its connectivity in the fifth dimension even though we are neither normally nor directly aware of either this connectivity or the new realm of physical reality that is the single field. Using reason and logic based on our perceptions and experiences in the four-dimensional world, we commonly impose a mathematical framework or artifice on the single field density variations to render them understandable in scientific terms. We can only abstract our four-dimensional perceptions to develop a general framework for the scientific analysis of our five-dimensional experience. Otherwise, we could explore our greater five-dimensional world intuitively as happens with NDEs and NDLEs. In either case, when conducting such explorations we interpret them as places. This dichotomy exemplifies the "Catch 22" situation that emerges from trying to realize something that is purely intuitional by utilizing reason and logic. NDEs follow this same dualistic pattern of knowing since the NDE represents the evolving conscious awareness of connectivity in the five-dimensional realm when we die based upon the four-dimensional interpretations of our 'mind's' pre-death knowledge of reality. Olaf and others interpret the fifth dimension as another very strange place as opposed to the four-dimensional world experienced during LIFE, never completely realizing that they are ultimately the same place as perceived from the two different points-of-view.

After reaching this point in progress of the MIND/CONSCIOUSNESS complex, CONSCIOUSNESS grows. Since all information of the death state of being comes from NDErs, it is not known how far this process can progress after death when the surviving complex finally enters the 'light' for good. For lack of pertinent information and reasonable observation, any knowledge of what lies next is partially speculative and partly scientific extrapolation. However, the SOFT model predicts the possibility that there is a still higher state of CONSCIOUSNESS that could be reached, even after death. Just as increasing knowledge of the world and new memories ultimately resulted in the evolution of CONSCIOUSNESS out of the raw 'stuff' of the MIND, we can assume that increasing knowledge and understanding of the greater five-dimensional universe should precipitate a new and even higher level complexity corresponding to a sixth dimension of space. Buddhist and other mystical traditions do tell of even higher states of consciousness, but these realms are not completely pertinent to this particular physical model of death.

Death refers only to the material body and physical LIFE. Upon death of the body, LIFE absorbs the disruption caused by death allowing MIND and CONSCIOUSNESS to move onto a separate extra-dimensional existence. Evidence of this new existence or level of being comes from observations made by people who have had NDEs. The physical realm that the NDErs report visiting is perfectly compatible with the five-dimensional model of our space-time continuum. So SOFT can be readily used to explain the physics (or paraphysics) of this new extra-dimensional extension of reality.

However, the actual existence of this extended reality cannot be demonstrated or verified by NDEs and NDLEs alone. Proof of the existence of a fifth dimension must come from the understanding of purely physical processes and scientific experiments. However, there are various bits of circumstantial evidence that render the existence of a fifth dimension quite likely. Perhaps the most significant evidence can be found in the testimony of a single NDEr, as reported by Raymond Moody, the first scientist to take the phenomena seriously and explore the testimonies of NDErs.

> Now, there is a real problem for me as I'm trying to tell you this, because all the words I know are three-dimensional. As I was going through this, I kept thinking, well, when I was taking geometry, they always told me there were only three dimensions, and I always accepted that. But they were wrong. There are more." And of course, our world – the one we're living in now – *is* three-dimensional, but the next one definitely isn't. And that's why it's so hard to tell you this. I have to describe it to you in words that are three-dimensional. That's as close as I can get to it, but it's not really adequate. I can't really give you a complete picture. (Moody, 26)

If this woman's observations are correct, and there is no logical reason to believe that they are not in light of the present emphasis in theoretical physics on higher-dimensional spaces, then the fifth dimension does indeed exist, it is being 'visited' by people during NDEs and NDLEs, and its formal 'discovery' will completely change the way we think and understand our universe. In this testimony and similar cases, science has direct observational evidence that space-time is not limited to the normally accepted four dimensions of our common experience. It would seem that death ends nothing, but actually represents a new beginning and existence in this higher-dimensional physical reality.

Otherwise, the seeming appearance of an evolutionary inevitability to the development of LIFE, MIND, CONSCIOUSNESS and death, especially with theoretical and observational evidence of the survival of MIND and CONSCIOUSNESS after material death, implies that some measure of purpose and possibly outright cause in a viable factor in the evolution of our universe. The very fact that individual CONSCIOUSNESS is so heavily entwined with the rest of the universe, should alone imply some type of conscious choice in the evolution of LIFE, MIND and CONSCIOUSNESS as well as what lies beyond. In the past, this implied sense of purpose has been interpreted as 'evidence' of the existence of some type of 'supreme being' that created and runs our world, sometimes with an iron fist. However, the religious interpretation is not completely necessary. The level and type of purpose that exists in our universe, if it exists at all, can only by decided by science.

CHAPTER 7
A UNIVERSE OF PURPOSE

The ideal of the Supreme Being is nothing but a regulative principle of reason which directs us to look upon all connection in the world as if it originated from an all-sufficient necessary cause.

—Immanuel Kant – 1781

Intelligent life on a planet comes of age when it first works out the reason for its own existence.

—Richard Dawkins – 1967

Everything that has a beginning has an end

Death is inevitable, a fact which no one can or would even attempt to deny. That which lives eventually dies. Nothing material is truly immortal, a fact that seems to be the rule in our universe, which is itself governed by the rule of ever-increasing entropy, which amounts to a growing chaos in the material content of the universe. Some philosophical arguments even claim that eons from now, the universe itself will enter a stage of death where the overall temperature of all things will dissipate evenly throughout the whole volume of the cosmos. Such a fate is called the "heat death of the universe." Everything will reach a final equalized temperature that will be close to absolute zero, and all life will freeze into oblivion. Although this particular scenario will unlikely never play out, the concept does provoke some thought in the minds of philosophers and scientists. Other more likely fates will decide the future course of the universe long before this particular fate has a chance to materialize.

Some things live short lives, gone in a whisper, but other living beings do not die for millennia. Humans normally live less than a hundred years on the average, which is old by some measures and young by others. Under these circumstances, it would be nice if everyone could live the full measure of their lives and die peacefully in their sleep in the twilight years of their old age. But that is seldom the case. So it seems that we automatically jump to the conclusion that we were somehow robbed of our expected immortality when we die and therefore look for the purpose of our deaths. Yet we are not really cheated out of our lives if we do not fulfill our overly optimistic expectations to live long and fruitful lives. It is inevitable that we die, but we see our death as if it comes unexpectedly knocking down our doors. And then we ask "why?"

Why do we die? Why do we die at such and such a time? We ask, why now of all times?, as if death is an some inconvenience that was just developed to harm us? We also ask, why did so-and-so have to die in such and such a manner, as if any other way of dying would have been better? Why this? Why that? Why? Why 'why'? The answer is easily found. We ask why because we have expectations of an afterlife and it does not matter whether our expectations represent a true reality or if they are imaginary, truly perceived or completely unrealistic. There is an old adage that claims we only ask questions if we already either suspect or know something of the answer before the questions are asked. For example, we can ask what time the sun will rise because we fully expect the sun to rise. If we did not expect the sun to rise, then we would not ask what time the event would occur. The root of the answer is always implied in the question. So what does the question 'why?' imply? We ask why because we anticipate a specific answer. We assume that any and every event will always have a cause, while not knowing the cause of a particular event is not acceptable as a reasonable or sufficient answer to our curiosity and stubbornness. We ask why we die because we assume that the cause of our death itself has a deeper, fundamental and more meaningful cause within the context of the whole universe of our being and existence. This would not be the mundane cause of any particular death such as old age, disease, a heart attack, a stroke, or a gunshot wound, but the ultimate underlying cause of the very concept of death. Why do all living beings die?

We sense a fundamental continuity of being and becoming in nature with every moment of our lives and every last bit of our intuition. Just try and think of a period of 'no time' between two different moments of time, or two different clicks of a clock. Or try and imagine a 'no-thing' between two consecutive points in empty space. Both are impossible because the mind and consciousness are primed and programmed by their very existence for a completely continuous universe. We are incapable of even imagining a discontinuity let alone sensing one. So after a person dies, the people who survive can still imagine and think about that person, almost as if the person were still real or alive. Thus we assume that something of that person survives outside of our mind and thoughts. We automatically apply continuity of being to life without any thought or hesitation and that continuity continues right through death, because it seems natural to do so. So we cannot accept death because accepting death as a final act of life would seem to represent a discontinuity in nature and the universe.

Since continuity rules the flow of time into the future, each effect (such as death) within the 'cause and effect' linked-chain that is physical reality becomes a cause for the next effect rather than an end in itself, ad infinitum. Within this context, death can be considered either the cause of the afterlife or just a point that momentarily occurs between the continuity of life and afterlife, just as it was an end to life, confirming our sensed continuity as a fundamental rule of the universe. Between the inevitability of death and our assumed continuity of existence and being through and beyond death, we sense some greater cause that rules the complete sequence of life, death and afterlife. It is a question of relativity, our continuous existence and being relative to the rest of the material universe and the flow of time. So the proper question is not why we survive death, but how we survive death, and that is a matter for science to deal with.

Before the scientific revolution in the seventeenth century, the philosophers asked 'why?' when they wanted an explanation of some event or phenomenon in nature, but alterations in our attitude and outlook toward nature changed during the sixteenth and seventeenth centuries such that science stopped asking 'why?' and began to ask 'how?'. The question 'why?' was relegated to either religion or the branch of philosophy called metaphysics. But that was a long time ago and attitudes are again changing. Since the scientific revolution, death and the possibility of an afterlife have only been considered seriously within the scope of a religious 'why?', but now they can be approached as a scientific 'how?'.

There are two different contexts for framing the question 'how' within science. We can assume that there is an actual physical reality independent of human consciousness and our physical laws of 'how' it works are accurate. In other words, any intelligent being that develop a science anywhere in our universe would derive the exact same laws that we have discovered. On the other hand, our physical laws explaining 'how' the universe works are not really part of nature or inherent in nature but are just a product of our sensual impressions and interpretations of nature and the natural order of the universe. The 'how' would then be an illusion created by the human mind and imposed by us on our physical environment. So cause and effect would be reduced to just the view of the world by the human mind. Cause and effect are just interpretations of our mind about nature as sensed through our five senses, so we can never know what reality truly is and we can never really now if that reality includes the survival of something after death. In this case, the universe could actually be something entirely different from what we sense. So the question of 'how' we survive death is irrelevant.

A similar view was actually proposed over a century ago by Ernst Mach, although it had nothing 'directly' to do with death and the afterlife, although Mach was reacting in part to the historical forces that affected both science and society, resulting in the rise of modern spiritualism. Mach believed that we could never really know physical reality and that our physical laws, the laws of nature that science had developed, were merely convenient ways of representing our sensations of the world around us in a language common to all scientific investigators. The physical laws were not actually part of nature, but part of our collective minds' interpretations of nature and the world around us.

To some extent Mach's view and the view presented above might be true, but Mach would never agree with the modern implications of his philosophical positions on science. What we normally sense as matter is really space-time curvature of the 'sheet' and the 'sheet' is just a density structure in the single field. We do not directly sense the structure of the single field, only its effects on us. But at the same time, the ideas are partially untrue because there actually is a single field, there is a 'sheet' and there is curvature and structure within the single field while the single field does follow specific rules of action and reaction, even if we do not perceive them in their truest forms. They form the basis of our laws of nature, as interpreted through our limited senses. So we may impose our limited interpretation on nature and the universe, but if our limited interpretation of the rules of the universe were very far out of synchronization with the universe's own rules, expressed in the universe's own unique terms, then our laws of nature would not be feasible and would just not work when we test them. We can do nothing that does not follow the rules of the universe within limits, even if we do not know those rules completely, because nature would fight back and tell us that our rules and laws were not compatible with the rules of the universe.

We may impose our limited interpretation on nature and the universe, in the form of physical laws, but there is really and truly a physical reality that is governed either by the laws that we have already discovered or by similar rules that we seek to find and define more precisely. Unfortunately, Mach's and similar arguments do not take into account that our brains and minds, just those entities that are interpreting what our sensations are telling us of the universe, are also subject to the same rules of the universe that they are sensing. So the brain and mind must still operate within the boundaries set by the universe's rules when interpreting the sensations of the world and establishing our physical theories and laws of nature. This fact alone would guarantee that our physical laws were compatible with the actual rules of the universe at some level of reality.

And this brings us to the second possibility. There is truly a physical reality and a universe that exists independent of our ability to sense it, a universe that follows specific rules that we interpret in our own fallible manner as best we can. Both possibilities lead to the same conclusion that there is a physical reality that follows universal rules, whether we able to discover and state those rules or not. LIFE, MIND, CONSCIOUSNESS and death must therefore be products of those rules even if we cannot fully and completely understand them. So if there is any purpose to death as well as a continuance of something that survives the death of the body, it must also be a product of those same rules of the universe. To answer the question 'how' some part of us all continues after death, is to try to discover the more basic rules of the universe that would govern life and death, which is to discover if there is a physical or natural purpose in the universe to which humans must be subject. This new type of natural purpose need not be the same as a human concept of purpose, just as our laws of nature may not be complete when judged against the universe's actual rules of action.

The word 'purpose' is normally associated with religion in our society and culture, rather than science, because religion attempts to answer the ultimate question of "why?" and discover the first cause instead of the immediate 'how?' that science seeks to answer. According to the most common view of the subject, religion assumes the existence of a supreme being who decides our future, which includes our deaths, and decisions or judgments regarding our future are based on a perceived morality of conduct while living. Otherwise, people assume some ill-defined concept of fate that drives our lives toward their inexorable ends. This fate is usually a non-specific non-religious but possibly spiritual concept. Yet there is no absolute necessity for either of these views. They only fill gaps in our knowledge of the universe that science has not yet filled. Purpose may be something altogether different than what we assume and it need not be associated with religion or any other form of spirituality that is not scientific. Yet purpose does imply some ethical and moral considerations that are not directly scientific. As little as science has to do with questions of ethics and morality, nothing other than religion has had so big an impact on ethics and morality as science.

The human concept of 'purpose' is one of those 'rules' that are corrupted by our sensations and interpretations of nature and physical reality that we attempt to impose over physical reality and the universe. We do have an intuitive sense of the natural purpose by which the universe operates, and it is upon that intuitive sense that our misinformed definition and concept of 'purpose' is based. But natural purpose is not what we humans sense or interpret as human purpose. Therefore, to ask 'why' according to the human concept of purpose is either wrong or irrelevant, but to ask 'how' after the scientific manner is correct.

Natural Purpose

Purpose is a rather strange concept, as concepts go, because the word purpose carries connotations that far exceed its own value as a simple word. The definition of purpose and the images that it evokes when applied are not always equivalent. In a simple sense, purpose means that the event to which the word is applied has a direct identifiable reason or cause, but usually any use of the word implies a still deeper and more fundamental cause beyond this obvious or immediate cause of the event that has purpose. This perceived implication is a product of the human 'quality' of 'freewill' in our choices.

We choose to follow one course or another and therefore our human actions have cause or purpose, the ultimate goal of our course of actions. We choose a specific course of events because we have a specific goal, objective or agenda in mind. Unfortunately, we also tend to impose this same idea of purpose on the world around us, giving the word purpose a power over us according to the whims and forces in nature which affect our lives, but over which we have no control. If a person dies in a hurricane from perfectly natural causes, we lament and ask why that person had to die. We know why but we still ask, because that person chose to go in harms way and paid the consequences. But we still ask why, referring to a deeper level of reality beyond both that person's choice and the hurricane that killed him. This transfer of purpose from our own human choices and actions to nature in general as well as physical reality's actions and influence over us raises a very important question, especially for science and philosophy: Does purpose exist in nature, independent of human action and influence?

Answering this question is extremely difficult since we don't know exactly what form purpose might take in nature and the natural world. Yet we assume that purpose, even in nature, implies a conscious choice over present and future natural events as well as nature's affect on us. People project their perceptions of human-like purpose on nature and the universe out of a sense of arrogance and the mistaken belief that we understand enough about the universe to decide on how the universe SHOULD work. However, purpose in nature might actually be quite different from what we, as humans perceive as purpose in our everyday lives. Perhaps there is no 'conscious' choice of events in nature, even though 'purpose' could still exist in nature and the universe in a different and less recognizable form than we would expect to find.

We, as humans, do not yet know enough about the universe to project our concept of purpose onto nature and the universe. Our own concept of purpose is ill defined at best, and then it is subject to change over time as well as changes from culture to culture. Yet there is still some general consistency in the concept of purpose, which would again seem to lead to a conclusion that there might be an absolute definable purpose inherent in nature. Any purpose inherent in nature would not necessarily change over the course of time or space (location) or even between different human interpretations. The universe follows its own time, not human time, and adheres to its own constant rules of operation, which may or may not be described exactly by human theories and laws of how nature works. Theories and humanly derived laws of nature are subject to change as we grow and learn more about the universe. We hope they become more exact and work toward that end, thus coming closer to nature's own rules. So we should not expect any form of a universal or natural 'purpose' to be the same as our own human concept of purpose.

In a very real sense, the universe is determining itself, its own outcome, because it must be consistent within its own initial parameters and rules. The universe can be nothing else. The laws of nature, as observed and developed by science, only represent our human interpretation of those parameters and rules in the most general, fundamental and common terms that we can find. So, when we apply the scientific method to discovering purpose in the universe at large, we need to find the most general, fundamental and common properties (results) of purpose in nature so that we can isolate the concept. Does this then imply that the universe has a specific end or goal toward which it is progressing? Possibly. Does the universe itself have a purpose or is purpose part of the structure of the universe, inherent in physical reality? Maybe. And, does the physical evolution or progress of physical events in time follow a predetermined path? Maybe they do, but not necessarily. The questions have definitely become more complex if not more specific, yet even if we can't answer these questions completely because of their complexity, we can at least indicate how they could be answered although we still hope to answer them more completely.

Physics is the science that generates our most fundamental view of reality, so it is to physics that questions whether purpose is inherent in the universe should be directed. Chemistry will not work in this regard and psychology is a purely human science studying the behavior of humans among humans, not between humans and nature, so it is only toward physics that we can turn. This presents a difficult problem for science since the quantum paradigm dominates current thought in physics so completely that a very significant portion of the scientific community could not possibly consider purpose.

The quantum theory renders the future state of the physical world indeterminate and completely unpredictable, especially at the smallest levels of that physical reality that deal with the motions and interactions of elementary particles and out of which our macroscopically sensed reality emerges. Knowledge of future physical states at this level seem to be governed by the Heisenberg Uncertainty Principle which guarantees that we cannot possibly know anything of physical reality below a certain level of measurement. So, we cannot really know the future physical state of any material system with absolute accuracy. We can only predict the future that is the most probable. Since absolute accuracy would seem to be a simple necessity for the case of a natural or universal purpose, it would seem that quantum theory rules out any possibility of 'purpose' within both nature and the universe. So physicists, whose thoughts and attitudes are shaped primarily by the quantum paradigm, would never even consider any possibility of purpose in nature and more than likely dismiss any attempt to do so by others.

On the other hand, Albert Einstein made convincing arguments in 1934 that the quantum theory is incomplete and that there is a knowable and determinable physical reality below the limits set on our knowledge and measurement by the Heisenberg Uncertainty Principle. To complicate matters further, quantum theory is incomplete on yet other grounds, independent of Einstein's arguments. The strictest interpretation of quantum theory claims that consciousness is necessary to collapse the 'wave packet' or otherwise determine which one of the infinite number of possibilities will become reality and determine the future state of the universe, single physical event by single physical event. So it would seem that consciousness must be something different from the physical system described by quantum theory, such that consciousness must exist outside of quantum reality.

Recent experiments have also determined that particle systems and physical events are somehow 'entangled' in a manner that defies the present laws of physics and theories of physical reality, including the quantum theory. Like consciousness, 'entanglement' lies outside of the limits to our knowledge of the universe as expressed by the Heisenberg Uncertainty Principle as well as more recent views of quantum reality. So, quantum theory is truly incomplete. Therefore, it has nothing conclusive, or possibly even relevant, to say regarding purpose within physical reality. Where then can science turn?

Fortunately, there is another fundamental theory upon which the structure of physical reality can be based, the theory of relativity. Relativity is completely deterministic in that the future does not evolve randomly, or probabilistically, as in quantum theory. Instead, the present and future states of the universe are determined by the past sequence of events. While this view does not yet approach the level where we can say there is purpose in the universe due to relativity, it at least leaves the possibility of purpose open at first glance, but only at first glance. Relativity theory simply does not imply purpose because every action or event is relative to all others, so there is no absolute future, while an absolute future is necessary to demonstrate human-like purpose in the universe itself. In other words, no 'conscious' goal-oriented choices in nature are implied by relativity theory. However, relativity theory does leave an opening for other interpretations and forms of purpose to evolve out of the structure of physical reality. Other interpretations need not be human-like purpose, but could represent 'purpose' of a more general and physical variety.

In relativity theory, space and time are not separate but linked together, continuous with one another, such that a change in one necessitates a change in the other. When this approach to physical reality is taken, a sequence of events through time follows a continuous path called a world line. The world line represents the position of the sequence of events in space-time or the history of the sequence of the events in question. Any sequence of events can be represented and mapped out in space-time as a world line. So all that is needed to demonstrate purpose in the universe in a scientific manner would be to determine or know the future or end-point of the world line for any sequence of events. In other words, the universe does not need to make conscious choices on an event-to-event basis to demonstrate the physical existence of purpose. Nor is there any need to invoke freewill on the part of the universe as a whole to show purpose in the universe

It is simply not necessary to know each and every point along the future portion of the world line, the moment to freewill moment choices which direct our immediate futures, or how the next moment after the present moment progresses toward the endpoint to show purpose. The universe does not need to micro-manage each and every event in the universe to move toward its goal or purpose because it has already set the rules in motion, and all action, by necessity, follows those rules even if science has not yet completely defined these rules.

The human mind and consciousness, along with the human penchant toward 'freewill', evolved in full accordance with the restrictions and the rules by which the universe operates and evolves, so freewill cannot do anything but help to bring about the goal of the universe, even if humans are unsuspecting of that goal. Humans evolved within the context of the universe and unwittingly adhere to the dictums of the universe, moving toward the endpoint of the universe as decided by the initial conditions and true rules by which the universe evolves in time. The universe, or rather the rules that the universe must follow, the rules that define the universe and nature at each and every point and moment of this existence, already limits the choices of action that can be made by human or any other freewill. When we make everyday common freewill choices, they affect the universe in such a way that the universe compensates by making minor adjustments toward its own world line and its own goal. So human freewill is not absolutely free, it is only freewill 'within the context of' or 'relative to' the rest of the universe. The universe offers choices according to its rules and then humans decide the immediate outcome, without altering the universe's long term goals, or purpose, whatever that purpose might be.

So questions about the existence of purpose in the universe should take a completely new form, such that we need to ask if the universe is goal-oriented or not, and this can be answered, to a certain degree, by science. Toward this end, we take into consideration the evolution of life itself, followed by mind and finally consciousness. If this evolutionary path is a general rule throughout the universe, and not unique to the earth, then it would seem, as science suspects, that consciousness is the goal of the universe, whereby the universe can know itself, or better yet its 'self'. It would seem that knowledge is the goal of the universe, knowledge of itself as discovered by humans and other sentient beings, and this notion is consistent with both relativity theory and the existence of a physical reality below the quantum limits to measurement.

From the special perspective of science and scientific insight, by which scientific thought is not limited by physical space-time, one can follow the history represented by the world-line from the past until the present and thus predict how a system might develop in the future if freewill doesn't later enter the system. Freewill as well as the physical rules that define the universe are already part of the previous history of the event sequence, as represented by the world-line, but the world-line cannot account for freewill and conscious choice along the future portion of any world-line.

If freewill could be suppressed, or otherwise rendered inactive in the future, then and only then, could the world-line be accurately extrapolated from the present into the future as nature fulfils the rules by which the universe operates. Even if freewill could not be suppressed in the future of a specific series of events, freewill acts only within the restrictions of the universe's system of rules of operation along the future portion of the world-line. So, to break the rules of the universe would be to create a discontinuity in the world line which would dissociate the sequence of events from their past, which would be impossible. Yet purpose would assume that the world line is progressing toward a certain known point, a goal, and attaining that goal would be the purpose of that particular sequence of events.

If we assume then, as science does, that the universe began its own physical evolutionary path in some singular spectacular event, which need not be defined at this time, then the universe has its own world line. Yet the universe represents all of space and time, so the universe is also the complete space-time diagram upon which the world lines of everything in the universe play out. The universe, therefore, has nothing against which it can distinguish itself, or rather; it can have no knowledge of itself outside of its own context, that is, if the universe were conscious to some degree.

But the universe has an internal context due to the initial conditions and rules as well as the rules that it follows as it unfolds through time, even if it doesn't have an external context to determine, know or, after its own manner, realize itself. Within the universe's internal context, at some point or even at many different points in place and time, the chemical and physical conditions became favorable for life to emerge out of some primordial ooze, or so science believes. Once the simplest form of life emerged, it began a natural evolutionary path, in a sense its own world line. Life grew and nurtured as it came into contact with a greater number and wider variety of chemical reactions and environmental extremes. Life, or rather life's world line, continued as life begat new life, its own progeny, and as it adapted and absorbed other chemical reactions that were either 'conducive to' or 'advantageous for' its continued growth and existence, LIFE evolved.

Eventually, individual organisms became differentiated within themselves by developing functional units to serve a variety of chores necessary to maintain each organism, such as skin for protection from environment and bones to maintain structural stability. As these functional groups became more complex, a new functional organ evolved within the organism to organize, coordinate and control the other functional organs so they would all work together toward the common goal – the maintenance and well being of the organism. That newly evolved organ was the brain. And with the brain came mind.

But the universe was not yet satisfied and evolution did not stop with just mind. Eventually electro-chemical reactions within the brain, corresponding to collected data from the organism's interactions with its natural environment, formed memories in the mind. First, the mind became aware that the environment continued beyond the immediate local range of the host organism's senses, and a concept of non-local space and spatial location emerged. Next, the mind came to know that there is change due to the passage of time, that there is a past, a present and a future. Mind became aware that time exists beyond the immediately sensed or experienced moment, that time is also non-local. Consciousness evolved through these steps by following the rules of the universe, as would any other evolving system within the universe. So human consciousness, and indeed all consciousness, must be natural to the evolutionary process of the universe and consciousness must be necessary in some manner to the maintenance and well being of the universe.

With the mind's new awareness of the sensed moment, relative to the passage of time, non-locality in both space and time emerged. Mind learned that it had a unique place and individual position within the space-time continuum. Mind thus distinguished itself within the continuum, the total physical environment, and consciousness emerged. In physics, a material object is distinguished by its unique position relative to other objects in space and time, that is the basis of relativity theory. So, when mind distinguished the unique position of its host organism in space and time, it actually realized its self as a separate entity, and awareness of self is the most basic distinction of consciousness. Consciousness thus emerged naturally as a more fundamental unit than mind within the overall complexity of life.

Now, science assumes that life is a natural evolutionary characteristic within the physical universe. To think otherwise, that life could somehow be unique to our earth, would be an admission that our position or place in the universe is unique, but all is relative and our position is not unique according to relativity theory. Scientists have actually predicted the number of stars that could have planets that are suitable to evolve and thus possibly harbor life of some form. But if life is so natural, a simple product of the parameters and rules by which the universe evolves over time, then mind and consciousness are also natural consequences of the universe's own evolutionary path. So individual consciousness has evolved within the context of the operating rules of the universe, it interacts with the universe according to these rules and is subject to these rules at all times even while it is continuing to evolve as an emergent quantity within the universe.

On the other hand, consciousness evolved from a greater knowledge of the environment of the organism, and thus the whole of the universe by extension, while the universe is continuous at its smallest and most fundamental structure, meaning that all things are connected through the continuity of the universe. Obviously then, the existence of a single conscious being in the universe must render the universe itself conscious, in a strict sense. As a natural consequence of the universe's rules of operation, this would mean that the universe must be striving internally to know its 'self'. The continuous development of consciousness through the gathering of information about itself, the universe's internal rules and parameters of operation and their consequences, the gaining of knowledge, must then be the goal of the universe, its purpose, the endpoint of the world line of the universe.

But this is not exactly a goal that is limited to the future of the universe. As individual consciousness grows and evolves, as should any consciousness, even beyond present human consciousness, the universe comes to know more of itself. Eventually, we could assume, the universe would finally come to know itself completely. This goal, this purpose, would not entail a human-like purpose that is made by conscious choices or otherwise freewill driven, unless it could be deemed a choice of the universe as a whole. Since freewill must also bend to the rules (the will) of the universe, then each and every moment is the product, either directly or indirectly, of the rules and parameters by which the universe unfolds. Human freewill can only operate within the theater and upon the stage of this universe. Instead, this purpose, for which our freewill is only a minor actor in a play whose purpose is to help bring about the 'desired' endpoint of the universe, a universal consciousness of self-realization, is an ever evolving purpose that is present in each and every point in space-time as well as a future purpose or goal. This means that the ultimate purpose, which is the physical goal of the universe, is self-realization by the universe at the same time as self-realization is the path followed by the universe toward its own self-realization.

We could never completely understand the concept that the universe is both being and becoming the awareness of its 'self'. The universe does not distinguish or differentiate the passage of time, as do humans. The universe does not care about the past, present and future – all are internal to the universe and internal is just internal, and nothing else. Everything is internal to the universe all at once in the eternal moment because there is nothing external to the universe by which to relativistically compare or distinguish the universe, although there is a probability that other levels of the universe exist, i.e., still higher dimensions, for which we have no knowledge at present. The universe is all of space, all of time and all of space-time simultaneously. The universe realizing itself must therefore be an internal process of the universe. The goal or purpose of self-realization, the primary property of consciousness, is both imminent in every point of space-time and transcendent beyond the mere collection of all the points in space-time, from the perspective of the universe itself. With the evolution of the first consciousness in the universe, the universe began to become aware of itself and that process of self-realization continues as any organism which has evolved consciousness then evolves toward greater consciousness. In this manner, each consciousness fulfills its own purpose or goal of gaining more knowledge about itself as part of the universe and the more knowledge about the universe as a whole in all of its varied manifestations. The path is knowledge and the goal or purpose of the universe is its own self-realization, which comes with a higher and vastly different consciousness than we, as humans, are yet aware.

Within this theater of the universe, on the stage of nature, we humans are but one actor. As such, we read the lines put to pen by the universe, but we play our role with our own interpretations, emotional content and unique perspective, thus expressing our freewill. Within this grand play we also have our own goal to reach toward, helping the universe attain its own goal of self-realization and the conscious awareness of itself. The human goal is merely part of the means by which the universe attains its own goal. True knowledge is the purpose of all sentient and sentient-to-be beings, so the search for true knowledge is the purpose of life itself. And true knowledge, whether it is gained by intuition or reason, emotion or logic, is knowledge that follows and works within the rules and parameters of the universe.

Scientists seek to learn the rules of the universe through logic and reason. They develop theories to explain observed phenomena and events, while religious mystics seek true knowledge through intuitive explorations and contemplations. And yet this is still not enough because the universe is one, it is not one part that can be accessed by intuition and a separate part that can be explained by reason and logic. The universe is one single whole in itself that does not differentiate between different methods of collecting data and the perception of that data. This was Mach's mistake and the mistake of all who separate MIND and matter or observer and observed. Such dualities are in our perception of the world, not in the universe itself. Well beyond both reason and intuition we still need know our individual places and roles in the universe relative to other living beings as learned through emotional relationships, if only to completely understand the limits of freewill and how freewill works within nature and our universe.

We live in a theater of reality where we are meant to grow more knowledgeable of ourselves (the actors), our relationship with nature (the stage), and to learn more about the universe (the theater) and how the universe works (natural laws). Only through such knowledge can each individual consciousness evolve as the universe evolves. Individual humans may have personal goals and relationships, which have nothing to do with our universal being, but we are infused with the universe within each and every point inside the volumes of space-time that we inhabit, as with all life. The universe does not care one wit about our little freewill choices between a red sports car and a blue convertible compact. These creature comfort choices are irrelevant to the real goal of expanding our knowledge. We are 'continuous with' and 'connected to' the whole universe, and exist accordingly, to act out our natural roles in this purposeful universe.

Within this context, it is the natural obligation of every human being to seek and learn true knowledge and thereby add to the overall level of consciousness of the universe just we are expected to form individual relationships and societies which enhance our search for knowledge, whether academic, social or emotional. But it is also more than a simple obligation. An obligation implies that we have a choice in the matter, but in this we have no choice. We live at the whim and fancy of the universe and we 'choose' not to do so at our own risk and pain. The type of knowledge that we need to increase our level of consciousness is unimportant so long as it is positive or true knowledge. True knowledge consists of facts, ideas and concepts that come closest to physical reality and the actual rules by which the universe operates. True knowledge does not violate the rules of the universe and does not harm any aspect of the universal consciousness. This knowledge can be theoretical, practical, emotional or experiential. These general varieties of knowledge expand the level of awareness of the universe and consciousness, thereby aiding the universe toward it goal of self-realization. However, there are some facts, concepts and ideas that are detrimental to the search for true knowledge. If there is good and evil in the universe, then what is good is that which helps the universe gain true knowledge of itself and that which is evil would be anything that disrupts the search for true knowledge as well as the development, support and dissemination of harmful false knowledge.

In some cases false knowledge will eventually lead to true knowledge and is therefore good for the universe. This occurs often in science where theories are sometimes adopted to advance science a small amount, but are later abandoned when a more accurate and truer theory is developed. On the other hand, religious bigotry and intolerance, social and cultural bias and racial prejudice are examples of harmful false knowledge because people who hold these views attempt to force their ideas on others, when in fact, no one should force ideas on another beyond presenting them and allowing each individual to make their own decision for which they and they alone will be responsible to the universe in the end. Anything that blocks other human beings and sentient beings in their own personal quests for true knowledge is wrong if not outright evil. It is upon this basis that the vast edifice of human morality is built and it is upon this basis that the MIND and CONSCIOUSNESS survive death to become a more integrated into a greater universe than we now either perceive or comprehend.

Some things never change and some things do

During the past, concepts of death and morality have been all but co-opted by religious thought and belief systems. Religions have long claimed the moral high ground in human affairs, deciding the difference between right and wrong as well as good and evil, in spite of secular legal systems that define what is legal or illegal in society as a whole. Religions have been concerned with the immortal souls of the individual, how humans relate to the rest of the universe both during and after life, while secular justice systems and governments have been concerned with whole societies and the actions of individuals relative to those societies. In most cases, science has been a passive if not willing partner in this deception since science has concentrated on a strictly objective view of nature and the universe. However, that attitude must change and has been slowly changing over the past century. Science has now progressed to the point where the subjective aspects of nature, the interactions of humans and nature as well as consciousness and physical reality, can no longer be ignored. As soon as subjectivity is completely accepted as a natural and integral part of nature, our concepts of morality and judgments of what is right and wrong will begin to develop along a physical rather than a spiritual basis, at least until science also accepts the fact that the concepts of death, afterlife and spirituality also have a physical basis and should therefore also be included in the physical basis of morality.

Before the advent of the quantum and relativity theories a century ago (1905), a scientist conducting experiments or observing nature was considered absolutely independent of the physical system studied. This belief represents the newest version of the Cartesian dictum that mind and matter are separate. Science now knows that such a separation of observer and observed is neither realistic nor necessary and it has become a priority for a few scientists to find the level of subtle interaction between the observer and the observed. Science is headed toward some type of accommodation between observers and observed as a way of saving the legitimacy of scientific inquiry into the most fundamental realms of physical reality. In other words, there is no such thing as a physical system that is isolated from the experimenter who is testing the system, a fact of nature that some scientists are now calling the 'experimenter effect'.

By accepting the dominant quantum paradigm in physics, scientists and scholars believe that consciousness is necessary to 'collapse the wave packet'. The 'wave packet', also called a state vector, is a mathematical equation that contains expressions for all of the various outcomes that could occur from a single action and thereby develop into a single event in our physical reality. When consciousness 'collapses the wave packet' it makes a choice between the possible outcomes and chooses one will become reality. This process is known as the measurement problem in quantum theory. The process occurs on a point-by-point basis throughout our cosmos to develop a single moment of material reality and then re-occurs each moment to create the flow of time. The combination of the material reality generated by consciousness in this manner and the flow of time create our total physical reality. So, if human consciousness is the only consciousness in the universe, which is an extremely egocentric point-of-view, then humans must be creating all of physical reality. However, there are extremely strong philosophical arguments against this possibility. This argument, no matter how outlandish it sounds, is accepted passively if not actively and vocally by many scientists and scholars and is known as the Copenhagen Interpretation. Other variations of this interpretation of simple quantum mechanics are not quite as extreme, but none seem completely satisfying and none are without philosophical if not practical problems. Yet this idea is implied in normal quantum theory, which is why it is so appealing to so many scientists.

On the other hand, in the past few decades the concept of quantum entanglement has become popular as an alternate way to either limit the possibilities contained in the 'wave packet' or state vector from which consciousness 'chooses' the particular outcomes or entanglement actually 'collapses the wave packet' itself to create our reality in keeping with the familiar guidelines expressed in our laws of nature. If entanglement does 'collapse the wave packet' then it does so independent of consciousness. Within this context an extreme interpretation of the entanglement concept would hold that entanglement is the real arbiter of physical reality such that the universe non-consciously 'chooses' its own course of being and becoming through entanglement. Whatever the case may be, physical reality exists somewhere between these two extremes.

However, it should be equally clear to anyone that both consciousness and entanglement are introduced from outside of the 'wave packet' or state vector that is used in quantum mechanics to create the scientific picture of our reality. So it is logical to conclude that consciousness and entanglement are both something extra beyond our material reality, as described by quantum theory, but are still physical since they act together to form our physical reality.

Quantum theory therefore assumes that consciousness and/or entanglement are required to make the quantum description of our world complete. Quantum theory would be incomplete without these two concepts. On the other hand, in the 1970s Alain Aspect and other experimenters confirmed that entanglement or some other yet unspecified mechanism can act non-locally, i.e., at a distance. In more common and simplified terms, if you kick a photon here, the photon's sister photon will feel the kick over there at the same time, even though no signal or warning could have passed between them. That result expresses the essence of entanglement and science does not yet know how it works. But then if you consider the accepted fact that consciousness 'collapses the wave packet' to create reality and feeling the kick over there at the position of the sister photon would be an example of 'collapsing the wave packet', then you must also believe the implication that consciousness can act non-locally. Consciousness acting non-locally is just a fancy and roundabout way of saying that consciousness can act on a material system paranormally. So quantum physics now implies the existence of the paranormal and serves as a powerful although unrecognized justification for accepting the reality of telepathy, psychokinesis and other psi phenomena. Yet physics is still unable to explain these phenomena because physicists still hold that the observer and the observed system are separate, isolated and completely independent systems, while the majority of physicists and scientists in general still refuse to accept anything paranormal in spite of the implications of their own acceptance of entanglement and consciousness as necessary elements in quantum theory.

The connection between the observer and physical reality is undoubtedly extremely subtle. In fact the connection is far too subtle to measure by normal material means. The connection is so subtle that it falls below the limits set on physical measurement by the Heisenberg Uncertainly Principle and thus beyond the capability of quantum theory to explain. So it should be no wonder that quantum theory is incomplete and cannot explain entanglement and consciousness, even though it must invoke both of them from outside of its own logical structure to look and sound complete. The wonder is that scientists do not yet see and accept this obvious fact. Entanglement is just a way of thinking about and applying the basis idea of relativity without talking about relativity, which would amount to an admission by quantum physicists that quantum theory is incomplete without relativity, as Einstein argued. In other words, explaining or accounting for relativity is beyond quantum theory, a fact that no quantum physicist seems willing to admit. Even this form of relativity is not exactly the relativity of Einstein's theories.

This form of relativity is far subtler and fundamental than either the special or general theories of relativity although it is the fundamental idea upon which they are formed. Einstein's theories of relativity are material theories that also happen to be physical because they are material. Einstein's theories do however imply a purely physical and non-material relative physical reality and there is evidence that Einstein intuitively suspected this fact. But he was unable to develop a purely physical theory of relativity without invoking a concept of matter, partly because he did not take into account the subtle connection between observers and observed or consciousness and physical reality. Einstein was a bit too Machian in his outlook on the natural world. He still assumed a separation between MIND (with CONSCIOUSNESS) and matter. It is this subtle connection that links MIND and CONSCIOUSNESS to the universe as a whole in the single field, provides both the basis and purpose of LIFE, MIND and CONSCIOUSNESS within the context of universal physical purpose and provides the continuity and survival of individual MIND and CONSCIOUSNESS after the death of the material body.

The survival of MIND and CONSCIOUSNESS beyond death of the material body and the subsequent disruption (decoherence) of LIFE is a natural part of the universe and therefore a result of the basic rules by which the universe operates and evolves. Death may be inevitable from the human perspective, but death also opens the door to the next step in the development of CONSCIOUSNESS. The next step in evolution could not be as effective as possible while the MIND and CONSCIOUSNESS are hampered by their connection to the material world by LIFE. So death is actually a celebration of LIFE because it is, in a sense, a fulfillment of LIFE and it should be accepted as such by people, if not welcomed by them when it comes naturally. While this is a radical point-of-view, there is indirect confirmation of this view by some people who have had NDEs.

NDErs who have 'sensed' or 'entered' the 'light' have reported that the experience is wonderful and beautiful beyond compare in the material world, so beautiful and comforting that they did not want to return to our normal world and LIFE. In those cases where they entered the 'light', they were actually yanked back by the physicians, medics or other care-takers who revived them, or told to come back by loved ones and friends that they met in the 'light'. These NDErs were told either that it was not yet their time to die or that they still had unfinished business to complete in the material world so they must return to LIFE. These stories imply both the subtle connection of CONSCIOUSNESS to the universe as a whole and a physical purpose to the NDErs existence as well as a greater physical purpose to the universe itself which must be fulfilled by each and every conscious being. These descriptions of events that occurred during NDEs supports the SOFT model of physical reality and its relation to the natural purpose of CONSCIOUSNESS and survival as described above.

However, knowing what awaits us upon death as well as the role of death within the context of the universe is a double-edged sword, full of misconceptions, misrepresentations, unforeseen consequences and choices as well as very grave responsibilities. If death is to be welcomed as an entrance to a beautiful and wonderful existence beyond the material world, then why should we not just die and be done with it? We cannot just commit ourselves to dying because that would be a tragically wrong, harmful and dangerous conclusion to reach concerning death. Every person must live as full and complete a life as possible. A simple misconception of the meaning and significance of death that could lead to a logic that would imply this question is totally false is completely at odds with the universe and represents a grave misunderstanding of the true nature of physical reality. To artificially end or take a life before its time, even our own, is against the rules of the universe and represents a disruptive element in the very fabric of physical reality. Also, concluding that we know better than the universe when our lives should end is obnoxiously egotistical and demonstrates corrupted thought processes in the MIND that render us completely out of touch with the reality presented to us by the universe.

Every person should live as long as possible so that they can learn as much as possible to develop their CONSCIOUSNESS and thereby prepare for what happens to them upon death. Since the 'purpose' of every living being is to seek, learn and disseminate true knowledge, ending the LIFE of oneself or another sentient being amounts to artificially and wrongfully ending a line of unique knowledge and experience of the universe. Taking another person's or sentient being's LIFE is a crime against the universe. For that reason and no other, murder is the first, foremost and most serious crime that anyone can commit.

Nor do individuals have the right to take their own lives by committing suicide. Human freewill is not so free as to completely ignore the rest of the universe and the special role of each and every one of us in the universe. The choice to commit suicide can only be made by a MIND and CONSCIOUSNESS that is so out of touch with its connectivity to the universe that it cannot ever 'see' or 'enter' the 'light' upon death. A suicide is so extreme an action, that the MIND and CONSCIOUSNESS of that person has little to no hope after death. Only in extreme cases where no more is to be gained by continuing a person's LIFE, such as a person on their deathbed with no more hope of gaining any worthwhile experience and knowledge to expand their CONSCIOUSNESS, can that person's LIFE be mercifully ended. By the same token, there comes a point in keeping a person alive by artificial means beyond which there is no more hope of gaining any worthwhile experience and knowledge to expand their CONSCIOUSNESS beyond which that person should be allowed to die their natural death.

The same rules apply to all murderers. A murderer has committed the gravest and most serious of crimes and has thereby created his or her personal 'hell' to experience after their death. We create moral and civil laws against such crimes because we intuitively sense the physical consequences of such acts, but the universe has its own way of punishing those people who commit murders after they die. When murderers die, they suffer the pain and grief that they have caused others. If the murderer has reached a high enough level of consciousness to experience a past-life review, and a murderer by the very nature of the crime could not have reached a higher level of consciousness needed to skip the past-life review, he or she would experience the emotional distress and anguish that they have caused others at that moment in their past-life review. So taking your own life or another person's life fixes the course of your own development after your death. Of course those who have committed mass murders and other terrorists will just have that much more to deal with after their own deaths. Anyone who uses death or the threat of death to invoke or promote terror is committing the most horrendous of crimes possible. Terrorists who use suicide to kill others have compounded their own distorted MIND patterns to such a high degree that they have little or no hope after death. They have chosen their fates by their material actions and crimes against others, but also by their crimes against the universe and the universe is not kind to those who bend and break its rules.

However, there are times when taking a LIFE is not completely wrong or in violation of universal rules and will not harm one's afterlife experience. Self-sacrifice is one of those instances. To save the LIFE of another is CONSCIOUSNESS elevating, so to lose your own LIFE while saving another does not harm your CONSCIOUSNESS in the afterlife. Self-sacrifice to save the LIFE of another is the sign of a highly evolved CONSCIOUSNESS. There are also times when a single person's MIND and CONSCIOUSNESS are so corrupted (evil) and distorted that society justly holds their LIFE forfeit through legal judgments or, in the case of political leaders, through wars. Every person involved in these acts has a responsibility to the universe and the unwritten rules of the universe, so judgments against others that result in their death (such as death penalties for murderers) must not be taken lightly. Wars are another matter and take far more justification then just a bad leader. If a leader or politician so thoroughly corrupts the people he leads then war might be the only option, but that choice should still be taken under extremely careful consideration. It is not a question of saving thousands of lives by killing many hundreds during a war, the numbers do not add up in that way. It is a question of a leader or a group in political power corrupting the MINDs of the populace as well as murdering people while posing a danger to others. Whenever a judgment is made and a course of action taken, as in the death penalty of a murderer or a war, those who made the decision carry the responsibility in their own MIND and CONSCIOUSNESS. So each person must be absolutely sure in his or her own MIND that the judgment was correct or they will pay their own price. The final judge in all such matters is the universe and the responsibilities of the individual are reflected in their connectivity to and within the universe.

Other crimes also damage a person's MIND and CONSCIOUSNESS to a greater or lesser extent, depending on the severity and the results of the crime. Crimes of violence are especially damaging because of the severe emotional responses and trauma that result in the victim's lives. These crimes damage the MIND and CONSCIOUSNESS of the victim, but the victim can overcome that damage and have a net gain in their own level of CONSCIOUSNESS if they do not dwell and fixate on the damage caused by the violence. The perpetrators of the violence have their own problems with disruption and chaos in their own MIND and CONSCIOUSNESS patterns, but they also have to deal with the negative emotional responses of their victims through the natural connectivity that we all share through the five-dimensional single field. The simple rule is the greater the negative emotional response to a crime, the more intense the memory patterns that are stored in the victims MIND and the greater the payment the criminal will face when the criminal dies and experiences his or her victims' pains through the past-life review.

There are still other actions that we take in this four-dimensional world that affect other people negatively to different degrees, but are not illegal. Even though legal, these other actions may otherwise be immoral or unjust so they still cause harm to other people that is reflected in each victim's memory patterns and MIND. The perpetrator of these smaller 'sins' against others is still responsible to the universe for their acts even if the acts go unpunished or unnoticed in the four-dimensional world at large. Examples of this run the gamut from a child torturing animals, to bullying or beating other children, to spousal abuse and rape. Some people just dismiss their responsibilities and think that they are getting away with their crimes, but that is not true. Nothing goes unpunished in the physical universe.

Each and every one of us is responsible for all of our interactions with other people and living beings, at all levels of contact, and each and every one of us will pay for our 'sins' and abuses against others in our past-life reviews, no matter how small or seemingly insignificant the initial act. The best advice is to do no harm, or better yet 'do unto others as you would have them do unto you'. This 'golden rule' or principle can be found in nearly all religions because it is a simple and precise rule of the universe regarding all interaction between sentient and conscious beings. The same rule also holds in the purely material domain of interactions where it is called Newton's third law of motion: For every action there is an equal and opposite reaction. Everyone would do well to heed these warnings and live by them as first principles for conducting LIFE.

So it is that human morality and ethics have a purely physical basis in nature and the universe since they govern 'how' we interact with others, just as our physical laws govern how material bodies interact in our four-dimensional material environment. Both are based on the unwritten rules of the universe by which the universe evolves and all LIFE evolves within the universe. All of our interactions with people, other living and sentient beings, influence and shape each of our individual LIFE, MIND and CONSCIOUSNESS patterns to prepare us for the continuity of existence that we experience at death and how well our MIND and CONSCIOUSNESS survive the material death of the body. All physical interactions bind us to the universe as a whole and determine how we experience our natural connectivity of MIND and CONSCIOUSNESS within the single field after death. Within this context, death is actually a celebration of LIFE and an eternal memorial to how each of us conducted our LIFE. In this sense, death should be welcomed as a new beginning through which we continue to evolve, but those that are left behind still grieve and feel the pain of loss upon the death of someone who touched and influenced their LIFE.

To live and let die

When a loved one, friend or highly respected person dies, we grieve, we feel anguish and feel a tremendous sense of loss as well as an emptiness in our 'hearts'. That loss is the result of chemical reactions in the Brain that are similar to the withdrawal symptoms associated with chemical dependency, drug addiction or alcoholism, but of a far more durable and lasting type because they are of a far more fundamental nature than artificially induced chemical states. While loved ones are alive, we communicate with them through all of our five senses. If we touch, that touch initiates chemical responses in our bodies that amount to pleasure when the sensation reaches the Brain. And we like that feeling. We grow to depend on that feeling in close relationships. The quality, intensity and variety of the sensation depend on whether the person with whom we shared the sensation is a lover (emotional response), relative (familial response) or just a friend. But in each case, a powerful signal of chemical response is sent to the Brain and is stored as memory in the MIND, strongly affecting the development of the MIND pattern in the single field.

The same is true for our other senses, although the sense of sight is particularly strong and intense. The mere sight of a person can send an emotional/chemical shock through our body to our Brain which can be interpreted as anything from desire to respect and back again to outright lust. And, of course, our sense of smell logs in some very strong markers. The perfume and cologne industry make a fortune by attempting to enhance this form of sensed and sensual contact between people. With the sense of smell, complex chemical markers are actually transported between people to stimulate sexual arousal. These markers are called 'pheromones'. When a person senses them, the chemical triggers are stored as memories in the MIND, and very intense memories at that. The strength and intensity of the effect of these various sense-induced memories is most evident in the stories told by NDErs who claim to 'see' the 'light' and also 'see' and 'meet' friends and loved ones who had already passed away before their experience. Many times they are greeted by these loved ones who then act as guides to help them along their journey toward the 'light'. The love that we feel for others is a special direct connection between the CONSCIOUSNESS patterns of different people that resulted from the intense chemical reactions associated with material love. The overwhelming intensity of the emotional patterns of these loved ones in the CONSCIOUSNESS and MINDs of the NDErs actually calls out to them by the process of 'pattern matching', bringing the greeters to the NDEr in or before reaching the 'light'.

Depending on the depth and intensity of any single relationship, we become addicted to that chemical stimulus in our Brain and MIND. That particular addiction is not necessarily a bad thing because it leads to improved connections in the single field through the individual connectivity to all living beings and process of 'pattern matching'. However, under other conditions of the MIND and CONSCIOUSNESS, that addiction can be quite harmful. Examples of this can be found in jealousy and obsession to name only a few. These are personality disorders stemming from pattern conflicts in the MIND that are unique to each person.

So when a loved one or friend dies, we realize that we will never again experience that pleasurable chemical reaction associated with the direct physical sensation of that person. We realize we will never again come into material contact with that person to arouse those chemical pleasures of contact in our Brains. We normally feel the loss of that chemical stimulus whenever a loved one is out of contact with us for a period of time, so it is very natural. That is why we miss people that we like or love when they are absent from our everyday lives as well as why we are so 'happy' when we see someone dear to us after a long absence. But the death of a loved one represents a whole new and magnified level of that feeling by which there is an acute chemical reaction in our Brain that directly affects our MIND and CONSCIOUSNESS. This acute reaction does not normally happen when a loved one or friend is just out of contact for a while. When this type of acute reaction does happen during the LIFE of that person it is a form of abnormal behavior and that person is confronted with insane jealousies and obsessions that may quickly get out of hand and become something more dangerous. But again, with the death of a loved one, the chemical loss is qualitatively different.

The strong memory patterns in the MIND that form as a result of these chemical/emotional responses complete a special 'resonance' or 'pattern matching' situation with another person that we call love. When the loved one dies and the MIND/CONSCIOUSNESS complex survives, the 'pattern matching' connection of love does not end but continues between the dead person and the living person. However, the chemical responses that brought pleasure to them both through natural contacts while they were both alive have ended. The living person's MIND is still receiving subtle input from the surviving CONSCIOUSNESS of the dead partner via the sixth sense, but receives no corresponding chemical signals via the normal five senses. This mismatching of input signals to the MIND, coming from two different 'directions', causes a 'dissonance' in the MIND from the eyes and inner ear. This situation is a mental analog to the physical disorientation and nausea called carsickness that occurs when the vision of a person moving in a car at sixty miles-per-hour indicates to the MIND that there is motion, but the equilibrium system in the inner ear of the person says that there is no motion. Disorientation and nausea then occur due to the mixed signals being interpreted by the MIND. A similar situation can also occur when a person listening to a beautiful musical piece by an orchestra feels the grating of dissonant chords by one instrument that is playing ever so slightly out of tune and out of time with the rest of the instruments. A similar 'dissonance' associated with death in the MIND is perceived as grief, anguish and the great sense of loss that we all feel when a loved one dies. It can also be interpreted as a chemical withdrawal at the material level of our being and existence, but one which does not completely correspond with the constant physical contact by our sixth sense sending signals of the intimate connection of the dead person to the MIND that now exacerbates the material/chemical withdrawal experienced in the Brain.

A person can deal with this situation in many different ways, depending upon his or her personality and perception of the events, but also depending upon his or her overall level of CONSCIOUSNESS. Mental, physical and emotional isolation from others can occur. Other severe mental and emotional problems might develop as other memory 'patterns' in the MIND are disrupted. The distraught MIND can also cause health problems in the body through disruption of its organization functions over LIFE. Overly intense grief and anguish resulting from the death of a very special loved one can lead to new chemical dependencies to replace the chemical stimulus loss, resulting in drug addiction or alcoholism. A person could just drown or anesthetize his or her sorrow in beer and then snap out of it or it could get worse and cause permanent damage of specific MIND patterns that make up the personality of the person. But these are extreme cases, for the most part people just hurt and then sometimes for a very long time. There are as many different reactions to death as there are people in the world, while the severity of the reactions is completely non-predictable for even a single person because each particular reaction depends on very deep-rooted characteristics of an individual's personality. Yet the severe problems are unnecessary because the 'essence' of the person that died still survives and our subtle connections to that person have not been completely broken, just altered to fit the new situation.

There is a simple rule in psychology: You cannot start to cure a mental problem until you have identified and accepted the problem. We can now consider the problem with death as identified, so we can begin to accept the problem to overcome our grief. By understanding what death really is and really means, by understanding that the person does not really die because the 'essence' of the 'person', what really makes that person a unique individual, the MIND and CONSCIOUSNESS of that person, survives material death of the body, we can start to deal with our grief, anguish and sense of loss. We do not need to be overly concerned or fear the loss of a loved one or person who is important to us, although some concern is justified. The knowledge of what happens, alone, will not cure the grief, but it helps us accept the loss, which is the first step in overcoming the negative impact of a death. The negative emotions that result from a death are real and any attempt to deny or mask them can cause a disruption within the MIND pattern. By just accepting the loss, the denial route is completely ignored. So questions of 'why?' are irrelevant and serve absolutely no purpose, while dwelling on them amounts to making excuses for not accepting what has happened, which is just as bad or worse than denial and non-acceptance. So the first step is to live and let die.

Each and every one of us must recognize and accept death for what it is, both the severity and the beauty of what has happened and continue to live our lives as normal while letting the dead go. Live and let die! Accept the death as what it is to us materially, a very long absence until we physically meet again on the other side. There is no need for radical or extreme actions on the behalf of those that live on, because the essence of the connection which is 'love' is still, and will remain for all time, intact. That connection has become part of the structure and 'stuff' of physical reality. The 'love' that people feel and their feelings for each other have been imprinted in the single field, added to the overall evolving CONSCIOUSNESS of the universe and will be part of that physical reality forever.

Although it may not be easy, according to the circumstances of any given case, the grief, anguish, sorrow and sense of loss can be largely minimized by anyone and perhaps completely ended by a few. We all have an intuitive sense that 'something extra' survives death because our sixth sense maintains subconscious contact throughout death and the afterlife with the person who died. That contact is important because the MIND/CONSCIOUSNESS complex of the dead person very often needs help orienting its new 'self' in its new environment. But that contact cannot be allowed to be so strong and intense as to bind the surviving complex to the four-dimensional material world. The living have to let the dead go for the sake of both the surviving complex and the CONSCIOUSNESS of the grieving person left behind. All that is needed is the conscious recognition that the 'essence' of the dead person, the MIND and CONSCIOUSNESS of that person, survives. That recognition is enough to maintain the 'pattern matching' contact in the MIND of the living with the still existing pattern of the dead 'person'. That contact and the conscious knowledge that the MIND and CONSCIOUSNESS have survived will also help to stabilize the disruptive 'dissonances' (perceived as grief and such) in the MIND of the living person who must carry on in the material world. This contact and the stability that it creates form a positive feedback system. The more stable it is at any given moment, the more stable and stronger it becomes with the passage of time. The stability grows exponentially. But if disruption in the MIND of the grieving person is too strong, it can also increase exponentially if it is not checked and halted in time. Such a disruption can form a cascading failure or emotional break in the MIND, which could become calamitous for the living person. That is why people sometimes 'go to pieces' when a loved one dies.

Once this stability of contact is established, the MIND can use its organizational controls over LIFE to minimize the damage and disruption that death causes through the chemical /emotional dependency of the Brain on material contact with the person who died. The MIND automatically follows this procedure if it is not disrupted in some manner, it just takes longer with some people and they will feel grief longer unless they can use the MIND to control it. This is a self-protective mechanism of the MIND and the higher the level of CONSCIOUSNESS of the person, the more stable and 'incorruptible' the MIND, thus the more impervious it is to such disruptions. The amount of time it takes to emotionally recover from a close death therefore depends upon each person's level of CONSCIOUSNESS and familiarity with their own intuitive powers and awareness of the sixth sense.

Since duration and the amount of grieving is proportional to the level of CONSCIOUSNESS that a person has reached, there is no need to continually mentalize or form mental images and thoughts about the dead person, that memory will not die because the natural five-dimensional connection will not end. Remembrance is good and helps us to cope, but not to the point of obsession. Artificially masking the pain and grief as well as denial only prolongs the problem. Masking by false measures, such as drugs and alcohol, abuse the body, the MIND and CONSCIOUSNESS, and add to the final intensity of the grief as a person spirals downward into guilt and other negative emotions. Accept the grief and pain for what they are and the MIND, LIFE and body will heal themselves in the proper time. This process will be substantially expedited by the knowledge of what had happened, what is happening and what will happen as described by the SOFT model. The comfort of other friends, family and other loved ones also helps, even though the chemical responses that they initiate within the grieving person is not the same as the loved one lost. Their contact, compassion and feelings, however, do help to naturally 'mask' the emptiness felt by the loss, a strictly chemical response in the body. The trick to recovery is just to realize that the pain, grief and sorrow are just natural chemical processes, which detracts nothing from the quality of the emotions, intensity or significance of the feelings felt for the dead person, which will continue and not end and will eventually be consummated in a new and wonderful way when we meet again on the other side of the equation.

CHAPTER 8
EPILOGUE:
(SOME THINGS NEVER CHANGE, BUT …)

All revolutionary advances in science may consist less of sudden and dramatic revelations than a series of transformations, of which the revolutionary significance may not be seen (except afterwards, by historians) until the last great step. In many cases, the full potentiality and force of a most radical step in such a sequence of transformations may not even be manifest to its author.

—I.B. Cohen – 1980

Some things do

Small-scale advances in science occur all the time. That is the nature of the beast as science inches forward. Some advances are large and effect science and culture to a greater or lesser extent. That is simply the evolution of human thought as science takes bigger leaps forward. However, occasionally an advance occurs that is so fundamental to the very nature of science and human thought that it affects all of science and culture and a revolution occurs. Revolutionary advances are much rarer occurrences than the others. Revolutionary changes in science, culture and the human worldview are so great that they require a new look at the history of science and human thought through new eyes. Everything changes in these instances, including how we view our own past, so revolutions necessitate re-evaluation and introspection. The realization of a fourth dimension of space, as proposed by SOFT, is a change of this last kind, so the acceptance of a four-dimensional spatial reality or a five-dimensional space-time necessitates a reappraisal of past science and how the current science developed.

It is not that the past has changed in any manner or is somehow different from what it was before the revolutionary changes occurred. The data and information that represent the past are still the same. However, ideas and problems that were thought to be irrelevant before the revolution may have gained new importance during the revolution. Old and well-known facts may appear different or have altered meanings and significance within the new scientific context, while facts thought important and historically significant before the revolution may actually have been unimportant or even misleading in the final analysis. When a person is on the inside of a building looking out, what that person sees and perceives may not be the same as a person on the outside looking out. As the theoretical context of historical events, facts and data changes, the relationships that we deem the stream of history and the passage of time will also change, so it is advantageous and necessary to reappraise history in the light of these new revolutionary theoretical structures.

Play it again Sam

The advance of science and the human intellectual progress that accompanies it are normally studied by counting the revolutions through which science has passed. To date, scholars, academics and others who are interested in the subject have only noted the passing and completion of two such revolutions, the Scientific Revolution of the seventeenth century and the Second Scientific Revolution at the beginning of the twentieth century. However, portraying scientific progress as only a result of these two revolutions is woefully inadequate, as admitted by many historians. First of all, there was a Zeroth Scientific Revolution and secondly there are overlapping periods of scientific evolution. And finally, given the various accepted characteristics of scientific revolutions, it is quite easy to show that science is presently experiencing yet another new pre-revolutionary period that will very probably end in a Third Scientific Revolution. Yet perhaps even this description of the progress of science is far too limiting given the phenomenal leap in progress that science is presently experiencing.

The Zeroth Scientific Revolution was characterized by the rise of Natural Philosophy and the resulting simple definition of physics as an attempt to explain nature on the basis of 'matter in motion'. It represented a revolution in human thought that started from the premise of mythological and religious explanations of nature and our world and ended with the first development of a comprehensive philosophy of nature at the hands of Aristotle in the fourth century BCE. Although Aristotle defined the path for science to follow in its future development, his work also ended a period of ambiguous inquiry into the material basis of nature that began with Thales in the seventh century BCE. The seventh century was an era of change that spanned across the major cultures that made up the civilized world at the time.

Almost simultaneously, Lao Tzu and Confucius initiated major philosophical movements in China, Siddhartha Gautama, the Buddha, tried to reform Hinduism in India, Zoroastrianism was introduced as the official religion of Persia and Cyrus the Great set the Israelites free from their Babylonian captivity. The Israelites added facets of Zoroastrianism to their Jewish beliefs, laying the foundations for Judaism as the first national religion and influencing the later development of both Christianity and Islam. These religious revolutions greatly influenced the major cultures of the world at nearly the same time as Thales and his students began the movement toward Natural Philosophy. Given these facts, it would seem that a major worldwide revolution in human thought had occurred, not just a revolution in Greece. However, natural philosophy as interpreted by Aristotle began humanity the road to true science only at the end of the revolution in Greece.

Aristotle never defined matter nor did he develop a theory of motion even though he correctly based his physics on 'matter in motion'. Philosophers from Thales to Plato, Aristotle's teacher, developed the notion that the nature of the world is change, but they could not decide on what 'exactly' was changing. Aristotle finally came to the conclusion that change in position is the common element that is shared by everything in nature, but he also realized that matter is also fundamental to everything in nature. Although Aristotle developed a physical theory of reality based upon the four elements of earth, air, fire and water, he also believed that the ultimate nature of matter was continuous. Aristotle called the continuity of matter the Plenum. Since matter was continuous at the most fundamental level of reality, there was no such thing as a void or vacuum, a portion of the universe without matter. He further theorized that all matter had a natural place that it strived to occupy and he used this idea to explain gravity and his geocentric model of the universe. Yet Aristotle had no abstract concept of space. Aristotle also had no theory of motion and no concept of speed, as we understand the concepts of motion and speed today. To Aristotle motion was merely the changing of position or place, without any special regard to time.

Aristotle's model of nature was opposed by the alternate hypothesis of Democritus and the atomists. To the atomists, discrete material particles bounced around in a void to give us the material world that we sensed. Even though they accepted the idea of a void, while we associate the void today with empty space, the atomists never developed an abstract concept of space. To the Greeks, a void was a simple 'nothingness' or a 'no-thing', while space would become a 'thing' simply by giving it a name. The ancient Greek concept of motion only went so far as to note a change of position (place) and a separate change in time that was completely independent of the change in position. Place and time were different 'things' so their passage could not be related to each other to develop a model or theory of motion based on the 'speed' of material objects. 'Speed' would be a different 'thing' than motion because it was thought impossible to change one 'thing' into a different 'thing'. However, the ancient Greeks did originate the basic idea that matter could be either continuous or discrete and this duality of interpretation of the material world is still found in the modern concepts of the continuous field, as expressed by relativity theory, and discrete material particles, as explained by the quantum theory.

True physical concepts of motion or speed were not developed until about 1300 CE at Oxford University in England. Medieval natural philosophers first used graphical analysis to define speed because there was no strictly philosophical basis for modeling speed as distance divided by time. While abstract models of speed and acceleration developed steadily from this time forward, the abstract notions of space and time took a few centuries longer. During the Dark and Middle Ages of European civilization, the Church conducted all education. So philosophical and the early scientific concepts of the time were intimately tied to religious beliefs. In fact, the first concepts of space were directly associated with GOD. Space was equated to GOD in many different ways. Both were thought to be infinite, indivisible, immutable, and continuous, to mention only of few of their common characteristics. GOD could not be reduced or otherwise analyzed, so real physical space was originally thought to be non-reducible for the sake of logical analysis. On the other hand, natural philosophers and religious scholars debated whether GOD was immanent or transcendent. If GOD were immanent, God would either 'correspond to' or 'exist in' every point of space and the universe, but if GOD were transcendent, then GOD would exist beyond the natural universe. Francesco Patrizi finally defined physical space as a philosophical abstract idea in the late 1500s by differentiating between absolute and relative space, thus solving the dilemma for early science. While GOD could be either immanent or transcendent, space could be either relative or abstract by similar arguments. The influence of the debate between immanence and transcendence can still be seen in the early versions of relative and absolute space, respectively.

Well before the Scientific Revolution, western intellectual discourse and natural philosophical thought had become so entwined with religion that it became necessary for science to separate itself from its own religious background and thereby distinguish between the supernatural and the natural worlds. The development of the abstract concept of space is only one example of how religion and natural philosophy were entwined. The task of separating them was not easy, that is why it took several centuries just to make a simple beginning at separating the new science and natural philosophy from religion and no more. Science then erected an intellectual wall around itself to protect it from all things superstitious and supernatural, but that wall also excluded those things that are paranormal (praeternatural), occult or otherwise exist in between the natural and the supernatural. While a basic theory was necessary for physics to fully develop, a basic philosophical approach to the separation of religion and science was also necessary for the development of science. Newton supplied the fundamental theory of physics only after Descartes supplied the fundamental philosophical basis for the separation of religion and science. Both events were crucial to the Scientific Revolution for different reasons.

Although his ideas were not completely new with Descartes, he did put them in the proper form and context for use in science and defined the issues of the duality between pure thought and the interpretation of nature in such a manner that they are still useful in science today. Descartes sat on the cusp of progress in human knowledge, like Janus facing in two directions at one time. He faced backward to scholasticism in that he still believed that philosophy defined and ruled over nature, as early scholastic philosophers thought. He believed that natural laws originated in the mind rather than through the observation and interpretation of nature, but he also faced forward to the new science in that he thought that specific natural laws could be derived to explain nature. The earlier scholastics did not look to nature and the world around them to explain phenomena, but interpreted events though the writings of even earlier philosophers and scholars, even when the earlier scholars were obviously wrong. Scholasticism has come to mean the acceptance of a fact as truth without testing the validity of that fact against reality merely because someone wrote it down. Today, many people believe something to be true if they see it on the evening news or if they are so informed by a computer, without realizing that the evening news offers facts about events not truths and computers are only as exact and accurate as the people who programmed them. In spite of the negative influence of his scholastic predecessors, Descartes still discovered the most fundamental criteria for choosing what in nature was scientific and what was non-scientific with his separation of MIND and MATTER.

With this duality, Descartes defined the fundamental breakdown of human knowledge in the mid 1600s. Descartes' distinction between MIND and MATTER literally forced scholars to distinguish the difference between the verification-based search for knowledge followed by science and a religious faith-based search for knowledge. This distinction not only allowed scholars to determine which phenomena in nature could be reduced for scientific analysis, but it also institutionalized a dilemma in science that science could only deal with at a later date and a higher level of knowledge. We could call this the Cartesian Dilemma. Descartes did not distinguish which is ultimately the truer basis of reality, the sensed world of matter or the mind that senses the world of matter. Neither Descartes nor his peers were able to answer this paradoxical question, so Descartes deduced that there were two realms of nature and reality, represented by MIND and MATTER. That is why we are still faced with a fundamental separation of consciousness and material reality today and some philosophers still argue whether consciousness of the material world is the real world.

Modern philosophers and scientists, who study consciousness and search for a theory of consciousness, are interested in the 'qualia' that are associated with consciousness. 'Qualia' are those characteristics of consciousness that cannot be so easily reduced for study and analysis by scientists. For example, 'redness' would be an example of a 'qualia'. Everyone can easily distinguish which colors are different shades of red without extensive analysis of the physical phenomena of being 'red', but scientists cannot determine how consciousness can determine which colors are red by their 'redness'. Consciousness must have some sense of 'redness' since it is able to determine the difference between different shades of red and shades of pink or other colors. When exactly does a color change from red to pink or from orange to red? Questions such as this are called the 'hard questions' of consciousness studies because they are so hard to answer within the present paradigms of science. At present, science can only ask and answer questions about 'things' that can be precisely defined, measured and subjected to careful analysis. Science can answer the easy questions about brain chemistry and function and their relationships to states of consciousness, the brain correlates of consciousness, but science cannot answer questions about why consciousness perceives one color as red and the other color as pink rather than a shade of red. In a sense, we can consider Descartes' distinction of MIND as opposed to MATTER as his way of dealing with the difference between the 'hard questions' about mind and consciousness and the 'soft questions' about the natural material world.

Descartes' realm of MATTER was no more or no less than the ordinary material world that humans commonly sense and he designated the MATTER as the realm of science. He viewed this realm as a complex mechanism that was defined and ruled by specific Natural Laws. To Descartes, the Natural Laws were to be studied and applied to the world mechanism by natural philosophers using reason and logic, but those laws could not be derived directly from nature. Natural Laws were given or revealed to natural philosophers and scholars by GOD through MIND. Descartes was well aware of Galileo's experimental methods, procedures and results as well as the early scientific work of others, but he thought that the purpose of experiments was merely to confirm the GOD given laws of nature, not to help derive and develop those laws. Experiments were clearly not part of the discovery or developmental processes of science in Descartes' original conception of MIND and MATTER, nor were they meant to model, expand or explore our observations of nature. Yet he did develop the idea that the material world of sensations was a mechanism and thus instituted the first mechanistic worldview for science.

On the other hand, Descartes gave equal if not more importance to the realm of MIND. After all, he was a philosopher in every sense of the word. MIND was the realm of spirit and therefore MIND was the realm of GOD as studied by religion. Descartes delegated the ultimate purpose and meaning of life, mind and consciousness to the realm of MIND, not to the scientific realm of MATTER. Humans could only know of MIND through emotion, revelation and intuition, never through reason, logic and direct observation. Intuition and emotion were the primary methods of exploring MIND, and MIND could only be confirmed by faith. While this separation, at least at some level, was completely necessary for the early development of science, the separation of MIND and MATTER also led directly to the Cartesian Error. So we have the Cartesian Paradox, the separation of MIND and MATTER, the Cartesian Dilemma in which one represents the ultimate reality in the long run, and the Cartesian Error that covers the application of his philosophical doctrine to scientific inquiry.

Descartes' own error arose from his belief that our knowledge of physical reality derived directly from MIND rather than observation of the world, which directly and adversely affected his attempt to develop the correct Natural Laws governing motion. Instead of observing the nature of real moving bodies colliding, Descartes based his theory of collision, and thus of motion, upon his philosophical biases. His failure to correctly describe collisions of real material bodies resulted from the mistaken belief that philosophy rules over nature rather than adopting a philosophy of nature that is derived from nature itself. Scientists and scholars still suffer from this Cartesian Error when they believe that nature must follow some particular philosophical or mathematical system that they insist is true without any direction or confirmation from nature. Some modern scientists still develop theories that look good on paper or seem 'aesthetically' pleasing, but have no basis in reality and are not 'falsifiable' or otherwise testable because the scientists did not take nature into account in their initial development. It is an unwritten rule of science that a theory must be 'falsifiable', that it must provide a way to test whether it is false or true. A strict belief in philosophy rather than nature, which is the basis of the Cartesian error, has proven bad for science. Philosophy can only suggest possible solutions for the problems that science faces; it cannot force nature to follow its suggestions.

The Cartesian Error implies that the Cartesian concept of MIND represents ultimate reality and our scientific concept of MATTER, in the form of theories, can be altered by our minds without reference to the external reality of MATTER. So MIND rules over MATTER. In reality, MIND can only discover MATTER through observation of our natural world, which is opposite to what Descartes thought. If science truly believes the idea that mind can only learn the truth of the material world through the direct observation of nature, then science will eventually be led to the possibility that MIND, or at least parts of MIND as defined by Descartes (such as mind, consciousness, life and the afterlife), are parts of the mechanistic world of MATTER. This does not mean that a belief in GOD or the soul is wrong. It only means that GOD, soul and similar concepts cannot be defined and are thus not amenable to scientific study.

In 1687, Newton published the Principia. This book is the crown jewel of the Scientific Revolution because it finally defined the correct laws of motion. Newton first equated the inertia that an object experiences when its state of motion is changed to the mass or amount of matter from which the object is composed, allowing the first true measurement of how much matter an object had. Descartes had earlier defined mass in a similar manner, but the restrictions of Descartes' philosophical system led him to the wrong theory of motion. Developing an erroneous theory of motion resulted from Descartes' failure to recognize and thus take into account the Cartesian Error. But Newton's ideas and concepts were based upon nature and direct observation, so he deduced the correct laws of motion and was thus able to give a quantitative definition of matter. He also correctly explained and defined forces relative to his definition of mass, but he needed an abstract concept of space to develop his theory of motion and forces so he adopted and adapted Patrizi's concepts of absolute and relative space to his needs. Quite literally, forces move matter around in space, between different positions in space, so a concept of space, not place, was necessary before physics could evolve as a true science.

Aristotle's idea that 'matter in motion' forms the basis of physical reality was not realized until Newton's new physics. It took a workable theory of how matter moved within space to develop the science of physics and make science happen as a successful intellectual pursuit. Newton's laws of motion thus became the fundamental theory upon which both physics and all of science were built over the ensuing centuries. The method of observing nature and deriving Natural Laws from nature became the scientific and experimental methods. Even today, the concept of 'matter in motion' remains the basic concept behind physics although a new 'political spin' and limitations have been placed on the concept by the quantum theory.

When Newton defined absolute and relative space and absolute and relative time, he also stated that only relative space is necessary for mechanics and the scientific explanation of nature. The existence of absolute space and time were only implied by observations of how the mechanical system of Natural Laws worked. So Newton deduced that absolute space is the container of our material world and does not normally affect the workings of nature or the universe. Newton equated absolute space to GOD, calling absolute space the 'Sensorium' of GOD. Doing so redefined the Cartesian Paradox and opened the door to a shift in the boundaries between MIND and MATTER that Descartes had established.

Matter was discrete, but the space that contained matter was continuous according to the Newtonian theory. Even then, only relative space was necessary to explain the interactions between bits discrete matter and other bits of discrete matter, which rendered matter subject to analysis, while a method needed to be developed to analyze motion in the continuous space. Problems arising from the opposing ideas of continuity and discrete were not so easily overcome. Absolute continuous space suffers from the fact that it cannot be reduced for analysis, but Newton circumvented this problem by allowing relative space to be reduced. Space as a whole remained continuous, but Newton solved the reducibility problem of relative space by the development of a new mathematical system called the calculus. Neither calculus nor any other form of analytical could be applied top absolute space, which made absolute space a perfect place to equate to GOD. So Newton escaped the philosophical bullet of the MIND versus MATTER dilemma that Descartes had faced, but did so by circumventing rather than solving the continuity versus discrete duality. He equated mass to matter, but never commented on the ultimate fundamental nature of matter just as he never explained the source of gravity that acted on matter without ever explaining the mechanism by which gravity acted on matter.

Nor did Newton say anything about life, mind or consciousness, so he delegated them by default to a middle region between MIND and MATTER if not directly to the realm of MIND. Newton was not wrong in doing so, he had to limit the material system of the world to what he could reduce and analyze to give science a start, but in conducting this self-limited version of science, Newton left questions and problems for later science to deal with. Newton's notion of the absolute soon came to define the role, place and boundaries of religion in the natural world, in a sense he legitimized religion from the scientific perspective, while Newton's other ideas defined and established the correct physics that carried forward into the future as Newtonianism, the first true science of physics that humankind had been able to develop. Life, mind and consciousness were not originally part of Newtonian science and could only considered from either the philosophical perspective of metaphysics or as questions by the religious community. Under these various limiting conditions, Newtonianism became the basis for all science and the whole of the natural world, not just physics.

Cartesian MIND		MATTER
Realm of Religion (and Metaphysics)		Realm of Science
GOD spirit-soul-ghosts consciousness-mind-life occult forces		PHYSICS
Angels afterlife Alchemy		Astronomy
Heaven & Hell morality & society creation		Cosmology

Original Boundary as of 17th century

In other words, Newtonianism was all of science in the beginning and physics was just one part, albeit the most important part, of natural philosophy. Yet even then, the study of nature was not called science, but rather natural philosophy as it had been called since the days of the ancient Greeks.

What goes around, comes around

Under these conditions, Newtonian science began to conquer and explain all of nature. A new view of the universe developed called the 'clockwork' universe. In this view, the universe was pictured as a clock mechanism that ran by specific rules, as defined by Newtonian physics. GOD had designed, created and built the clock according to Natural Laws, wound it up and left it to run on its own. Humankind could study the clock mechanism (nature) to learn the Natural Laws. GOD would only step in from time to time to either adjust the clock or wind it, events which we call miracles. A new religion called 'Deism' was even developed along this line of thought. As time passed, Newton's laws became so successful that philosophers began to believe that society must follow similar Natural Laws, so they began to develop social laws and study the relationships between people and societies as well as societies and their rulers, influencing the course of the American (1776) and French (1789) Revolutions. As the successes of Newtonian physics spread, areas of nature that had originally been off-limits to MATTER fell under the growing Newtonian influence.

Life became associated with the chemical reactions that sustain living organisms, just as alchemy matured into chemistry. John Dalton developed the basic theory of chemistry in 1803, the atomic/molecular theory. Charles Lyell established the basic theory of geology, uniformitarianism, in 1833. In 1828, Friedrich Wohler synthesized the organic compound urea using non-organic chemical methods, changing the scientific perspective on life. This fostered the attitude and belief that there was nothing special about life with respect to the scientific point of view. Life was nothing more than an extremely complicated chemical process, so life should be amenable to scientific reduction and analysis. Michael Faraday discovered magnetic field induction in 1831 and developed the concept of physical fields over the next few decades while James Joule conducted his experiments equating heat to mechanical energy in 1841. These and related advances in science forced the demise of Natural Philosophy as the home of all scientific thought.

The end of Natural Philosophy did not initiate or result from a new revolution in science in the classical sense of the word because no overturning of scientific ideas and ideals was involved. This evolutionary episode of science continued along the same path that science had been following since the last revolution; so no revolving of thought and attitudes was involved. Instead, the chemical element of heat, caloric, was found to be a kinetic or mechanical process in physics, precipitating a change in perspective that allowed the different sciences to emerge as independent academic disciplines from within natural philosophy. The time for this to happen was so ripe that William Whewell declared in 1840 that scholars conducting professional work in the sciences should be called 'scientists'. Whewell later came to be regarded as the first philosopher of science, yet he could not have been regarded as a philosopher of science as long as science was regarded as part of natural philosophy. It would have been silly to think of Whewell as a philosopher of (natural) philosophy, directly indicating that scientists and scholars of that era believed that science had separated from natural philosophy.

On the other hand, no science can be developed without a fundamental theory upon which it can grow. By the 1840s, all of the basic sciences as we know them today had begun to grow around their basic theories so they split apart from Natural Philosophy and Newtonian physics. Aristotle had set the whole system in motion by defining 'matter in motion' as the fundamental basis of physics (nature), but a true science of physics was not feasible until Newton institutionalized Aristotle's discovery by developing the laws of motion that govern 'matter in motion'. The other sciences followed when their own basic theories were discovered and applied successfully to explain natural phenomena.

The notion of progress in science developed in conjunction with the successes of Newtonian physics and the idea of progress in science then influenced the idea of progress in human development. Toward the end of the eighteenth century Jean-Baptiste Lamarck and Erasmus Darwin put forward simple theories of evolution based upon the inheritance of acquired traits. During the 1830s, a naturalist by the name of Charles Darwin, the grandson of Erasmus, traveled around the world on the HMS Beagle. As a naturalist, Darwin was able to scientifically observe, classify and compare different species of flora and fauna from all over the world. Putting together the information and knowledge that he gained from that voyage, Darwin concluded that living beings progressed through the transmutation of species.

By 1838, Darwin developed his first notions of 'natural selection', the mechanism of evolution. However, Darwin did not publish his book *The Origin of Species* and publicly announce his theory of evolution until 1859. The delay was caused by his desire to support his theories with ample evidence because of the important groundbreaking nature of his discovery. Darwin's theory of evolution clearly demonstrated that the development of life and mind were just natural mechanical processes without regard to religious concepts of design. With this discovery and the evidence that Darwin had compiled, Newtonian science completed a very important phase of its inroads into the formerly forbidden Cartesian realm of MIND, precipitating a sequence of events in the history of science that have been grossly misunderstood by scholars, scientists, academics, both religious philosophers and the general public, and culture in general for over a century and a half.

As the many successes of Newtonian science accumulated, both he role of science within culture and the realm of science within nature expanded. These successes led to four different but related development sin science after the middle of the nineteenth century: (1) As the successes of Newtonian science weakened the position of religion within culture and society, Modern Spiritualism evolved to take the place of religion. Science then reacted to the Modern Spiritualism movement by developing the first scientific studies of the paranormal; (2) Reacting to the same historical 'forces' and influences, the Cartesian MIND/MATTER boundaries rapidly shifted so a new philosophical basis for science evolved, based upon the work of Ernst Mach and others, called positivism; (3) Newtonian science was so successful that it directly challenged the concept of mind and consciousness, leading to the birth of a new science of mind called psychology; (4) Successes of Newtonian science in different areas of physics led to conflicts between those different areas, which resulted in the second Scientific Revolution.

In each of these cases, the same historical and social forces were at play, while different sectors of culture reacted to the forces in different ways. However, this view of the history of science indicates that the development of paranormal science is as much a part of natural science as any other of the natural sciences, which is at odds with the normal views of scholars. The normal view of historians and other scholars is that the paranormal is not a part of real science, the afterlife is a 'non-question' and of no consequence to science, Modern Spiritualism was an aberration and the Second Scientific Revolution was caused by specific crises in science rather than the successes of Newtonianism. Yet the historical evidence does not completely bear out the normal views of historians and scholars.

Newtonian influences also created a crisis for religion that came to a head with the development of evolution theory. Evolution then enhanced and expanded the sense of crisis within society that had been growing as science grew in its influences on society. Both science in general and evolution in particular replaced religion as the sole authority on nature and the world at large. When the Fox sisters reported the rappings of a dead salesman's spirit in their cabin in Hydesville, New York, in 1848, they triggered the development of Modern Spiritualism. However, the rapid spread of Modern Spiritualism throughout the world occurred because Newtonian successes had prepared society for such an event. The event with the Fox sisters only set a new cultural phenomenon in motion. The Fox sisters did not cause the rise of Modern Spiritualism as many claim. At present, science does not want to take credit for the rise of Modern Spiritualism and is willing to allow that the misinterpretation of historical facts remains intact without challenge regarding the role of the Fox sisters. Many other earlier influences played direct and indirect roles in the development of the movement. The list of historical factors that influenced the rise of Modern Spiritualism includes witchcraft, early forms of spiritualism, Mesmerism, phrenology, earlier forms of superstition and alchemy, but is not limited only to these factors.

In the end, no matter what its origins were, Modern Spiritualism was essentially a cultural movement by non-scientists to render the afterlife scientific. It was also an effort to develop a personal form of non-doctrinal religion based upon the paranormal that arose in the face of advancing science and the growing reliance of society and culture on the new technologies and industries that were fed by scientific advances, but the effort went astray as mediumship was reduced to showmanship as practiced in many cases by prestidigitators, charlatans and con-artists. Still, a few scientists took some of the phenomena associated with the new form of spiritualism seriously enough to investigate the phenomena and found scientific societies such as the Society for Psychical Research and its American version to study spiritualism and the claimed phenomena. These scientists documented many different and new varieties of paranormal phenomena, but were never able to confirm or verify either the afterlife or communication with spirits. However, they did establish a precedent for the scientific study of the paranormal and developed techniques, standards and methods for its later scientific study. In the meantime, the same historical forces that influenced the scientific study of the paranormal had other effects on the progress of science.

Quite frankly, there were two different but related backlashes to Modern Spiritualism that were every bit as much responses to the shifting boundaries of the MIND/MATTER split as was Modern Spiritualism itself. They were also a direct response to the successes of Newtonian physics as it spread into and encroached upon the formerly forbidden realm of Cartesian MIND. In general, there are three ways or directions in which science can expand its realm of influence. The theories of science can predict new phenomena that are then discovered and incorporated into the structure of science, such as the discovery of wavelengths of light beyond normal vision that were predicted by the electromagnetic theory. But science can also discover or observe new, previously unsuspected events or phenomena that science must then incorporate into its accepted theories or create a crisis for science. Examples of this way of expansion range from the discovery of the photoelectric effect, to the discovery of electrons, x-rays and radioactivity. And finally, science can incorporate well-known phenomena that it had never before considered within the realm of science. In other words, science could also expand progress by expanding its range into the Cartesian realm of MIND by accepting the reality of paranormal phenomena and the afterlife, but only when science is ready and able to do so.

In the late nineteenth century, science did find itself ready to expand into the realm of MIND by considering the role of mind and consciousness in science. Given recent events, science was secure enough and assured of its own infallibility enough to tackle mind. Evolution theory implied that life would eventually be conquered y science, which further implied that mind and consciousness would also fall to scientific analysis. At first, a philosophy of mind began to develop as scientists and scholars considered the role of mind in the development of Natural Laws and theories. New questions on the role of mind in science formed a direct challenge to the old Cartesian boundaries between MIND and MATTER, as the definition of Cartesian MATTER changed to fit new advances in science. Of the different philosophical schools that developed, the work of Ernst Mach became the best known and the most influential in further advances of science. Mach thought that the human mind could never know physical reality directly. The mind was only capable of interpreting human sensations of material reality, not of experiencing reality without the interference of sensations. So he determined that our Natural Laws reflected nothing more than a convenient and efficient description of human sensations of physical reality. The Natural Laws of science were not inherent in reality itself, but were the product of human senses and would reflect any faults within the senses. Mach's approach was thus a middle-of-the-road approach to the boundaries of science between MATTER and MIND. The philosophical system that he helped found is called logical positivism or empirical positivism, even though he claimed that he was not a positivist.

On the other hand, the scientific community as a whole came to realize the necessity of dealing with questions regarding the nature of mind and the role of mind in physics and science. Scientists did not wake up one morning and decide to develop a new science of mind. There was no single father of psychology with the status of an Aristotle or a Newton. The whole of the scholarly community came to the realization that a new science of mind had become necessary, so several notable scientists made their own special contributions to the origin of psychology. In this respect, the philosophy of mind evolved into a new science of mind called psychology. The simple fact that nearly all if not all of the scientific community came to the same conclusion regarding the need to develop a new science of mind is easy to demonstrate because the new science developed from several directions simultaneously. Within the span of just s few years, Gustav Fechner developed psychophysics (physics of the brain), Wilhelm Wundt developed the experimental wing of psychology, William James worked on the concepts of consciousness, the philosophical foundations of psychology and the paranormal aspects of mind, Sigmund Freud developed psychiatry and the medical aspects of the mind and, of course, Mach developed the philosophical foundations of the relationship between mind and nature. All of these men, and still others, were the parents of the science of mind, psychology.

So, at the end of the nineteenth century, science was faced with the birth of psychology, the birth of positivism and Modern Spiritualism. In the end, psychology, the rise of positivism and Modern Spiritualism, along with the recent developments of thermodynamics and electromagnetic theory, all added to the long record of successes for Newtonian physics and science and fostered an attitude that science was nearly finished in its total explanation of nature. Newtonian science had reached an unparalleled level of success, and the scientific community was fully aware of its successes. The same historical forces and successes that caused these changes in science led to a general feeling that Newtonian science could explain everything, even new phenomena as they were discovered, but the same Newtonian successes also led to clashes between electromagnetism and Newtonian mechanics. Solving these crises led to the Second Scientific Revolution, but scientists and scholars did not consider them crises at the time because everyone assumed that Newtonian science would overcome their inherent problems. No one so much as suspected that new ideas and attitudes toward nature would be needed to solve the crises and new discoveries that were made at the end of the nineteenth century, and these new ideas emerged in the forms of the quantum and relativity theories. It is simple to see from this discussion, that the Second Scientific Revolution that incorporated the quantum and relativity theories was a direct result of the successes of Newtonian science as much as it was due to the failures of Newtonian science, a view that runs directly in the face of normal historical accounts of the new revolution. But then a better understanding of where science is moving toward in the near future, the incorporation of the paranormal and the explanation of the afterlife, renders this new interpretation of the last revolution and all for the history of science more obvious.

And goes around again

Science does not progress by proposing new theories to replace old theories that are successful unless there is some compelling reason to replace the older theories. Then, the new theories must incorporate and explain the successful applications of the older theories before they can replace those older theories. The reason for developing the newer quantum and relativity theories only became obvious after the new theories were developed. They were not obvious before they were developed or during the time that they were being developed, and then they seemed to come in answer to specific crises in physics. Max Planck solved the blackbody radiation problem in 1900 and Albert Einstein explained the photoelectric effect in 1905, together leading to the founding of quantum theory. Einstein also explained Brownian motion in 1905, which led to experiments that verified the existence of atoms for the first time in just a few years. The birth of quantum theory marked a distinct change in the relationship between physics and nature, but many aspects of the quantum theory follow directly from the Newtonian past of physics. So the quantum theory grew faster and more immediately than relativity theory since it did not offer any fundamental breaks with Newtonian physics with regard to space, time, and matter, the most fundamental quantities of physics. Now this is the catch: Quantum theory prides itself on the fact that it proposes a model of nature that is indeterministic even though the Newtonian model of nature is deterministic. So scientists presently and falsely believe that quantum theory offers a very large philosophical break with its Newtonian past. But the quantum theory still regards 'matter in motion' as the basis of physics as well as a more Newtonian concept of space and time. Within this context, quantum theory is continuous with Newtonian physics even though it is indeterministic.

Still other problems led Einstein, again in 1905, to propose the special theory of relativity. Relativity deals with a completely different set of problems than the quantum theory, so it forms the other leg of the revolution. Yet relativity theory offers a far larger break with Newtonianism than quantum theory, so it has taken longer to assimilate the changes that it wrought in science. In neither case did scientists ever even dream that a new revolution in science had begun. That idea was only realized more than a decade later. In fact, the majority of scientists paid no attention to these advances at first, and science carried on with its duties as normal, not yet relying on the fundamental changes that were being wrought upon their fundamental scientific beliefs. We can thus accept as a basic rule of thumb that the more fundamental the change, the greater and deeper and more revolutionary the effect, also the harder it is for science to belief that a change has occurred and accept the new science. Relativity theory offered so simple yet radical a change with Newtonianism that special relativity was still largely unknown until after 1910 and was then only adopted across the broad range of science after general relativity was developed. But even then, general relativity was largely irrelevant and impractical for decades, allowing quantum theory to dominate physics during the following decades.

The material effects of relativity theory only appear at extreme speeds (special relativity) or near extremely large masses (general relativity), so relativity was impractical for common purposes until the space programs of the 1960s and the technology boom that the space program brought. Since then, both relativity theories have been incorporated into common scientific calculations. In the special theory of relativity, Einstein unified space and time to yield a new space-time continuum that is essentially different than the Newtonian concepts of a separate space and time. In so doing, Einstein demonstrated that science only needs relative space-time and needs never refer to absolute space or time, which are superfluous in science. Within this context, the very concept of time changed drastically as moving matter was found to dramatically affect the passage of time experienced by the moving object relative to stationary clocks.

The new space-time was also found to share a special relationship with matter itself, such that Einstein derived a new energy and matter equivalence based on his famous formula $E = mc^2$. So relativity changed the basic Newtonian relationship between space, time and matter, those very quantities that are basic to our whole concept of physical reality. However, special relativity implied the existence of more general rules that would cover accelerated material objects. Einstein then developed general relativity in 1916, which redefined the law of gravity and introduced the concept of space-time curvature using the non-Euclidean geometry of a Riemannian manifold. The scientific consequences of 'curvature' in our natural universe are only being realized today and a great deal of work still needs to be done on this subject. For example, the SOFT model of physical reality begins with the concept of 'curvature' as expressed in general relativity. While the quantum and relativity theories together formed the foundations of the Second Scientific Revolution, the revolution also included fundamental advances in the other sciences. Important work was conducted in psychology, medicine, biology and the other sciences, all of which added to the breadth of the revolution, but the main thrust of this second revolution, like the past two, came in physics and physics alone. And that physics, in turn, affected the other sciences.

Both relativity theory and the quantum had specific effects on the world of science, all of which are documented in any good history of the subject. But both theories also influenced the development of science and human thought in other lesser-known ways. Both had unintended consequences in science and human intellectual development that have influenced science in the long run rather than the short run that has been studied by historians, philosophers of science and other scholars. For example, relativity theory demonstrated that absolute space and time are superfluous, i.e., they are totally unnecessary in science. That fact is well known. However, it is not so well known that the loss of the absolute in physics tossed religion to the wind. The absolute had been related to GOD and relegated to religion by Newtonian doctrine, so when relativity theory totally dismissed the absolute as irrelevant, physics relinquished any say on any of the subjects within the old Cartesian realm of MIND. According to relativity theory, MIND became irrelevant to physics through its association with the absolute. Religion was still busy debating evolution theory and never noticed the shift in Cartesian boundaries caused by relativity theory and they still have not figured it out. Psychology was already in place so it picked up the slack, at least until it dropped the ball by ignoring consciousness.

In the case of quantum theory, the unintended consequences are far more troubling because they have been totally ignored by science. Quantum theory has so mesmerized the intellectual community that almost everyone has accepted inconsistencies in the fundamental concepts of quantum theory as unimportant aspects of nature even though they are actually among the most important aspects of nature. The Heisenberg Uncertainty Principle actually separates space and time from each other, much in the same manner as Newtonian physics, but quantum theory always assumes a combined space-time background for uncertainty as proposed in relativity theory, causing a new unrecognized paradox for science. Einstein and his colleagues addressed this inconsistency in 1935 and argued that quantum theory was incomplete, but the scientific community still considers quantum theory complete and thereby propagates fundamental errors into newer versions and advance sin quantum theory. Quantum theory cannot consider space and time separate for the sake of calculating uncertainties while they assume a space-time background without introducing at least a philosophical error if not quantitative errors. On the other hand, consciousness is needed to 'collapse the wave function' and create material reality, which is a standard interpretation of the quantum model of reality. If material reality does not exist until 'the wave function collapses', then consciousness must exist before material reality or it could not cause (or choose) the 'wave function' to collapse. Therefore, consciousness must lie outside of the quantum model, an idea that is completely new and wholly unknown in science. Quantum scientists and scholars assume in general that consciousness will ultimately be explained in the context of the quantum theory, but consciousness can never be explained by the quantum hypothesis because consciousness is prior to the 'collapse of the wave packet' and thus prior to the quantum created reality.

If this is not bad enough, quantum theory introduces yet another paradox that must be associated with the quantum model of reality: the creation of a Quantum Cartesian Paradox. Whereas Descartes distinguished between MIND and MATTER, quantum theory has altered that paradox to one between 'consciousness' and the 'collapse' that creates material reality. Yet this new dualism has passed totally unrecognized by everyone. It is not mentioned anywhere in either the scientific or the supporting literature. Instead, scientists and scholars have been mistakenly focusing on the determinism (classical)/indeterminism (modern) split. Evidence of this concern is found throughout the literature on modern science. Science will never progress to a higher understanding of nature until it can recognize and address the real problems in nature, i.e., solving the Quantum Cartesian Paradox and its precursor the classical Cartesian paradox. Unfortunately, scientists who follow the quantum theory believe that there is no Cartesian Paradox to solve. They are apologists. Quantum theory has forced the scientific community off on a tangent to the really relevant questions about nature and the nature of reality and they still have not figured it out. Those people who support the quantum paradigm as if now exists and is interpreted are not only barking up the wrong tree, but they have placed science in the wrong forest by obfuscating the issues with the wrong philosophical approach to the major 'classical' problems that physics has faced in the past and now faces. Science needs to return to the original fundamental problems out of which science evolved and solve the division between MIND and MATTER.

The solution of the problem of MIND might seem more a matter for psychology than physics, but psychology is actually unprepared for tackling the concepts of mind and consciousness, let alone their relationship to physical reality. After the Second Scientific Revolution began, psychology lost consciousness and did not regain consciousness until the 1970s, if at all. Prior to 1910, psychology was all about mind and consciousness, but scientific attitudes changed. Psychologists looked to behavior as the fundamental concept to drive their new science forward rather than seeking to discover and understand either mind or consciousness directly. Behaviorism took over the new science of mind due to positivistic influences, such as the ideas behind the quantum theory. In other words, psychology followed Mach's middle-of-the-road approach to MIND. In the behaviorism point-of-view, mind and consciousness cannot be directly known so psychology studies the mechanisms of interaction between different minds or between individual minds and the world around them. In a sense, psychologists have reduced humans and their interactions with each other to lab rats and statistical methods, demonstrating the influence of quantum theory on their science. Lab rats wandering through mazes are the living equivalents of material particles (without mind) bouncing around in the void in a random chaotic manner. So behaviorism belittles the Cartesian Paradox and reduces it to a difference between the objective and subjective 'in' nature rather than the mental (MIND) and material (MATTER) aspects 'of' nature. Psychology assumes the reality of mind and consciousness, but it does not seek to understand them or study them directly.

The subjects of mind and consciousness might have disappeared from science altogether if it were left to psychology as practiced for the first two-thirds of the twentieth century, but it did not disappear precisely because there is some physical truth to the proposition that consciousness is a 'thing-in-itself' and/or a quantitative variable in nature and is therefore at least semi-independent of the material brain and body of living organisms. The very fact that questions about mind and consciousness have become popular once again in both culture and the sciences other than psychology, but only in the past few decades, would seem to indicate that mind and consciousness are real 'things' that science must take into account in its studies of nature and the natural world. The more science comes to understand nature, the more mind and consciousness reappear in the basic scientific models of physical reality. Mind and consciousness cannot be glossed over and forgotten, nor can they be circumvented or worked around, because they are real parts of nature and the physical world. If and whenever science ignores consciousness as it did for the first six decades of the twentieth century, nature will again show the workings of consciousness and mind in the natural world and force science to reconsider their reality. As science grows closer to describing and understanding the natural world, as it really exists, in all of its most fundamental details, science must deal directly with mind and consciousness and their natural roles in our physical existence.

The same is true for the paranormal because the existence of mind and consciousness as 'things-in-themselves' implies that there can be direct communication between them or between them and other material bodies without the intervention of material brain or the common five material senses. Their semi-independent existence also implies the possibility that the mind and consciousness can survive the death of the material body and brain. As far as science was concerned, Spiritualism was out after about 1910, but the possibility that other paranormal phenomena could be verified in nature was still accepted by a few scientists. Following other normal trends in science, paranormal phenomena migrated into the laboratory as a new branch of science called parapsychology. Parapsychology evolved as research on the paranormal moved out of the Victorian parlors of scientists and into the laboratory in the 1930s, primarily due to the efforts of J.B. Rhine in North Carolina. The development of parapsychology as a lab science based upon the new psychology was essentially the positivistic answer to Spiritualism. Scientific study of the paranormal was reduced to the study of human 'lab rats' and statistics. The first decades of the new science represented the data-gathering phase of parapsychology. Already by the middle of the 1940s, common characteristics for different varieties of phenomena became apparent from the data, resulting in the development of the concept that a single common mechanism, called 'psi', could explain all paranormal phenomena.

The influence of quantum physics on parapsychology is quite apparent in the choice of the word 'psi' to represent that mechanism o paranormal communication because the 'psi function' is the physical quantity that is studied in quantum and wave mechanics. Otherwise, the first several decades of work in parapsychology were marked by debates whether 'psi' was a purely mental or partially physical quantity. In this particle debate between the mental and physical nature of 'psi' it is easy to discern the influence of the MIND and MATTER duality first enunciated by Descartes. Was Cartesian Paradox was alive and well at some subliminal level even though it had been emasculated and minimized by modern science that thought that they had solved it. The mentalists in parapsychology dominated the field for several decades and only two physicists worked as professional parapsychologists before the 1960s.

All in all, psi is the closest thing that parapsychology has ever had to a fundamental concept. They have never had a fundamental theory on which to base the science of the paranormal. The very existence of psi represents an assumption that a separate consciousness exists, beyond the normal material functions of the brain, but questions about consciousness and its role are completely missing in parapsychology. Instead, parapsychology is presently a study of the communication between different consciousnesses (ESP phenomena) or between consciousness and matter (PK phenomena). Nor does psychology have a specific fundamental theory upon which it is based. Psychology is merely based upon an assumption that mind and consciousness exists and no more. A theory of psychology would necessarily include statements about the nature of mind and consciousness, but psychology does not deal with them directly so it has no fundamental theory on which to progress beyond its present status. At best, psychology has a working hypothesis based upon behaviorism. Beyond this, parapsychology, psychology and quantum physics have many common characteristics. In particular, they all assume the existence of consciousness, but say nothing about what consciousness might be. None of them tries to define mind or consciousness. In fact, psychology, parapsychology and quantum physics all deny the existence or relevance of the Cartesian Paradox because recognition of the Cartesian Paradox would imply the existence of mind and consciousness. It would seem for them that the easiest way to explain something you don't understand is to ignore its existence, if not totally deny its existence and/ or relevance, at least until it comes back to slap you in the face and wake you up to its reality. They all ignore the existence of mind and consciousness at the practical level of interaction and deny its existence at the level of the Cartesian Paradox.

During the 1960s, a minor cultural revolution occurred in the United States and elsewhere. That revolution in attitude was accompanied by a major shift in scientific attitude. The shift was partially due to the space program, which expanded the human view of its role in the universe. But the space program also rendered general relativity relevant and practical for the first time. This change of events was then followed by rapid technological changes that brought relativity theory right into our homes. Ironically, Einstein had searched for a unified field theory based upon general relativity and the continuity the space-time for the last three decades of his life, while many physicists who were interested only in the quantum concept of a discrete particulate reality considered Einstein an eccentric. Yet the boundaries between the older Cartesian realms of MIND and MATTER did not shift. As scientific attitudes changed without a concurrent shift in the boundaries of science, new stresses began to develop. The stresses so placed on the scientific and academic communities as well as the paradigms upon which the scientific community based its attitudes were thus seriously challenged, but the scientific community did not realize that it was so challenged. The strain experienced by the community was not yet great enough to cause a revolution.

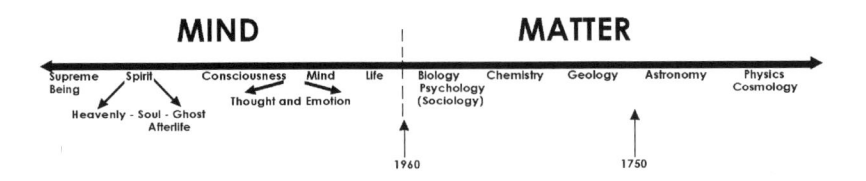

```
         MIND                    |           MATTER
<--------------------------------|----------------------------------->
Supreme   Spirit   Consciousness  Mind  Life | Biology    Chemistry  Geology   Astronomy   Physics
Being            Thought and Emotion         | Psychology                                  Cosmology
      Heavenly - Soul - Ghost                | (Sociology)
              Afterlife                      |
                                             ↑                       ↑
                                           1960                    1750
```

The only change in the boundary between MIND and MATTER from 1840 to 1960 came from the addition of psychology to science after the 1890s

The changes in attitude that occurred in the 1960s brought with them a wider and more lenient view of the world that incorporated the views of Einstein and his search for unification in physics. Direct challenges to the strict interpretation of quantum theory known as the Copenhagen Interpretation, where there is no reality until the 'wave collapses', began to appear and physics seemed to rediscover the concept of consciousness and its relevance to physical reality. The attitudes of the scientific community changed so radically during the 1960s that parapsychology became a legitimate science in 1969 as recognized by the acceptance of the Parapsychology Association into the American Association for the Advancement of Science. But parapsychology still had no basic theory. It would seem that a new period in the progress of science was about to commence.

As the 1970s drew nearer, different trends began to come together and show evidence that they would eventually converge. It is not normally the task of the historian to think about the future or predict the future, because the historian merely records and analyses the past. Writing, studying and understanding history on its own merits is hard enough without using history to predict the future. The history of science is much harder because it deals with science, which is itself a difficult subject. But trying to use history to predict future history, especially in the case of science, is the most difficult and frustrating tack of all. However, if you understand what science is and how it got to its present state, some good educated guesses can help to tell its probable future, at least what direction science is heading in. There is a point where everything fall\s into place, and once that point has been reached the future becomes inevitable. This is the point that has been explored in this book.

Solving the Universe

In a rather offbeat if not a historically perverse sense, at least when compared to normally accepted historical interpretations, the Second Scientific Revolution could be interpreted as only offering compromises that solved the immediate problems and crises in physics at the expense of avoiding the long-term and more fundamental problems of science. This interpretation is made from the point-of-view of one who has already experienced the attitude shift that others will experience withy the Third Scientific Revolution, so this interpretation is relative to the issues and trends that are most significant to the next revolution. The new changes in science wrought by the Second Scientific Revolution did not include solutions of the very most fundamental problems that are inherent in science as a whole, as represented by Cartesian Paradox of MIND and MATTER. Scientists left important issues unresolved only to crop up later in new forms in anticipation of the revolution that is even now emerging. The Second Scientific Revolution was hijacked and left incomplete because two different strands of development were inaugurated in physics (the discrete and continuous natures of reality) rather than unifying the physics that already existed in a single theoretical structure and completing the revolution, once and for all. In retrospect, the band-aid solutions that defined the new post-revolution science of the twentieth century, behaviorism and a philosophically incomplete quantum theory as Einstein argued, were never more than partial solutions, although they were more than likely necessary at the time.

Science took the quantum theory and its standard philosophical underpinnings far too seriously to heart, overemphasizing discreteness at the expense of continuity, and mistakenly interpreted the quantum reality as a final solution to all the problems in the natural world. Yet this strict emphasis on the quantum aspects of reality has failed to solve all of the problems in physics after nearly a century of trying. The development of 'quantum field theories' and the later 'standard model' have only perpetuated the fundamental errors of the quantum approach to reality. Unfortunately, science has ignored or perhaps even avoided the truly fundamental changes in science proposed by Einstein with the relativity theories. Criticizing quantum theory in this manner is tantamount to scientific heresy if not outright treason within the present scientific establishment because of the strong faith that the scientific community places in the quantum theory. However, expedient solutions of the type offered by the quantum theory, which are not really final solutions in spite of the good intentions of theorists, are not new to science. Although revolutionary at the time, such theories still offer a valid path of progress for science after their own manner, but the quantum also unnecessarily limits science as the time to take the next step beyond them becomes more and more evident. Like the original Scientific Revolution, which sowed the seeds of the next revolution, the development of the quantum theory carries forward the seeds of the next revolution.

Aristotle finalized the Zeroth Scientific Revolution by establishing 'matter in motion' as the fundamental principle in physics, but he also set up the changes that would bring about the Scientific Revolution of the seventeenth century by failing to go any further and define either speed or matter. The first Scientific Revolution was set up by Descartes' differentiation between MIND and MATTER, but failing to completely and accurately define them set-up the changes made by Newton that would eventually lead to the Second Scientific Revolution. Planck, Einstein and others initially solved the problems that led to this last revolution, but in so doing they also established a basic duality of mutually incompatible ideas (reality is either continuous or discrete, but not both) that were bound to lead to a new and far greater revolution, the one that is to come in the near future. So the Second Scientific Revolution was hijacked by the overemphasis of science on the quantum theory and behaviorism before it could run to completion. That sense of completion in our overall worldview of nature will not occur without a Third Scientific Revolution.

Science only solved the problems of MATTER presented by the crises in physics in 1900 by circumventing new and more fundamental problems such as the new concept of the space-time continuum, repeating the absolute/relative space (non-reducible continuum as opposed to a reducible continuum) problem that was solved at a lower level of reality by Patrizi and Newton four centuries earlier. MIND was finally approached directly by the newly emergent science of psychology shortly before the beginning of the Second Scientific Revolution, but events and changes in the fundamental nature of science during the revolution effectively evolved as the stopgap compromise of behaviorism to the problem of scientific understanding of mind and consciousness. By failing to solve either the Cartesian Paradox of MIND and MATTER or the conceptual relationship between the scientific concepts of mind and matter and the split between the discrete and continuous nature of reality, the Second Scientific Revolution has unwittingly defined the central problems that will bring about the Third Scientific Revolution. It would thus seem that the seeds of the next revolution have always been found in the successes of the last revolution, and such can thus be accepted as a general historical rule.

From this perspective, we can peel away even more layers of historical topsoil to discover a more accurate picture of the development of science by taking the point-of-view of a post-third revolutionary in the near future. In this new view, scientific research into the paranormal becomes an integral part of the natural evolution of science rather than an aberration or historical anomaly. In a still broader context, the Zeroth Revolution resulted from a major shift from myth to natural philosophy by the ancient Greeks. Humans first started to think and use their minds to interpret nature and the world around them rather than blame everything on anthropomorphic Gods. Even the concept of God became more abstract and less human-like, while natural philosophy began to consider the realms of the paranormal and afterlife legitimate philosophical fodder. Between the Zeroth and first revolutions, natural philosophy reverted and was co-opted by religion, but the religious connection was doomed to failure by further abstractions and the growth of rote knowledge. The scientific portion of natural philosophy quickly outgrew the antiquated attitudes of religion even though religion was abstracted to the cause of natural philosophy. This natural evolution of human thought guaranteed the course that the next major revolutionary changes would take.

The resulting Scientific Revolution in the seventeenth century offered a major shift within natural philosophy from religion to science, during which the major differences between the internal mind that perceives the world and the cold hard external world of matter being perceived were uncovered and popularized by Descartes. Religion and spirituality were relegated to MIND and science was firmly ensconced in MATTER. The next change came with the 1840 splitting of natural philosophy into the individual scientific disciplines, but this was only a minor shift or correction in the boundaries between MIND and MATTER so there was no revolution. MATTER, the realm of science, was becoming ever more fractured by specialization within the Newtonian framework. Yet this shift prepared science for the next revolution by opening the human mind to new possibilities and freeing it from earlier religious restrictions.

The Second Scientific Revolution ended with a false shift between those areas of nature that were traditionally included in Cartesian MIND and MATTER. It was characterized by attempts to alter and replace the fundamental problems of science by convenient middle-of-the-road solutions to the Cartesian Paradox, such as indeterminism and 'collapse of the wave packet' instead of a fundamental concept of a single physical reality shared by both MIND and MATTER.

Quantum theory has managed to redefine the primary duality on its own terms and thereby obfuscate the central paradox for science

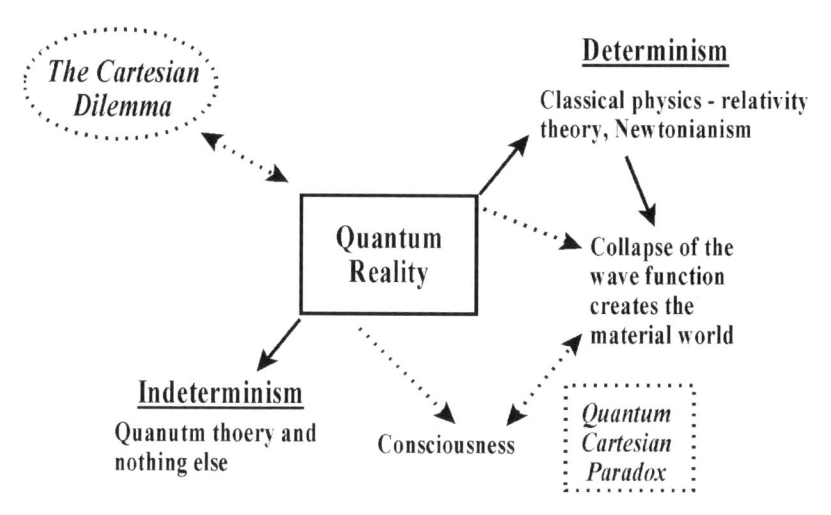

The Cartesian Dilemma

Determinism
Classical physics - relativity theory, Newtonianism

Quantum Reality

Collapse of the wave function creates the material world

Indeterminism
Quanutm thoery and nothing else

Consciousness

Quantum Cartesian Paradox

Science subconsciously realized that the Cartesian Dilemma must be resolved, but was not yet up to the task, so compromises were made and these compromises propelled twentieth century science forward for a short while. Scientific progress was made within the new limitations that these compromises placed on MIND (psychology) and MATTER (quantum physics), so all was not lost, but these middle-of-the-road solutions guaranteed the advance of a new revolution as science delved deeper into the fundamental nature of physical reality.

Indeed, a great deal of progress has been made under these conditions. In fact, science has suffered from an unbelievable bout of progress that has carried it to new heights of knowledge and understanding that would seem to indicate that the compromises were necessary for science to progress in spite of the limitations of these compromises. However, the progress under these compromises has also made it that much harder for science to accept that they were just 'temporary' measures, while the advances made by our modern science, compromised as it is, can only go so far before it becomes corrupted by its own limitations and that point has already been reached. In both physics and the science of mind, the superficial solutions to fundamental problems that seemed to define the Second Scientific Revolution so completely also tended to distract science from the possibility of more fundamental solutions, at least until nature more recently forced science back to considering the more fundamental 'hard' problems that science must eventually face, solving the MIND/MATTER and the discrete/continuity problems. A good example of this rediscovery is the rapid rise of 'consciousness studies' as an interdisciplinary science since the late 1980s.

In each successive 'revolving' of human thought and attitudes toward nature that we call scientific revolutions, the historical situation has become all the more complicated as rote knowledge grows and new data accumulates. Both more and a greater variety of people have become involved in the development of science so more opinions are added to the fray, while other evolutionary trends in science complicate the picture still further. All of these have become important factors that have influenced the course of events during the present pre-revolutionary period. In other words, each successive revolution in science becomes all the more complicated by the addition of new historical factors beyond the common crises in science that science must solve to progress as well as the fact that science is considering ever more fundamental questions about nature, the natural world and the nature of physical reality as it progresses. However, the choices to be made for science to progress and the changes to come after the coming Third Scientific Revolution have also become more transparent to historical analysis as the new revolutionary period draws nearer, at least to those who have some understanding of the new science that is coming. Thomas Kuhn identified the basic characteristics of scientific revolutions in the early 1970s, so historians, philosophers and scientists can now complete limited historical analyses on the go, as science is changing.

In general, the direction that the new post-revolutionary science will take depends upon the pre-revolutionary trends in science and human (cultural) thought as well as the fact that the new science must incorporate all of the correct but not quite accurate science that has preceded the revolution. New theories only replace old theories when it becomes necessary to adopt the new theories in favor of the old theories. This would of course seem natural because humankind is coming closer to the 'truth' of nature's inherent rules of operation with its own evolving physical laws, so it cannot discard the older solutions to problems already solved by science unless it has new and more accurate solutions. Taking all of this into consideration, past trends clearly indicate the direction that future science will take. The major trends that have so far been identified are the unification of MIND and MATTER as well as a parallel unification in physics (MATTER) that will depend upon a higher-dimensional space-time. A unification of sorts will also occur in MIND, as psychology and parapsychology merge in a concrete model and theory of consciousness that will become the fundamental theory of the first true and complete science of mind. The overall trend is toward unification and we are moving toward a single science that fits all after a long period (since 1840) of ever-increasing separation and specialization of the various branches of science.

The problem with the scientific study of MIND is that both psychology and parapsychology presently operate at only the same low level that Aristotle reached for physics more than two-thousand years ago, looking in the right direction at the correct concept, but not having developed a basic working theory. Psychologists accept the reality of mind and consciousness, but they do not address them directly in their normal research, so they have no fundamental theory upon which to base their science. Parapsychologists also accept the reality of mind and consciousness with no theoretical basis for the concepts, but they also accept the reality of a mechanism of interactions between different minds called psi. However, parapsychologists know nothing of psi directly. So both psychology and parapsychology are searching for a theory consciousness upon which they can build models of the interaction of minds as well as mind and matter. So while physicists search for a single unified view of MATTER and its interaction with consciousness, scientists of the mind are searching for a single concept of MIND and its interaction with consciousness and matter. Since major trends in science in the two different areas of MIND and MATTER have now been identified, heading in the same direction in lock step with each other, it is easy to see the coming revolution that will undoubtedly result in a unification of MIND and MATTER that will force major changes in both science and culture.

To understand and predict the shape that the next revolution will take, or rather the basic tenets of the theories that will emerge, a person need only define the problems of the present paradigm of science and correctly identify the most important, significant and fundamental trends of change in science that are presently occurring. In this instance, the present trends show a growing interest in the study and understanding of consciousness by all areas of science and a growing interest in the paranormal and the afterlife within common culture, if nor in science, both of which are related to consciousness and fall within the realm of MIND. On the other hand, the major trend in MATTER is toward the unification of the quantum and relativity theories and the subsequent unification of the four basic natural forces (or interactions); gravity, electromagnetism, the strong nuclear and weak nuclear forces. Yet there is also an overwhelming and inherent merging of MIND and MATTER within this pattern of trends and that is where paraphysics comes onto the scene. The unification process in physics must account for role of consciousness and mind in our interpretation and perception of physical reality or that unification will be as incomplete as past unifications and theories. In such a case, the unification process will devolve into just another compromise, perhaps more fundamental compromise than the last, but still just a compromise and temporary solution to the Cartesian Paradox.

Our science has become so accurate and fundamental that the next step forward, represented by the next revolution in thought, will amount to 'solving the universe' as best we can from our limited human perspective on Earth. Science is experiencing simultaneous revolutions in both MIND and MATTER; even as the boundaries between them are blurring and they are converging toward one another. There is no doubt that a flashpoint is nearing whereby a new paradigm will be in the offing for human thought, rather than just the sciences and academic studies. In particular, increased education and rapid advances in technology, especially in the electronics available to individuals, has pushed the envelope in other areas of psychic research. Common stories of apparitions, ghost sightings and hauntings have brought research down to the personal level. The interest of the general population in the paranormal is only surpassed by the public interest in science and the universe as a whole. These interests should force science to seriously consider the paranormal instead of dismissing anomalous phenomena outright and out of hand, as it has largely done in the past. Scientists must go beyond normal science to consider new problems and answer questions that were previously ignored, avoided or passed over as irrelevant, if science is to make any meaningful advances.

On the other hand, there is still a lot to be said for normal science. For example, we have derived a great deal about the inner workings of our universe and nature through a careful application of the scientific method. Science seeks the logical truth in our universe and nature. It relies on facts and theories whose truth can be verified. The scientific method has brought us a long way and it will take us a good deal further, but it is not unlimited in that science is not everything to everybody. Science and the scientific method are not absolute and no scientist would ever claim that they are, or else science would have become a broken system of knowledge long ago. Science is and should be immanently adaptable. The scientific method is also adaptable although some scientists act as if science and its methods are rigid and absolute by claiming that studies of the paranormal are and always will be beyond the scientific method and thus not part of valid scientific inquiry. The scientific method is not fixed and the role and scope of science are constantly expanding outward, incorporating new phenomena as well as old. Science, its methods and tools evolve and change as humans further explore their experiential world, and while we logically experience our material world we intuitively sense a vaster world beyond the merely material world. Phenomena once thought impossible are now extremely probable, including many paranormal phenomena, causing new stresses on the present paradigms of science.

These new stresses share a great deal in common with the stresses faced by science during the period immediately prior to the Second Scientific Revolution. In fact, the similarities between the two periods are astonishing. In a very fundamental sense, the period of Modern Spiritualism is repeating itself in a more modern form as exemplified by a growing interest in parapsychology and relevance of paraphysics. In the last two decades of the nineteenth century three theories dominated physics; Newtonian mechanics, thermodynamics and electromagnetism. They did not work well together in those areas where their application overlapped, precipitating a crisis in physics and science. The same is true for relativity theory and quantum mechanics today. Relativity and the quantum are mutually incompatible even though there is a growing sense that they must be combined into a single whole if science is to continue to progress. Meanwhile, the crisis facing the scientific world by the existence of Dark Matter and similar unexplained phenomena easily parallels the crises in physics that precipitated the Second Scientific Revolution.

In the late nineteenth century, culture was ripe for changes as education reached new levels and the educated questioned old loyalties to cultural institutions such as the church. Today, cross-cultural influences are rapidly changing attitudes as our intellectual heritage collides with the information age, resulting in changes in our concepts of spirituality. The concept of mind was elevated to a science of mind in the late nineteenth century, while the concept of consciousness is now pushing to the forefront of scientific inquiry. Undoubtedly, there are significant stresses on both our cultural and scientific beliefs and these stresses will soon cause irreparable strains on our scientific and cultural worldviews, if they have not already appeared. The only possible conclusion is that science has entered a pre-revolutionary period of development in which the paranormal, including the scientific study of consciousness and the survival of consciousness, will play an important role.

The pre-revolutionary period began about 1970 as attitudes in general changed; at least 1970 is a convenient date to mark the emergence of the new revolution. Ironically, or perhaps providentially, paraphysics began its first phase of development in the late 1960s and 1970s. Paraphysics emerged from the realization that 'psi', whatever it would ultimately prove to be, would at least prove to be physical in nature. In any case, a small group of scientists began to look seriously at phenomena 'beyond' those of normal physics and thus 'para' physics emerged. Earlier debates to decide whether 'psi' was ultimately a mental phenomenon or at least partly physical in nature seemed to ease. They dwindled to insignificance and just ended without being resolved. Their end came just as physicists became interested in the paranormal, shifting the weight of opinions in the direction of a physical nature of psi. This 'coincidence' was probably due more to an expansion of physics and changing attitudes within the physics community as much as it was due to any research that parapsychologists had completed or evidence that they could have presented.

A growing interest in unification and the possibility of a TOE also emerged within the physics community during the same time frame, as did renewed interest in consciousness and its role in the material world. In particular, Eugene Wigner, Henry Stapp and other physicists offered new interpretations and began a series of debates concerning the relationship between quantum and consciousness. Otherwise, only two physicists worked in the field of parapsychology prior to the 1960s, Joseph Rush and R.A. McConnell. After 1970, many more physicists and physical scientists began research in the field of parapsychology, including Hal Puthoff, Russell Targ, Edwin May, Elizabeth Rauscher, Evan Harris Walker, Robert Jahn, Brenda Dunne, Jack Sarfatti, Beverly

Rubik, Edgar Mitchell and others. Within the next decade, other physicists became interested in the physics of consciousness, which provided a safe haven for physicists to talk about consciousness and related matters such as the paranormal without having to suffer the stigma of being associated with parapsychology. These changes marked the new attitude of scientists and their renewed interest in mind, consciousness and related subjects, just as 'psi' began to emerge as a matter for physicists to contend with.

Popular books on these subjects also began to appear from the perspective of physics as well as the other sciences. Within the span of a few years during the early 1970s, Raymond Moody published his seminal work on 'Near Death Experiences', *Life after Death*, and Robert Ornstein effectively challenged science to once again consider the concept of consciousness directly on its own merits by publishing his popular book *The Psychology of Consciousness*. Ian Stevenson began his groundbreaking research in reincarnation, implying that our reality extends far beyond any that science had considered for many centuries. Fritjof Capra published *The Tao of Physics*, an extremely popular book offering a comparison between eastern philosophy and modern physics, while Lawrence LeShan compared mystical philosophy in general with both modern physics and the findings and impressions of mediums in *The Medium, the Mystic and the Physicist*. The astronaut Edgar Mitchell also edited and published a collection of essays under the title *Psychic Exploration: A Challenge for Science*.

Mitchell's book included articles by physicists, including a foreword by Gerald Feinberg, who claimed that it was time to end efforts aimed at proving the existence of psi and paranormal phenomena. It was now time to begin investigating the physical properties of psi so that a theory could be developed. Two other articles in Mitchell's book clearly announced the emergence of the new science of paraphysics, but Mitchell's 'challenge' was never answered or acted upon by the scientific community. In the meantime, Puthoff and Targ publicly investigated 'spoon bending' and remote viewing at Stanford Research Institute while Jack Houck invented 'spoon bending parties'. The CIA and other government organizations initiated a secret spy program for remote viewing called Stargate, run by Puthoff, but Puthoff and Targ still published reports on their research in more public and open forums. This spate of books, publications and initiatives are indicative of a healthy evolving science as well as characteristic of a pre-revolutionary period.

A host of scholarly and semi-scholarly professional organizations were also founded at this time. The development of such organizations has long been associated with pre-revolutionary and revolutionary periods in history. For example, new scientific societies for the study of psychical phenomena like the SPR and the ASPR were founded in the late 1800s, just prior to the Second Scientific Revolution. These professional societies were founded to answer questions and concerns about the growth of mediumistic and psychic phenomena that scientists were beginning to raise. In the 1970s and following decades, a number of new societies to deal with the paranormal also emerged. Societies such as these only develop if there is a need for them in the scientific community. The Society for Scientific Exploration (SSE), the Academy of Religion and Psychical Research (ARPR), the Casey Foundation, the International Association of Near Death Survivors (IANDS), the International Remote Viewing Association (IRVA), and the US Psychotronics Association (USPA) were formed, to mention only a few.

In England, a brief attempt was made to spin a paraphysics group off of the SPR, but that attempt failed from political squabbling and questions regarding the scope and mission of the proposed organization as the 1980s neared. International organizations for consciousness studies such as the Association for the Scientific Study of Consciousness (ASSC) also emerged, as did independent professional peer-reviewed periodicals such as the *Journal of Consciousness Studies*. Annual international conferences such as "Toward a Science of Consciousness" and "Quantum Mind" (held every four years) have also been held since 1990. These various events are not aberrations in science, but part and parcel to a growing, evolving and healthy science of consciousness and psi. They are all part of the same revolutionary trends in science.

All of these changes are characteristic of a pre-revolutionary period in science even as the rise of interdisciplinary consciousness studies and the growing body of paranormal evidence serves as a direct challenge to normal psychology. Yet spirituality, which is associated with the paranormal by psychologists, is still considered a disease as listed in the *DSM4*. The *DSM4*, or the fourth edition of the *Diagnostic and Statistical Manual of Mental Disorders*, is the clinical bible for the psychological community. Mystical experiences, psychic experiences, visionary experiences, meditation and spiritual practices, Kundalini awakening, near-death experiences, possession experiences, UFO encounters, and Shamanic crises are all listed as spiritual problems in the *DSM4*. So it would seem that psychologists officially treat spirituality as a mental disease or disorder even though everyone has a spiritual part to their conscious nature.

It would thus appear that psychology cannot live with MIND or consciousness even though it cannot live without MIND or consciousness. This viewpoint is at complete odds with events that are occurring in science in general, so science is faced with incompatible worldviews in the science of MIND as well as competing paradigms in the science of MATTER. We also have an open competition between fundamental theories in physics for the first time in over a century. Quantum theorists are trying to quantize the continuous gravitational field while some physicists are trying to explain the discrete quantum based upon a continuous relativistic interpretation of physical reality. The discrepancy between the two theories has existed for a century, but it is only within the past three decades that the relativistic and continuous basis of reality has become a viable candidate for the next fundamental theory. Every present trend in science points to the inevitability of a new revolution in science and all indications are that science entered a specific well-defined pre-revolutionary period a little more than three decades ago.

From the evidence thus provided by history, it would seem that the coming theory that will define the next revolution would be a physical theory of unification that must necessarily account for both mind and consciousness. By historical and logical necessity, the theory would define the mechanisms by which consciousness interacts with the world of matter, both locally and non-locally, and thus include the mechanisms of both behavior and psi. All evidence indicates that the new physical theory of unification will be hyper-dimensional, but not necessarily a superstring or brane theory. The 'single field theory' or SOFT is the only theory that fits all of these criteria, so at the very least, SOFT will soon be considered a primary contender for the basic theory of science leading to and/or emerging from the next or Third Scientific Revolution.

Buddha, Jesus, Human Enlightenment and the dawn of the Mysphyts

When the scientific community accepts the paranormal as perfectly and completely natural, science and what we now regard as religion will merge together to a large extent. The afterlife will become part of the scientific realm and subject to verification, no longer subject to religious distortion based upon faith alone. Science and religion will never merge completely because GOD is not a subject for scientific analysis, but both religion and humanity will come closer to knowing what GOD is by knowing what GOD is not, as science discovers more about our physical reality. Science will define what GOD is not and leave the question of what GOD is to what remains of religion.

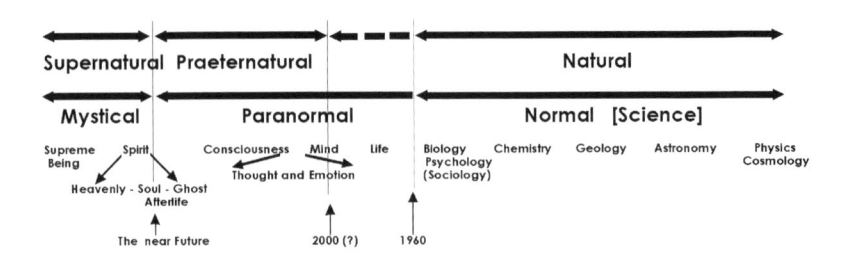

```
←——————————→ ←————————→ ←— — —→ ←————————————————————→
  Supernatural   Praeternatural              Natural
  ←————————→   ←————————→        ←————————————————————→
    Mystical      Paranormal              Normal [Science]
  Supreme   Spirit   Consciousness  Mind   Life   Biology    Chemistry   Geology   Astronomy   Physics
   Being                                         Psychology                                   Cosmology
         Heavenly - Soul - Ghost   Thought and Emotion    (Sociology)
                  Afterlife
                    ↑                 ↑                ↑
              The near Future      2000 (?)          1960
```

The still shifting boundary between Cartesian MIND and MATTER

Under these circumstances, religion and science will change to the extent that they will travel the same road toward conscious awareness and knowledge. There is, after all, only one reality even though there is more than one way to interpret that reality, just as there is only one human mind to interpret that reality even though there are two different ways for the human mind to mentally explore that reality, intuition and reason.

The paranormal and the survival of consciousness (the afterlife) will become a more respectable part of science, leading religion to deal with the concepts of GOD and soul and little else. So it would seem that the classical Cartesian view of MIND and MATTER are converging, offering a unique new view of the coming revolution. The new revolution, like past revolutions, will be triggered by crises in science, but it will emerge from the trends and successes of science rather than from any specific failures of science. Closer and more accurate observations of nature will present science with new problems that cannot be resolved within the dominant paradigms or by the old theories, and crises will thus emerge. But the solutions of these crises are already inherent within past trends that have already been evolving in science from the seeds that were sown in the shortcomings of the last revolution. The closest that science will come to failures will exist in the shortcomings and nearsightedness of the older theories and attitudes toward MIND and MATTER that will fall by the wayside of history in the light of future science. Given the expected changes that this new revolution will force upon science and culture, it is constructive to once again reinterpret the very notion of a scientific revolution. We will then discover that the number of revolutions does not change to three with the addition of the new revolution, but to four.

Within the context of this new analysis, it can be shown that the new scientific revolution will emerge from a change of the fundamental basis of science from 'matter in motion' and all of its implications to the curvature of space-time in a higher dimension (or dimensions). Matter and motion are actually our way of sensing, interpreting and understanding the curvature in a higher dimension of space-time than we normally associate with material reality. But humans somehow instinctively know that there is a higher reality than just our material world and our intuitive knowledge of that higher reality is the basis of all religion. Unfortunately, it is also the physical basis of all superstition and a few mental disorders. Physical reality does not proceed from our human sensations of a material four-dimensional reality to an extension in a higher dimension, but from a higher-dimensional reality to our material perception of a four-dimensional material reality. All life forms are really higher-dimensional beings than we presently perceive because life, mind and consciousness are really five-dimensional field structures within space-time.

Yet we still have the ability to learn about that higher-dimensional reality through either intuition (mystic reality) or reason (scientific reality) and accept it as our common birthright. Plato once used the metaphor of a shadow on a cave wall to explain this concept. Humans sense the shadow and believe that it is the reality, but the form or structure casting the shadow is the true reality. Learning and accepting this simple truth will force scholars and academics to reinterpret the changes wrought by the earliest development of Natural Philosophy, from Thales to Aristotle, out of which our search for a scientific and reasonable explanation of physical reality first developed. Just as Aristotle first defined the fundamental search for reality on the basis of 'matter in motion' and thus initiated the development of science, science will surely overturn 'matter in motion' as the fundamental basis and instead adopt the superior reality of 'curvature and variations in that curvature' in the next revolution. This change in our perspective of reality will elevate the period of Greek history from Thales to Aristotle to the status of a Zeroth Scientific Revolution from the hindsight of the Third Scientific Revolution.

The Zeroth Revolution marked a change from the purely mythical and supernatural interpretations of nature to newly derived mystical (intuitive) and logical interpretations of nature and the world in which we live. In the seventh century BCE, human beings started to use their minds for the very first time to think about and interpret their world of reality and thus undertook the development of the very first paradigm shift in the written history of the human race. In the Far East, that interpretation was introspective and the intuitive part of mind became dominant in culture, while the Greek tradition of logic and reason won the day in the West. But the single human mind did not just end with this split decision. The philosophical seeds for the next revolution were cast through a misunderstanding of the differences between intuition and logic as expressed in medieval Natural Philosophy by religion and science, respectively. On the other hand, the practical seeds for the next revolution were already cast by Aristotle's inability to define either matter or motion in any meaningful or logical manner that could lead to practical accurate applications to nature even though he chose them as the basis of logical science and explanation.

The twin functions of mind, in the form of intuition and logic, were unrecognized at first, such that Aristotle arrived at 'matter in motion' as the logical basis of nature by intuitive methods. But Aristotle did not have the logical or conceptual background necessary to adequately define either matter or motion, so the new science that emerged was a combination of both mystical (intuitive) and philosophical (logical) thought, allowing religion to reemerge and dominate Natural Philosophy in its new abstract monotheistic form until the scientific revolution in the seventeenth century. Aristotle unwittingly planted another seed for the next revolution by allowing a place for both religion and science in Natural

Philosophy, but the religion that evolved was not the same as it was before Thales. Religion became a more logical practice rather than a purely mythical institution. The development of different logical concepts of an abstract God or Supreme Being thus emerged from the pantheon of mythical gods with more human attributes and came to dominate thought in Natural Philosophy.

Under these conditions, Natural Philosophy progressed by the accumulation of new knowledge about nature and the world within its own newly established paradigm. Natural Philosophy became so successful in the pursuit of knowledge that discrepancies emerged between the logical interpretation of nature and the facts of nature. These crises were most evident in the logical concept of motion and direct astronomical observations, while the Natural Philosophical perception of reality slowly ceased to fit the physical and natural world as experienced. The Scientific Revolution that emerged in the seventeenth century partially corrected some of the shortcomings that Aristotle incorporated into physics, such that it at least defined matter and motion to the extent that these concepts became useful in explaining natural phenomena. Motion was better defined while matter was merely associated with inertia, which allowed the practical measurement of the amount of matter and the forces that move matter. However, the ultimate nature of matter, to a large degree, was not yet discovered.

Meanwhile, the Cartesian split between MIND and MATTER better defined the roles of intuition (religion) and logic (science) in the new world order set in place by the Scientific Revolution. So, while Newton defined and applied the science of 'matter in motion' for all practical purposes, the seeds for the next revolution were still inherent in Newton's work because matter was only measured and not completely defined. Newton further modified the Cartesian split of MIND and MATTER by instituting his concepts of relative space and time within the realm of MATTER, but left the question of the meaning of life and mind out of science, thereby relegating them to MIND and religion, which he associated with absolute space and time. Newton's negligence in this case planted other seeds of discontent for science to deal with later.

The Second Scientific Revolution represented the first attempt to develop a science of mind. This development was spurred on by the natural expansion of Newtonian science into the realm of MIND as well as Darwin's development of evolution theory. By the nineteenth century, Newtonian science had advanced to the point where scientists and philosophers needed to learn how minds interact with both other minds and matter (the environment) to form the natural laws of science. Science needed to correct newly found discrepancies between Descartes' concepts of MIND and MATTER since they defined the realms of intuition and logic, as relegated to religion and science. On the other hand, the successes and substantial growth of Newtonian Physics eventually turned up new discrepancies between our knowledge of the world and the Newtonian picture of the nature.

The pre-revolutionary period for the Second Scientific Revolution had begun by the end of the 1840s when natural Philosophy split into the more specialized sciences of biology, geology, chemistry and physics. This evolutionary step was followed by the development of the Darwinian theory of evolution, which forced the differences between life, mind and the lack of an understanding of them to the forefront of science. Evidently, the seeds of the next revolution that had been set by Newton finally began to bloom, and quite rapidly at that, by about 1850, resulting in the crises in physics that triggered the Second Scientific Revolution in 1900. Yet when science was confronted with new fundamental questions about the very nature and meaning of matter, in the form of specific crises dealing with the interactions of electromagnetism and matter at the end of the nineteenth century, science found it more advantageous to retreat to the very basic duality of matter that had originally been defined in the Zeroth Revolution: Is the basis of matter discrete (atomic) or continuous (Plenum or field)? Paradoxically, the answer to this question was yes: the wave/particle duality. Thus a new split, a serious fracture in science that was held over from Ancient Greeks, formed within the Cartesian realm of MATTER: Continuity in the form of the field, relativity and waves fell on one side of the chasm, while discreteness in the form of the atom, quantum theory and particles raised their ugly heads on the other side, even as science delved more carefully into the fundamental nature of material reality.

With quantum theory in general and the Copenhagen Interpretation of the quantum in particular, whereby reality does not exist until consciousness collapses the wave function, science began to follow the path of least intellectual resistance rather than considering the more difficult solutions to the problem of finding a more fundamental and permanent understanding of the true nature of matter. However, this does not mean that the quantum theory was completely wrong, just that it was "incomplete" as Einstein had argued. Quantum theory rapidly became the dominant paradigm in physics, in large part because it retained the Aristotelian-Newtonian notion that 'matter in motion' was the fundamental basis of nature, rather than something more fundamental. Although 'matter in motion' did indeed form the fundamental nature of the new Aristotelian/quantum paradigm, quantum theory still sacrificed determinism and changed the rules of normal causality in physics in order to accomplish the physical compromise that it proposed and thus changed the priorities and direction of physics away from its long standing search for the true nature of matter.

Physicists and philosophers no longer thought that a search for a more fundamental nature of matter and reality was even relevant. In the new quantum theory, the normal basis of 'matter in motion' and thus physical reality did not emerge until the 'collapse of the wave packet' by consciousness, which placed the true nature of matter beyond knowing, rendering it irrelevant in the quantum paradigm. The quantum paradigm and its interpretation of the nature of reality were at complete odds with relativity theory from the very beginning since relativity theory ultimately implied the reduction of matter to space-time curvature. In the very simplest case possible, two or more bits of matter determine relative space and time regardless of the presence or action of consciousness or the collapse of a wave function. The very concept of space-time curvature evolved from this simple notion of relativity. As such, relativity went beyond the limits to the physical world set by the quantum theory from the very outset. And thus the seeds of the next revolution were sown during the early days of the Second Scientific Revolution. Yet these seeds fell directly within the Cartesian realm of MATTER.

Still other seeds were sown in the realm of MIND where a similar compromise or 'trade off' was instituted to continue the development of psychology, the new science of mind. The new scientific concept of mind was minimized by the emphasis of science and psychology on behaviorism at the expense of consciousness. As of the early decades of the twentieth century, science was not quite ready to tackle the concepts of mind and consciousness directly, just as it was not quite ready to directly tackle the conceptual basis or nature of matter. The role of mind in nature was thus reduced to a study of the interactions between individual minds if not the relationship between minds and the material environment rather than the thing interacting, i.e., neither mind nor consciousness was defined or even directly investigated. So science lost consciousness for the next half-century or more. Neither Cartesian MIND nor MATTER was adequately redefined during the Second Revolution, and science slid into a moribund trend of evolutionary progress under false pretenses.

Under these circumstances, a vast number of phenomena were discovered and explained by attempting to redefine all of science within the new quantum paradigm. The quantum paradigm has been highly successful in its effort to support its own relevance and philosophical worldview, but still limited to work only within its own worldview and not expand beyond that worldview as it is now trying to do. Of this there can be no doubt. Meanwhile, the traditional scientific search for the true nature of physical reality was reduced to a pathological pursuit of knowledge that was unnecessarily limited and fraught with inherent and built in problems, discrepancies and paradoxes. An unrealized and unspoken Quantum Cartesian Error has ruled the quantum paradigm even though quantum theory has been highly profitable in searching for its own version of the fundamental nature of reality. In this search, the vast majority of physicists have tried to force the nature of physical reality into a box built by the quantum paradigm, rather than trying to determine whether or not nature was truly discrete at the most fundamental level of reality. Too many scientists just assume discreteness in nature without ever questioning that basic hypothesis. But nature cannot be forced into a pre-set or pre-determined philosophical mold. Just attempting to do so has created new cracks within the intellectual edifice that is modern science.

On the other hand, relativity theory represents an attempt to redefine matter as 'curvature' and motion as 'variations in curvature' from the very onset. So, one would think that relativity theory is an advance over quantum theory. Clifford first recognized the significance of 'curvature' and the 'variations in curvature' in 1870, but his theories and ideas were premature and are only now being realized in five-dimensional physics. Perhaps general relativity was also premature. After relativity was formulated, scientists retreated conceptually to treating curvature as an intrinsic property of space rather than an extrinsic property of space that would have required higher-dimensional embedding. In this manner, the theory was artificially limited and offered its own paradox: How could curvature be an intrinsic property of the continuum? In a manner of speaking, relativists decided on a form of curvature without any real curvature to it, creating yet another conceptual paradox. Then scientists and scholars claimed that the word curvature should never have been used to describe the concept because it was not truly curvature in the normal sense of the word. But they did not try to find a better word to fit the property of space that they were trying to describe because they ultimately and subconsciously knew that it was still common curvature necessitating a higher-dimensioned space. To complicate matters further, the adoption of the new quantum paradigm by science further minimized relativity theory and the concept of continuity, just as science in general minimized intuition and consciousness.

The new science of the early twentieth century was just not strong enough to overthrow the older Aristotelian priorities based on our normal senses and gross perceptions of the world. It just twisted them around a bit into new forms and forced them into the misshapen artificial box that it constructed for them. In so doing, science unwittingly established the relevance of consciousness and curvature in a higher dimension as the seeds for the third revolution by ignoring their relevance and fundamental nature in the second revolution. Yet, nature will never be denied in the face of human philosophical biases. So it would seem that our ever expanding and more accurate study of nature has now led science back to the reality of consciousness and curvature even though quantum and psychological logic took over science by overlooking and working around them. In a very real sense, the revolution that is now in progress has become Einstein's revenge by default, but still goes far beyond Einstein's earlier work. It brings consciousness of MIND and the continuity of MATTER back together again. In keeping with the convergence of concepts offered by these more recent trends in science and elsewhere, which demonstrate a merging of our intuitive and logical pictures of reality, this newest revolution in science will be characterized by the birth of the MYSPHYT. The MYSPHYT or *mys*tical/*phy*sicist will probe nature and perceive the nature of reality with the whole mind, directly accessing both intuition and logic in the search for true knowledge.

But how could this come to be? Mystics seek to intuitively and directly experience the whole of reality in a single moment rather than study it philosophically or logically. Yet mystics must also try to reduce their direct experience of the totality and connectedness of reality to philosophical and logical terms in order to apply that knowledge to normal life as well as inform others and guide them along their own paths toward experiential enlightenment. The physicist, on the other hand, seeks to make logical sense of reality and probe nature with reason. But that is not enough. Even scientists must occasionally make intuitive leaps in order to advance science. Intuition fills in when logic and reason outpace the basic theories that science has developed to further explore the totality of natural reality. In other words, both intuition and logic are incomplete without the other, a simple fact which leaders in both areas of endeavor have failed to realize or take advantage of in the past. Realizing this simple truth, both mystics and physicists must endeavor to develop both halves of mind and consciousness to truly understand reality.

Both groups must realize that intuition, like reason, is not static and can be developed through increased knowledge, at least through true or universal knowledge. Wrong knowledge, such as misconceived theories in science, can only lead us away from enlightenment while deluding us with a false sense of progress. So the closer that science comes to true or universal knowledge, the more probable it becomes that people who understand reality and nature through science will become intuitively enlightened rather than just philosophically enlightened. Mystics in the past who have studied their 'craft' from a philosophical but non-scientific perspective have long recognized this fact. So mystics and enlightened religious leaders should embrace science and scientists could then embrace mystic practices. They should add a scientific component to their studies to better understand the whole of the universe as both a connected single whole and separate parts at the same instant. The universe is internally connected, but also separated into individual parts, even while it is both and neither, all at the same time.

Mystic teachers have long utilized the philosophical knowledge of experiential enlightenment as a tool to move toward that experience, but prized the reality of the experience over the philosophical understanding of the experience. This could be called the Hui Neng effect, harking back to the demise of the Ch'an Buddhist sect with the advance of Hui Neng to the position of the sixth patriarch. The young servant and novice Hui Neng was elevated to the patriarch position over the older and wiser Monk, precisely because the Monk had only a perfect philosophical understanding of the experience of enlightenment and not the true experience of enlightenment as did Hui Neng. Yet even a perfect philosophical knowledge and understanding of enlightenment is not enough because enlightened mystics have no maps to follow in their explorations of the newly discovered reality when they become enlightened. An experientially enlightened person can be his or her own map of the new reality they have discovered, albeit a limited map without a scientific knowledge of the true nature of that reality. Only a science based upon a true knowledge of the universe can provide such maps and prepare mystics to take advantage of their newly experienced reality, possibly leading to an opportunity to transcend even this new reality, because the experienced enlightenment is no more than a direct physical contact with the greater physical reality of the single field in five-dimensional space-time.

Intuition is knowledge that is gained without any logical cause or method of gaining that knowledge. Intuition is 'knowing' without knowing how. It is a by-product of the direct connection between consciousness and the rest of the universe through the single field in five-dimensional space-time. In other words, intuition is a product of our sixth sense as it interacts with our physical extensions into the higher dimension of space. When a mystic experiences this connection directly, even though the mystic does not realize the science behind the connection, the mystic reaches enlightenment. The experience of connecting with the whole of reality, or otherwise experiencing the connection between the single field in the fifth dimension of space-time, alters and enhances the consciousness pattern of the individual self in the single field. Being able to utilize this intuitive connection would greatly enhance the scientific search for knowledge as well as the perceptions of reality experienced by an enlightened person. In other words, science has as much to gain from mystical experience, as mystics have to gain from scientific knowledge.

A small number of people in history have reached enlightenment spontaneously while a few others have succeeded through their own meditative efforts. The founders of the major religions would be the first among this group of beings. Siddhartha Gautama became the Buddha through spontaneous enlightenment under the

Bo tree, but only after shedding his philosophical efforts to do so. Jesus was at least an enlightened being, a fact that no one would argue with, although many believe that he was far more than just an enlightened human being. Mohammed also reached his own level of enlightenment under his own circumstances. The same is true for Lao Tzu, Zoroaster and others. Only the degree, extent and quality of the enlightening experience of these people should be in question, not whether they were enlightened. The same is true for many other religious leaders as well as some non-religious scholars and thinkers. But each of them reached a different level of enlightenment and then interpreted that enlightenment according to their particular individual and unique cultural backgrounds. Each person understood and reacted to their experience within their own cultural context. Many then tried to teach others the path to enlightenment, although they may not have called the end toward which they led their followers, enlightenment. Modern religions later evolved from those attempts. Religious salvation, in a sense, is the heir to their enlightenment. The Buddha taught compassion and Jesus taught love, both of which are material correlates of the intimate connection afforded by an experience of the connection with the universe as a whole during experiential enlightenment. In other words, both religious leaders used and taught emotional love as the path to enlightenment because emotional love is the closest three-dimensional chemical response to the five-dimensional experience of enlightenment.

With enlightenment, an individual experiences the higher-dimensional reality of space-time directly and intuitively rather than logically. Logical understanding of the next higher dimension comes through mathematics and science, not direct experience. But the logical scientific knowledge of that reality still enhances the intuitive experience, so science should not be ignored as a method on the path to enlightenment. The truly enlightened individual need not recognize the higher dimension of space-time as such during this process, but still gains the experience of the connectivity of all things through the single field in the higher dimension. However, knowledge of the logical science behind the experienced connection would greatly enhance an enlightened teacher's abilities to understand his or her own experiences and guide others toward enlightenment as well as enhance and broaden the teacher's own experience. With the next revolution in science, scientific logic will have caught up with mystical intuition. Science is now beginning to understand the nature of the single field and higher dimensions logically, just as an enlightened mystic understands his or her experience of the connection to the universe afforded through the single field intuitively. When scientists understand enough about nature and physical reality to experience them in their entirety in the present moment or instant of time, then the true MYSPHYT will have been born.

Logically speaking, it stands to reason that it will become more likely that a greater proportion of the human populace will reach enlightenment as time passes, at least the intuitive experience known as enlightenment. It also stands to reason that the proportion of the human populace hat is truly enlightened will grow even larger as logical knowledge, understanding and acceptance of the reality of enlightenment grows. As rote scientific and true knowledge of our universe increases, in all of their aspects as scientifically understood, more humans will experience enlightenment. So, true knowledge offers a path for humanity as a whole to approach experiential enlightenment. The human race should eventually reach a point of saturation or shear weight of knowledge where enlightenment becomes a natural phenomenon for all human beings collectively and perhaps even simultaneously. The human race should eventually experience enlightenment as a whole, not as individuals. This point could well represent the next step in human evolution, just as life, mind and consciousness evolved as humans came into a more comprehensive contact with their three-dimensional world and time.

This physical model of reality indicates that we are not just gaining new knowledge, but are actually evolving ourselves by understanding the five-dimensional nature of our physical reality. So the human race could be nearing a super-convergence of MIND and

MATTER that could directly influence the physical and biological evolution of our species within the not-to-distant future. The convergence of mysticism and physics during the coming revolution could well ignite the next biological step in human evolution as well as the next scientific/cultural revolution of the human race. Impending changes may not represent just a simple scientific revolution or a simple cultural revolution, but something far greater. Humanity has come full circle and will return to its roots with a scientific revolution that redefines and reduces matter in motion to curvature and redefines human beings in the process. Just as mind was used by the Greeks and their contemporaries to analyze the world around them for the first time, two and a half millennia ago, turning superstition and mythology into abstract concepts of GOD and science, the MYSPHYT will evolve as a normal intellectual being when we begin to properly use our consciousness in its entirety: We will evolve into new humans.

When this occurs, what we have regarded in the past as religious truth at one level at an earlier time will be reduced to an earlier metaphor for the higher reality that we once upon a time sensed intuitively, at least, if not more fodder for exploration in science at the next higher level and superstitious at a still higher level. Evolution will carry the human race beyond its religious biases. This conclusion is undeniable. We are becoming MYSPHTS within our own skins, bodies, brains and material realities. And, as our perceptions of reality change, so will our biological and mental capabilities evolve. There will come a point where scientific theories become facts and facts become universal truths. The search for universal truth is, after all, the goal of science and the scientific method. Humans will experience reality just as they understand reality. In the end, the MIND reflects the universe and science reflects the MIND reflecting the universe, while CONSCIOUSNESS affords the purest contact between the MIND and the universe.

The Kaballah—a Warning

Once upon a time, there were two friends, drawn together by their common interest in the unknown nature of physical reality and its relations to the hidden recesses of the human mind. The first friend was a scientist, a MYSPHYT. The second friend was a self-made mystic, looking for answers in advanced science. She had had visions earlier in her life and she was looking for an explanation to what she had seen. She was very Jewish, so she first sought answers to her visions from her Rabbis. Even though she could not formally study the Kaballah, because she was a woman, the Rabbis taught her all that they could of the Kaballah. They taught her the real Kaballah, not the Kaballah of pop culture.

She eventually learned that the Kaballah did not supply all of the answers, so the Rabbis told her to consult the scientists. However, they did not direct her to pharmacists, neuroscientists, psychologists, doctors or psychiatrists; because they knew that she had had visions of a higher reality. So they sent her to seek answers from physicists. She studied modern physics as best she could from some of the best scientists in the field, but she sensed that that was still not enough. She eventually learned of a scientist who worked outside of the normal channels of science in the physics of consciousness and she contacted him. She found that he alone among the scientists understood her visions and attempted to explain the new reality that she seemed to have sensed. This scientist was a MYSPHYT, the friend mentioned above.

They shared ideas for a while, but finally drifted apart. One afternoon the scientist told her that the human race would evolve intellectually to learn of death and the concept of the afterlife. It would become possible to communicate with the consciousnesses that had survived after their own deaths. In fact, he wished to teach people not to fear the death of loved ones and friends, or even their own deaths, because the afterlife (or survival of consciousness) was a scientific fact that would eventually be verified. In fact, common communication with the dead was not beyond the realm of possibilities when everyone came to know and accept our greater reality and its nature. But the woman became scared.

She had been taught that knowing the dead and the Kaballah strictly forbade communicating with them. So the MYSPHYT asked her if the Kaballah told of human progress to a time when that would no longer be true. She confirmed that the Kaballah spoke of such a time, but it was neither the beginning of a new era for humanity nor the next step in human evolution, but rather the end of humanity that would accompany the return of the Messiah. She became so scared of this possibility that she no longer spoke or associated with the scientist. She did not realize that each of us is a messiah or rather each of us has a Buddha-nature. The return of the Messiah will occur when the human race as a whole realizes that each and every individual is a messiah. Their friendship ended and she carried on her search for answers elsewhere.

In reality, all ends of one kind are beginnings of another. That is the nature of scientific revolutions, at least when such changes occur with regard to human thought. Unfortunately, every revolution has its own set of losers, whether it is a revolution in human thought, a revolution in the human interpretation of nature, a political revolution, or an evolutionary step in human biology. All are not without a price to pay.

If a vote were taken on our next evolutionary step, or even on the next scientific revolution, many would surely vote on remaining ignorant with the present status quo. There are always those who think that change of any kind is frightening, especially changes as radical as those wrought during scientific revolutions. Some of those people that oppose change under any conditions, even when they are told that the changes are for their own benefit, will oppose the changes because they do not understand the changes. And sometimes, the opposition becomes violent. Even now many people are beginning to intuitively sense the coming changes and are opposing those changes by trying to bring back the 'good old days' of the Middle Ages, before science came to influence society and culture was still based largely upon religion and the prejudices that accompanied various religions. Glorifying the past as some kind of a 'golden age' is counterproductive and wrong when it leads to failing to move toward the future. Trying to return to or even rebuild the past is a direct denial and violation of the universe. It is not wrong to fear the future because the future is unknown, but it is wrong and dangerous to deny the future by seeking shelter in the past because it is known and familiar. The past can never be reconstructed. Reviving the past is impossible. The past is truly dead and one can only find death in the past, not in the future. Only knowledge, at least true universal knowledge, can overcome all of our fears of the future. People who oppose the future due to ignorance and bias will never be successful in their opposition to the coming changes because evolution, whether evolution of thought or biological evolution, are inevitable. Many children are afraid of growing up, but they still grow up.

The Boy Cried

> *The investigation of the truth is in one way hard, in another easy. An indication of this is found in the fact that no one is able to attain the truth adequately, while, on the other hand, no one fails entirely, but everyone says something true about the nature of things, and while individually they contribute little or nothing to the truth, by the union of all a considerable amount is amassed. Therefore, since the truth seems to be like the proverbial door, which no one can fail to hit, in this way it is easy, but the fact that we can have a whole truth and not the particular part we aim at shows the difficulty of it.*
> *Perhaps, as difficulties are of two kinds, the cause of the present difficulty is not in the facts but in us.*
>
> —Aristotle – *Metaphysics* – c.350 BCE

The boy that cried when his father died is grown up now. Yet he still cries. Now he cries alone and in private. He cries because he knows too much. He knows the truth of the nature of reality and thus he knows the future: Not the minute to minute or day to day details of the future, but the future end toward which humankind is progressing and evolving. He cries alone because no one else knows and he understands what it will take to make humanity understand as he does. In the words of the philosophers, he knows both being and becoming. They are the same. He knows both the path that humankind must travel and is traveling upon as well as the problems that humankind will face along that path. In his MIND, he weeps for humanity.

> *The test of a theory is its ability to cope with all the relevant phenomena, not it's* a priori *'reasonableness'. The latter would have proved a poor guide in the development of science, which often makes progress by its encounter with the totally unexpected and initially extremely puzzling.*
>
> —John Charlton Polkinghorne – 1981

BIBLIOGRAPHY
RELIGIOUS AND PHILOSOPHICAL TEXTS
(IN CHRONOLOGICAL ORDER)

(c. 2500-1500 BCE) *The Book of Coming Forth by Day.* (1240 BCE) *The Papyrus of Ani.* A group of books commonly know as "tomb texts" or collectively as *The Egyptian Book of the Dead.* (1998) Translated by Mnata A. Ashbi. Miami, FL: Cruzian Mystic.

(c. 2500-1500 BCE) *Vedas.*

(c. 2000 BCE) *The Epic of Gilgamesh.*

(c. 1800-1500 BCE) *Rig Veda.*

Moses (c. 1500 BCE). *The Pentateuch.*

(c. 1400-200 BCE) *Mahabharata.*

Zarathustra (c. 1000-600 BCE). *Zend Avesta.*

(c. 1000-600 BCE) *Upanishads.*

Lao Tzu (c. 600 BCE). *Tao Te Ching.*

Confucius (c. 600 BCE). *Analects.*

Daniel (c. 600-200 BCE). *The Book of Daniel.*

(c. 500-200 BCE) *Bhagavad Gita.*

Plato (c. 360 BCE). *Timaeus.*

Aristotle (c. 350 BCE). *Physics.*

Valmiki (c. 250 BCE) *Ramayana.*

Matthew, Mark, Luke, John, et al. (c. 40-100 CE) *The New Testament.*

Mohammed (c. 620 CE). *Qur'an.*

Padma Sambhava? (c. 700-800 CE). *Bardo Thodol* (*The Great Liberation upon Hearing in the Intermediate State*). Commonly known as *The Tibetan Book of the Dead.* (1992). Translated by Francesca Fremantle and Chogyam Trungpa. Boston, MA: Shambhala.

GENERAL REFERENCES

Edwin A. Abbott [A. Square] (1884). *Flatland: A Romance of Many Dimensions,* 2nd edition. Oxford: Blackwell, and London: Seeley.

L.A. Amos and A. Klug (1974). "Arrangement of subunits in flagellar microtubules". *Journal of Cell Science* 14: 523-549.

William Barrett (1926). *Death Bed Visions – The Psychical Experiences of the Dying.* London: Methuen.

Morey Bernstein (1956). *The Search for Bridey Murphy.* New York: Pocket Books.

Susan Blackmore (1993). *Dying to Live: Near-death experiences.* New York: Prometheus Books.

Robert T. Browne (1919). *The Mystery of Space: A Study of the Hyperspace Movement in the Light of the Evolution of New Psychic Faculties and An Inquiry into the Genesis and Essential Nature of Space.* New York: Dutton.

Fritjof Capra (1975). *The Tao of Physics: An Exploration of the Parallels Between Modern Physics and Eastern Mysticism.* Berkeley: Shambhala.

David J. Chalmers (1995). "Facing Up to the Problem of Consciousness". *Journal of Consciousness* Studies 2 (3): 200-219.

William Kingdon Clifford (1870). "On the Space-Theory of Matter". Read 21 February 1870, *Transactions of the Cambridge Philosophical Society* 2 (1866/1876).

Charles Darwin (1859). *The Origin of Species.*

Radin, Dean (1997). *The Conscious Universe: The Scientific Truth of Psychic Phenomena.* New York: HarperEdge.

Albert Einstein and Peter Bergmann (1938). "On a Generalization of Kaluza's Theory of Electricity". *Annals of Mathematics* 39 (3): 693-701.

Albert Einstein, Peter G. Bergmann and Valentine Bargmann (1941). "On the Five-Dimensional Representation of Gravitation and Electricity". *Theodor von Karman Anniversary Volume.* Pasadena: California Institute of Technology: 212-225.

Gustav Fechner (1860; Revised 1889). *Elemente der Psychophysik.* Leipzig.

Herman Feifel (1959). *The Meaning of Death.* New York: McGraw-Hill.

Gerald Feinberg in Edgar Mitchell, editor (1974). *Psychic Explorations: A Challenge for Science.* New York: Putnam's Sons: 1-15.

Douglas C. Giancoli (1998). *Physics: Principles with Applications*, 5th edition. New Jersey: Prentiss Hall.

Bruce Greyson and Charles P. Flynn, editors (1984). *The Near-Death Experience: Problems, Prospects, Perspectives.* Springfield, IL: Charles C. Thomas.

Bruce Greyson (1985). "A Typology of Near-Death Experiences". *American Journal of Psychiatry* 142: 967-969.

Stewart Hameroff and Roger Penrose (1996). "Conscious events as orchestrated space-time selections". *Journal of Consciousness Studies* 3 (1): 36-53.

Stephen Hawking (2001). *The Universe in a Nutshell.* New York: Bantam.

Charles Howard Hinton (1884). "What is the Fourth Dimension?" Originally published as a pamphlet in 1884; Republished in *Scientific Romances*, two volumes. London: Swann and Sonnenschein, Volume 1: 1-32.

Charles Howard Hinton (1888). *A New Era of Thought.* London: Swann, Sonnenschein.

Robert D. Jahn and Brenda J. Dunne (1987). *Margins of Reality: The Role of Consciousness in the Physical World.* San Diego: Harcourt Brace Jovanovich.

William James (1890). *The Principles of Psychology.*

William James (1902). *Varieties of Religious Experience.*

Theodor Kaluza (1921). "Zur Unitätsproblem der Physik". *Sitzungsberichte der Preussischen Akademie der Wissenschaften* 54: 966-972.

Robert Kastenbaum and Ruth Aisenberg (1972). *The Psychology of Death*. New York: Springer.

Oskar Klein (1926). "The Atomicity of Electricity as a Quantum Theory Law". *Nature* 118 (2971): 516.

Elisabeth Kübler-Ross (1969). *On Death and Dying*. New York: MacMillan.

Lawrence LeShan (1975). *The Medium, The Mystic, and the Physicist: Toward a General Theory of the Paranormal*. New York: Ballantine.

Clarence Irving Lewis (1929). *Mind and the World Order: Outline of a theory of knowledge*. New York: Scribner's Sons.

John Locke (1690). *An Essay Concerning Human Understanding*.

Ernst Mach (1883; Reprint 1897). *Die Mechanik in ihrer Entwicklung: Historisch-Kritisch Dargestellt*. Leipzig: Brockhaus; (1974) *The Science of Mechanics*. Translated by Thomas J. McCormack. LaSalle: Open Court; Reprint of the sixth American edition.

Ernst Mach (1906) *Analysis of Sensations*, 5th edition. Reprint translated by C.M. Williams and revised by Sydney Waterlow. New York: Dover.

John E. Mack (1995). *Abduction: Human Encounters with Aliens*. New York: Ballantine Books.

Anthony J. Marcel and Edoardo Bisiach, editors (1988). *Consciousness in Contemporary Science*. Oxford: Clarendon Press.

James Marchand (1975). *A.R. Wallace: Letters and Reminiscences*. New York: Arno.

Raymond Moody (1975). *Life After Life*. New York: Bantam.

Melvin Morse with Paul Perry (1990). *Closer to the Light*: Learning from the Near-Death Experiences of Children. London: Souvenir Press.

Frederic W.H. Myers (1903). *Human Personality and the Survival of Bodily Death*, 2 Volumes. London: Longman, Green.

Thomas Natsoulas (1978). "Consciousness". *American Psychologist* 33: 906-914.

Simon Newcomb (1902). "The Fairyland of Geometry". *Harper's Monthly Magazine* 104: 249-252.

Isaac Newton (1687). *Philosophiae Naturalis Principia Mathematica* (The *Principia*). Translated by Florian Cajori. Berkeley: University of California Press, 1934.

Russell Noyes (1972). "The Experiences of Dying". *Psychiatry*: 178.

Donna Olendorf, Christina Jeryan and Karen Boyden, editors (1999). *The Gale Encyclopedia of Medicine*. Detroit: Gale.

Robert O. Ornstein (1972). *The Psychology of Consciousness*. San Francisco: W.H. Freeman.

Karlis Osis and Erlendur Haraldsson (1977; 1986 Revised). *At the Hour of Death*. New York: Avon; New York: Hastings House.

J.E. Owens, E.W. Cook and I. Stephenson (1990). "Features of 'Near-Death Experiences' in Relation to Whether or Not Patients were Near Death". *Lancet* 336: 1175-1177.

Karl Pearson (1892). *The Grammar of Science.* London: Walter Scott.

Frank Podmore (1902). *Modern Spiritualism: A history and a criticism.* London: Methuen.

Kenneth Ring (1980). *Life at Death: A Scientific Investigation of the Near-Death Experience.* New York: William Morrow.

Kenneth Ring (1984). *Heading Toward Omega: In Search of Meaning of the Near-Death Experience.* New York: William Morrow.

Kenneth Ring (1992). *The Omega Project: Near-Death Experiences, UFO Encounters and the Mind at Large.* New York: William Morrow.

Kenneth Ring and Evelyn Elsaesser Valarino (2000). *Lessons from the Light.* Needham, Massachusetts: Moment Point Press.

Barbara Rommer (2000). *Blessing in Disguise: Another Side of the Near-Death Experience.* St. Paul, Minnesota: Llewellyn Press.

Michael Sabom (1982). *Recollections of Death.* New York: Harper Row.

Gary Schwartz (2002). *The Afterlife Experiments.* New York: Atria.

Mary Shelley (1818). *Frankenstein.*

Society for Psychical Research. *Homepage.* WWW. Accessed 30 December 2004. <www.spr.ac.uk>

Johann Bernhard Stallo (1881; Revised 1884) *The Concepts and Theories of Modern Physics*. Edited by Percy W. Bridgman. Cambridge, Massachusetts: The Belknap Press of Harvard University Press, 1960; Reprint of the second edition of 1884.

Ian Stevenson (1974). *Twenty Cases Suggestive of Reincarnation*, 2nd revised edition. Virginia: University Press of Virginia.

Ian Stevenson (1997). *Biology and Reincarnation: A contribution to the etiology of birthmarks and birth defects*. Westport, CT: Praeger.

G.F. Stout (1899). *A Manual of Psychology*. London: University Correspondence College Press.

P.G. Tait and Balfour Stewart (1875). *The Unseen Universe: or Physical Speculations on a Future State*. London: Macmillan.

R.H. Thouless and W.P. Weisner (1942). The Present Position of Experimental Research into Telepathy and Related Phenomena". *Proceedings of the Society for Psychical Research* 47: 1-19.

R.H. Thouless and W.P. Weisner (1947). "The psi process in normal and "paranormal" psychology". *Proceedings of the Society for Psychical Research* 48: 177-196.

Jessica Utts (1991). "Replication and meta-analysis in parapsychology (with discussion)". *Statistical Science* 6 (4): 363-403.

Jessica Utts (1996). "An Assessment of the Evidence for Psychic Functioning". *The Journal of Scientific Exploration* 10 (1): 3-30. Also at <http://anson.ucdavis.edu/%7Eutts/air2.html>

Eugene Wigner (1961; Reprint 1967). "Remarks on the Mind-Body Problem". In *Symmetries and Reflections: Scientific Essays of Eugene P. Wigner*. Bloomington: Indiana University Press: 171-184.

Wilhelm Wundt (1873-1874). *Grundzüge der physiologischen Psychologie*. Leipzig.

Wilhelm Wundt (1896). *Grundriss der Psychologie*. Leipzig.

Johann Karl Freidrich Zöllner (1881). *Transcendental Physics*. Translated and abridged by C.C. Massey. Boston: Colby and Rich.

ISBN 1425161677

Printed in Great
Britain
by Amazon